Application of Geophysical Methods for Hydrogeology

Application of Geophysical Methods for Hydrogeology

Guest Editors

Alex Sendros
María del Carmen Cabrera Santana
Albert Casas Ponsati

MDPI
Basel • Beijing • Wuhan • Barcelona • Belgrade • Novi Sad • Cluj • Manchester

Guest Editors
Alex Sendros
Mineralogy, Petrology and Applied Geology
Universitat de Barcelona
Barcelona
Spain

María del Carmen Cabrera Santana
Instituto Universitario de Estudios Ambientales y Recursos Naturales (i-UNAT)
Universidad de Las Palmas de Gran Canaria
Las Palmas de Gran Canaria
Spain

Albert Casas Ponsati
Mineralogy, Petrology and Applied Geology
Universitat de Barcelona
Barcelona
Spain

Editorial Office
MDPI AG
Grosspeteranlage 5
4052 Basel, Switzerland

This is a reprint of the Special Issue, published open access by the journal *Water* (ISSN 2073-4441), freely accessible at: http://www.mdpi.com/journal/water/special_issues/ZC47RT0HAR.

For citation purposes, cite each article independently as indicated on the article page online and as indicated below:

Lastname, A.A.; Lastname, B.B. Article Title. *Journal Name* **Year**, *Volume Number*, Page Range.

ISBN 978-3-7258-4352-7 (Hbk)
ISBN 978-3-7258-4351-0 (PDF)
https://doi.org/10.3390/books978-3-7258-4351-0

Contents

About the Editors

Alex Sendros

Alex Sendros is a tenure-track lecturer at the University of Barcelona in Spain. His research interests are focused on the application of near-surface electrical and electromagnetic methods for environmental and hydrogeological studies. His research includes projects integrating geophysical, geochemical and hydrogeological methods in managed aquifer recharge and nature-based solutions for wastewater treatment. Dr. Alex Sendros has authored many high-impact SCI papers and is active as a peer-reviewer for several scientific journals.

María del Carmen Cabrera Santana

María del Carmen Cabrera-Santana is an Associate Professor at the University of Las Palmas de Gran Canaria in Spain. Her research is focused on Hydrogeology, especially on Volcanic islands, including recharge, aquifer characterization and hydrogeochemical studies. She has experience as a Principal Investigator in National and International R&D projects. Dr. Cabrera-Santana has authored several SCI papers and is an active peer-reviewer.

Albert Casas Ponsati

Albert Casas-Ponsatí is an Emeritus Professor at the University of Barcelona in Spain. His research interests are focused on the application of near-surface electrical and electromagnetic methods for environmental and hydrogeological studies. He has extensive experience in leading projects integrating geophysical, geochemical and hydrogeological methods in different countries worldwide. Dr. Casas-Ponsatí has authored many high-impact SCI papers and is active as a peer-reviewer for several scientific journals.

Preface

Recent and unique works in the field of applying geophysical methods to hydrogeology are compiled in this book. When examining the integration of different electrical, electromagnetic, and seismic geophysical techniques alongside the combination of geophysical techniques and direct method data (which include geotechnical soundings logs, geochemical tracers, and physical parameters) to minimise interpretation ambiguity and validate the geophysical models, readers will find the contributions both inspiring and interesting.

In order to implement numerical tools for the modelling of the dynamics of groundwater quantity (flow) and quality (salinity and pollution) and obtain tools for the sustainable management of groundwater, it is necessary to characterize hydraulic properties, some transient groundwater features, and aquifer geometry and vadose zones. The findings and methods presented in these original contributions aim to be of interest for some of the issues associated with achieving a holistic strategy. The editors hope that scholars and practitioners will find these contributions interesting and that they will additionally aid in identifying areas for future research.

Alex Sendros, María del Carmen Cabrera Santana, and Albert Casas Ponsati
Guest Editors

water

Editorial

Application of Geophysical Methods for Hydrogeology

Alex Sendrós [1,2,*], María del Carmen Cabrera [3] and Albert Casas-Ponsatí [1,2]

1 Department of Mineralogy, Petrology and Applied Geology, Universitat de Barcelona, 08028 Barcelona, Spain; albert.casas@ub.edu
2 Water Research Institute, Universitat de Barcelona, 08001 Barcelona, Spain
3 Instituto Universitario de Estudios Ambientales y Recursos Naturales (i-UNAT), Universidad de Las Palmas de Gran Canaria, 35017 Las Palmas de Gran Canaria, Spain; mcarmen.cabrera@ulpgc.es
* Correspondence: alex.sendros@ub.edu

1. Introduction

Groundwater is considered essential in the world's current water supply. Subsurface water also supports ecosystems and is more resilient than surface watercourses to the negative effects of climate change and human activity. In many places, aquifers are the only source of water. However, there has been an increase in the intensive use of groundwater and a growing number of reports of degradation [1–3].

The characterization of the subsurface and its hydraulic properties is essential for an appropriate groundwater and surface water management [4], but they are often difficult to evaluate from traditional borehole drilling, water well and piezometer pumping tests, or soil sampling techniques [5,6]. Traditional soil sampling methods and the use of devices to obtain hydraulic properties in the field, such as permeameters or humidity sensors, typically only provide localized data from the upper layers, and borehole drilling is a destructive and expensive technique which requires accessible areas to place heavy machinery and only delivers spotted information [7]. The integration of geophysical data into direct hydrogeological measurements is a challenging issue that could be used to characterize, monitor, and investigate hydrological parameters and processes in the vadose zone and aquifers at different resolutions and over many spatial scales in a minimally invasive manner [8,9].

For the purpose of characterizing groundwater, electrical, electromagnetic, and seismic geophysical techniques are frequently employed [10]. While the latter is primarily used to deduce aquifer geometry and certain steady aquifer hydraulic parameters, the first two are usually used to infer aquifer geometry and certain transient groundwater features like the piezometric level, freshwater–saltwater interface, characterization of groundwater flow, and pore water conductivity [8,11]. The interpretations become clearer when several methodologies are combined and used on conductive structures and pore-filling fluids (both natural and man-made) that are subjected to the temporal dynamics of dissolved ions and water content. Integration can also include utilizing direct data (such as physical parameters, geochemical tracers, and lithological logs) to enhance and/or validate the geophysical models. There are several scientific software systems available with user-friendly interfaces, robust data inversion techniques, and tools for dealing with uncertainty analysis [12].

In this broad hydro-geophysical framework, this Special Issue aimed to attract specialized researchers using geophysical prospecting techniques for groundwater research and for gathering the advances and challenges associated with the use of geophysical methods. The special focus of this Special Issue is on case studies demonstrating the potential to improve our understanding of hydrogeological parameters in vadose and non-vadose zones used to modelize groundwater flow, study the transport of substances, and, therefore, improve our aquifer knowledge and manage many important processes such as contamination and saltwater intrusion. The accepted papers included (i) geophysical prospecting surveys as a part of the holistic strategy for aquifer conceptualization and modeling, (ii) integrated large and detailed scale near-surface geophysical prospecting techniques and time-lapse approaches to reduce the ambiguity of hydrogeological interpretations, (iii) experimental field and numerical operational designs, and (iv) case studies surveying saturated and unsaturated media for methodological and conceptual purposes. Other papers contributed to understanding the state of the art of geophysical techniques through specific study cases covering (i) hydrogeological environments such as polluted sites and urbanized areas in different countries; (ii) aquifer typologies in coastal and inland areas such as Paleogene and Neogene sedimentary rocks and Quaternary detrital sediments; and (iii) climate settings including humid, sub-humid, and semiarid to arid. The used techniques were (i) electrical, such as electrical resistivity tomography (ERT), vertical electrical sounding (VES), resistivity well logs, self-potential (SP) measurements, and induced polarization tomography (IPT); (ii) electromagnetic, such as ground-penetrating radar (GPR), the time-domain electromagnetic method (TDEM), and Global Navigation Satellite Systems reflectometry (GNSS-R); and (iii) seismic, such as Seismic Refraction Tomography (SRT), multichannel analysis of surface waves (MASW) and microtremor recordings elaborated with the horizontal-to-vertical seismic ratio (HVSR) technique.

2. Contributions

Twelve manuscripts have been accepted for publication since the March 2023 announcement of the call for papers. The manuscripts have been accepted for publication following a rigorous review process. To achieve a better insight into the Special Issue, we present brief highlights of the published papers below.

The authors of the paper "Hydrogeophysical Investigation in Parts of the Eastern Dahomey Basin, Southwestern Nigeria: Implications for Sustainable Groundwater Resources Development and Management" conducted geoelectrical resistivity measurements (VES and ERT) in five locations within the eastern portion of the Dahomey Basin (Nigeria). The geophysical results were integrated with the borehole logs to generate the spatial distribution of the subsurface lithologies and to estimate the hydraulic parameters (porosity, hydraulic conductivity, and transmissivity) of two highly productive local aquifers. However, it is crucial to consider the presence of sub-vertical faults in the study site, as these faults can significantly affect water wells' productivity and ultimately influence the overall water availability in the area.

The authors of the paper "Geophysical Constraints to Reconstructing the Geometry of a Shallow Groundwater Body in Caronia (Sicily)" analyzed and reinterpreted geoelectrical data, allowing for the construction of a preliminary 3D resistivity model. This initial modeling was subsequently integrated with a geophysical data campaign to define the depth of the bottom of the shallow Caronia Groundwater Body and the thickness of alluvial deposits. Finally, a preliminary mathematical model flow was generated to reconstruct the dynamics of underground water. The results show that the integration of multidisciplinary data represents an indispensable tool for the characterization of complex physical systems.

The authors of the paper "Coupled Geophysical and Hydrogeochemical Characterization of a Coastal Aquifer as Tool for a More Efficient Management (Torredembarra, Spain)" integrated hydrogeological, hydrogeochemical, and electrical resistivity subsoil data to establish a hydrogeological model of the coastal aquifer of this area. The obtained results could be used as a support tool for the assessment of the most favorable areas for groundwater withdrawal, as well as enabling the control and protection of the most susceptible areas affected by saltwater intrusion.

The authors of the paper "Identification of Breaches in a Regional Confining Unit Using Electrical Resistivity Methods in Southwestern Tennessee, USA" applied electrical resistivity and borehole data to delineate lithostratigraphic boundaries and image the geometry of confining-unit breaches in Eocene coastal-plain deposits to evaluate inter-aquifer exchange pathways. The results underscore the efficacy of the ERT method in identifying sand-rich paleochannel discontinuities in a low-resistivity regional confining unit, which may be a common origin of breaches in coastal-plain confining units.

The authors of the paper "An Integrated Approach for Saturation Modeling Using Hydraulic Flow Units: Examples from the Upper Messinian Reservoir" characterized and predicted the change in reservoir water saturation (SW) with time, while reservoir production life is based on the change in reservoir capillary pressure. The study introduced an integrated approach, including the evaluation of core measurements, well-log analysis covering cored and non-cored intervals, neural analysis techniques, and permeability prediction in non-cored intervals. The empirical formula was predicted for direct calculation of dynamic SW profiles and predicted within the reservoir above the fluid contact and free water level based on the change in reservoir pressure.

The authors of the paper "Geometry, Extent, and Chemistry of Fermentative Hot Spots in Municipal Waste Souk Sebt Landfill, Ouled Nemma, Beni Mellal, Morocco" aimed to detect and characterize fermentative hotspots in municipal waste dumps as well as the leachates that form within them using SP measurements. Despite the small size of the hotspots generating the leachates, the accumulation of leachates in ponds and the low soil permeability limits the percolation rate, resulting in moderate but permanent groundwater pollution.

The authors of the paper "A Real-Time Prediction Approach to Deep Soil Moisture Combining GNSS-R Data and a Water Movement Model in Unsaturated Soil" proposed a real-time prediction approach to deep soil moisture combining GNSS-R data and a water movement model in unsaturated soil. The approach was validated in a study area in Goodwell, Texas County, Oklahoma, USA, and validates the feasibility of the proposed procedure, which has the potential to play a crucial role in agricultural production, geological disaster management, engineering construction, and heritage site preservation.

The authors of the paper "Environmental Monitoring of Pig Slurry Ponds Using Geochemical and Geoelectrical Techniques" evaluated the relationship between electrical values and geochemical parameters and the risk of lateral contamination of pig slurry stored in a pond using ERT and geochemical analysis. The infrastructures dedicated to managing pig farm by-products are necessary to prevent environmental pollution and eutrophication of groundwater, and this non-invasive method provides detailed information on the distribution and characteristics of the fluids.

The authors of the paper "Coastal Groundwater Bodies Modelling Using Geophysical Surveys: The Reconstruction of the Geometry of Alluvial Plains in the North-Eastern Sicily (Italy)" reconstructed the pattern and extent of two groundwater bodies, located in populated and industrialized coastal sectors of north-eastern Sicily, through the integrated analysis and interpretation of several geoelectrical (VES, ERT, and IPT), seismic (active and passive seismic), and geological data. The procedure followed allowed them to recognize the areal extension and thickness of the various lithotypes and define the depth and the morphology of the base of the groundwater bodies and the thickness of the filling deposits.

The authors of the paper "Assessing and Improving the Robustness of Bayesian Evidential Learning in One Dimension for Inverting Time-Domain Electromagnetic Data: Introducing a New Threshold Procedure" applied Bayesian evidential learning in one dimension for stochastic TDEM inversion with a threshold approach on field data collected in the Luy River catchment (Vietnam) to delineate saltwater intrusions. Their results show that the proper selection of time and space discretization is essential for limiting the computational cost while maintaining the accuracy of the posterior estimation. Moreover, the selection of the prior distribution has a direct impact on fitting the observed data and is crucial for realistic uncertainty quantification.

The authors of the paper "Characterization of a Contaminated Site Using Hydro-Geophysical Methods: From Large-Scale ERT Surface Investigations to Detailed ERT and GPR Cross-Hole Monitoring" presented the results of an advanced geophysical characterization of a contaminated site, where a correct understanding of the dynamics in the unsaturated zone is fundamental to evaluate the effective management of the remediation strategies. Large-scale surface ERT was used to perform a preliminary assessment of the structure in a thick unsaturated zone and to detect the presence of a thin layer of clay supporting an overlying thin perched aquifer. Therefore, a deep trench was dug upstream of the site and a forced infiltration experiment was carried out and monitored using ERT and GPR measurements in a cross-hole time-lapse configuration. The results emphasize the contribution of hydro-geophysical methods to the general understanding of subsurface water dynamics.

The authors of the paper "Dynamics of Saltwater Intrusion in a Heterogeneous Coastal Environment: Experimental, DC Resistivity, and Numerical Modeling Approaches" conducted experimental, numerical, and geophysical field campaigns to assess the saltwater intrusion phenomena in coastal aquifers. Direct Current (DC) resistivity sounding data were collected using a laboratory physical model to determine the depth of the freshwater–saltwater interface, a finite element analysis was employed to generate numerical models based on experimental feedback and for validation purposes, and ERT data were acquired from the seacoast and an aquaculture area. The alignment of the experimental, numerical, and geophysical data suggests that this integrated approach could be valuable for studying saltwater intrusion and can be applied to different geological settings, including tidal flats and alluvial plains.

3. Future Prospects

The Guest Editors envision that practitioners and scholars will find the published papers in this Special Issue interesting and useful in identifying areas for further research in the use of geophysical techniques applied to groundwater. Applications such as obtaining reliable three-dimensional and time-lapse hydro-geophysical models, working with uncertainty reductions at greater depths, improving the uncertainty quantification, obtaining more robust correlation among hydro-geophysical and hydrochemical parameters, transitioning from homogeneous to heterogeneous subsurface models, and dealing with the integration of geophysical information with routine environmental matrices monitoring as defined by community regulations can benefit from the papers included in this issue. Additionally, we hope that readers will find this Special Issue's contents to be both educational and motivating as they investigate geophysical techniques for hydrogeology. The methods and conclusions offered in this compilation of publications add to the growing interest in the application of geophysical methods in groundwater studies.

Author Contributions: Conceptualization, A.S., A.C.-P. and M.d.C.C.; methodology, A.C.-P.; formal analysis, M.d.C.C. and A.S.; writing—original draft preparation, A.S.; writing—review and editing, A.S., A.C.-P. and M.d.C.C. All authors have read and agreed to the published version of the manuscript.

Funding: This research received no external funding.

Acknowledgments: The authors of this paper, who served as the Guest Editors of this Special Issue, would like to express their gratitude to the journal editors, all authors submitting papers, and the referees who contributed to revising and improving the twelve published papers.

Conflicts of Interest: The authors declare no conflicts of interest.

List of Contributions:

1. Oyeyemi, K.D.; Aizebeokhai, A.P.; Olaojo, A.A.; Okon, E.E.; Kalu, D. V; Metwaly, M. Hydrogeophysical Investigation in Parts of the Eastern Dahomey Basin, Southwestern Nigeria: Implications for Sustainable Groundwater Resources Development and Management. *Water* **2023**, *15*.
2. Canzoneri, A.; Capizzi, P.; Martorana, R.; Albano, L.; Bonfardeci, A.; Costa, N.; Favara, R. Geophysical Constraints to Reconstructing the Geometry of a Shallow Groundwater Body in Caronia (Sicily). *Water* **2023**, *15*, 3206, doi:10.3390/w15183206.
3. Pavoni, M.; Boaga, J.; Peruzzo, L.; Barone, I.; Mary, B.; Cassiani, G. Characterization of a Contaminated Site Using Hydro-Geophysical Methods: From Large-Scale ERT Surface Investigations to Detailed ERT and GPR Cross-Hole Monitoring. *Water* **2024**, *16*.
4. Sendrós, A.; Cubides, I.J.; Himi, M.; Lovera, R.; Urruela, A.; Tapias, J.C.; Rivero, L.; Garcia-Artigas, R.; Casas, A. Coupled Geophysical and Hydrogeochemical Characterization of a Coastal Aquifer as Tool for a More Efficient Management (Torredembarra, Spain). *Water* **2023**, *15*.
5. Hasan, M.R.; Larsen, D.; Schoefernacker, S.; Waldron, B. Identification of Breaches in a Regional Confining Unit Using Electrical Resistivity Methods in Southwestern Tennessee, USA. *Water* **2023**, *15*.
6. El-Gendy, N.H.; Mabrouk, W.M.; Waziry, M.A.; Dodd, T.J.; Abdalla, F.A.; Alexakis, D.E.; Barakat, M.K. An Integrated Approach for Saturation Modeling Using Hydraulic Flow Units: Examples from the Upper Messinian Reservoir. *Water* **2023**, *15*.
7. El Mouine, Y.; El Hamdi, A.; Bousouis, A.; El Jarjini, Y.; Touzani, M.; Valles, V.; Barbiero, L.; Morarech, M. Geometry, Extent, and Chemistry of Fermentative Hot Spots in Municipal Waste Souk Sebt Landfill, Ouled Nemma, Beni Mellal, Morocco. *Water* **2024**, *16*.
8. Luo, X.; Yin, C.; Sun, Y.; Bai, W.; Li, W.; Song, H. A Real-Time Prediction Approach to Deep Soil Moisture Combining GNSS-R Data and a Water Movement Model in Unsaturated Soil. *Water* **2024**, *16*.
9. Capa-Camacho, X.; Martínez-Pagán, P.; Acosta, J.A.; Martínez-Segura, M.A.; Vásconez-Maza, M.; Faz, Á. Environmental Monitoring of Pig Slurry Ponds Using Geochemical and Geoelectrical Techniques. *Water* **2024**, *16*.
10. Capizzi, P.; Martorana, R.; Canzoneri, A.; Bonfardeci, A.; Favara, R. Coastal Groundwater Bodies Modelling Using Geophysical Surveys: The Reconstruction of the Geometry of Alluvial Plains in the North-Eastern Sicily (Italy). *Water* **2024**, 16.
11. Ahmed, A.; Aigner, L.; Michel, H.; Deleersnyder, W.; Dudal, D.; Flores Orozco, A.; Hermans, T. Assessing and Improving the Robustness of Bayesian Evidential Learning in One Dimension for Inverting Time-Domain Electromagnetic Data: Introducing a New Threshold Procedure. *Water* **2024**, *16*.
12. Tiwari, P.; Rupesh, R.; Sharma, S.P.; Ciazela, J. Dynamics of Saltwater Intrusion in a Heterogeneous Coastal Environment: Experimental, DC Resistivity, and Numerical Modeling Approaches. *Water* **2024**, *16*, doi:10.3390/w16141950.

References

1. Dao, P.U.; Heuzard, A.G.; Le, T.X.H.; Zhao, J.; Yin, R.; Shang, C.; Fan, C. The impacts of climate change on groundwater quality: A review. *Sci. Total Environ.* **2024**, *912*, 169241. [CrossRef]
2. Custodio, E. Education and science to safeguard the human right to water: The role of groundwater. *Boletín Geológico Min.* **2020**, *131*, 745–755. [CrossRef]
3. Fernández-Ortega, J.; Ulloa-Cedamanos, F.; Barberá, J.A.; Batiot-Guilhe, C.; Jourde, H.; Andreo, B. A common framework for the development of spring water contamination early warning system in western Mediterranean karst areas: Spanish and French sites. *Sci. Total Environ.* **2024**, *956*, 177294. [CrossRef] [PubMed]

4. Duque, C.; Nilsson, B.; Engesgaard, P. Groundwater–surface water interaction in Denmark. *WIREs Water* **2023**, *10*, e1664. [CrossRef]
5. Ronayne, M.J.; Gorelick, S.M.; Zheng, C. Geological modeling of submeter scale heterogeneity and its influence on tracer transport in a fluvial aquifer. *Water Resour. Res.* **2010**, *46*, W10519. [CrossRef]
6. Sendrós, A.; Himi, M.; Estévez, E.; Lovera, R.; Palacios-Diaz, M.P.; Tapias, J.C.; Cabrera, M.C.; Pérez-Torrado, F.J.; Casas, A. Hydrogeophysical Assessment of the Critical Zone below a Golf Course Irrigated with Reclaimed Water close to Volcanic Caldera. *Water* **2021**, *13*, 2400. [CrossRef]
7. McLachlan, P.J.; Chambers, J.E.; Uhlemann, S.S.; Binley, A. Geophysical characterisation of the groundwater–surface water interface. *Adv. Water Resour.* **2017**, *109*, 302–319. [CrossRef]
8. Binley, A.; Hubbard, S.S.; Huisman, J.A.; Revil, A.; Robinson, D.A.; Singha, K.; Slater, L.D. The emergence of hydrogeophysics for improved understanding of subsurface processes over multiple scales. *Water Resour. Res.* **2015**, *51*, 3837–3866. [CrossRef]
9. Capizzi, P.; Cellura, D.; Cosentino, P.; Fiandaca, G.; Martorana, R.; Messina, P.; Schiavone, S.; Valenza, M. Integrated hydrogeochemical and geophysical surveys for a study of sea-water intrusion. *Boll. Geofis. Teor. Appl.* **2010**, *51*, 285–300.
10. Vereecken, H.; Binley, A.; Cassiani, G.; Revil, A.; Titov, K. (Eds.) *Applied Hydrogeophysics;* Springer: Amsterdam, The Netherlands, 2006; p. 383.
11. Flores Orozco, A.; Steiner, M.; Katona, T.; Roser, N.; Moser, C.; Stumvoll, M.J.; Glade, T. Application of induced polarization imaging across different scales to understand surface and groundwater flow at the Hofermuehle landslide. *CATENA* **2022**, *219*, 106612. [CrossRef]
12. Alcalá, F.J.; Paz, M.C.; Martínez-Pagán, P.; Monteiro Santos, F. Integrated Geophysical Methods for Shallow Aquifers Characterization and Modelling. *Appl. Sci.* **2022**, *12*, 2271. [CrossRef]

water

MDPI

Article

Dynamics of Saltwater Intrusion in a Heterogeneous Coastal Environment: Experimental, DC Resistivity, and Numerical Modeling Approaches

Prarabdh Tiwari [1,*], R Rupesh [1], Shashi Prakash Sharma [1] and Jakub Ciazela [2,*]

[1] Department of Geology and Geophysics, Indian Institute of Technology Kharagpur, Kharagpur 721302, West Bengal, India; rupeshgeokuk13@gmail.com (R.R.); sps.kgp@gmail.com (S.P.S.)
[2] Institute of Geological Sciences, Polish Academy of Sciences, Twarda 51/55, 00 818 Warszawa, Poland
* Correspondence: pt.tiwari75@gmail.com (P.T.); j.ciazela@twarda.pan.pl (J.C.); Tel.: +91-8950776466 (P.T.)

Abstract: Saltwater intrusion (SWI) is a critical concern affecting coastal groundwater sources due to natural and anthropogenic activities. The health of coastal aquifers is deteriorated by excessive SWI, mainly caused by the disturbance of the freshwater–saltwater equilibrium due to the escalating population, climate change, and the rising demand for freshwater resources for human activities. Therefore, gaining insight into the dynamics of SWI is crucial, particularly concerning the various factors that influence the intrusion mechanism. The present study focuses on the experimental simulation of saltwater in freshwater aquifers, considering boundary conditions and density-dependent effects. Two geological scenarios within coastal environments were investigated: First, a uniform, homogeneous case consisting of only sand, and second, a heterogeneous case in which layers of sand, clay, and sand mixed with pebbles are used. During the experiment, DC resistivity sounding data, as part of a widely recognized geophysical method, were collected and subsequently inverted to determine the depth of the freshwater–saltwater interface (FSWI). A finite element analysis was employed to generate numerical models based on experimental feedback. Further, for validation purposes, electrical resistivity tomography (ERT) data were collected from two distinct locations: near the seacoast and an aquaculture area. The ERT results show the presence of salinity intrusion in the study area, attributed mainly to groundwater overpumping and fish farming practices. The experimental findings indicate that the advancement of saltwater is affected by the geological properties of the media they traverse. The porosity (ϕ) and permeability (k) of the geological layer play a crucial role during the passage of saltwater flux into freshwater aquifers. The FSWI deviated along the clay boundary and hindered the easy passage of saltwater into surrounding layers. The alignment of experimental, numerical, and geophysical data suggests that this integrated approach could be valuable for studying SWI and can be applied in different geological settings, including tidal flats and alluvial plains.

Keywords: coastal aquifers; DC resistivity; numerical modeling; saltwater intrusion; sustainable water management (SWM)

1. Introduction

In recent years, an exponential increase in population has created enormous pressure on coastal groundwater resources. According to the high end forecast scenario, the global population residing in low-elevation coastal zones (LECZs) could increase by more than 50% between the base year of 2000 and 2030 [1]. The exponential growth of coastal populations imposes a significant strain on groundwater resources. Coupled with this, the elevated living standards prevalent in these hotspot areas exacerbate groundwater depletion. The extensive extraction of groundwater intensifies the threat of saltwater intrusion, particularly in coastal regions, emerging as a critical environmental concern. Researchers worldwide have reported salinity problems in their respective areas [2–7].

Anthropogenic and catastrophic events can also affect the dynamics of coastal aquifers; therefore, research must focus on analyzing and understanding the behavior of saltwater intrusion along coastal margins. Numerous efforts have been made to understand the mechanism of saltwater intrusion, with Henry's problem emerging as the benchmark for density-dependent SWI models [8]. Various experimental data sets were used to validate the analytical solution for groundwater flow, thereby establishing correlations with the freshwater–saltwater mixing conditions. Many simulation models have been created to address various aspects, including the impact of water levels, tidal effects, seawater intrusion concentrations, the freshwater–saltwater interface's migration rate, and the three-dimensional variable-density advection–dispersion model. Several laboratory experiments have been conducted to replicate the behavior of seawater mixing under controlled conditions [9–12]. Most of these studies concentrate solely on the progression of the saltwater wedge as a result of salinity contour distribution through a homogeneous geology. Thus, it is advantageous to trace the flow path of saltwater in various geological settings, utilizing geophysical methods that have been proven to be the most effective tools. Among these techniques, DC resistivity is particularly efficient in delineating saline boundaries in coastal aquifer environments [13–17]. Since the migration of saline water depends upon the geological characteristics of the host region, resistivity methods can help characterize various geological layers, such as clay, sand, silt, and shale. These layers exhibit different responses in terms of porosity and permeability based on their composition. Therefore, it is crucial to map the depth and thickness of different subsurface layers accurately. Resistivity methods have been previously used to characterize the subsurface geology for mapping saline zones [18–20].

The primary focus of laboratory simulations is to replicate the conditions under which salt diffusion arises due to the concentration disparity between fresh and saline water. However, flow velocity is also crucial for the advancement of the saline contours [21]. Several attributes have been incorporated into these experiments to address the dynamics of the freshwater–saltwater interaction, including an uncertainty analysis in fractured aquifers [22], saltwater up-coning [23], the pumping effect [24,25], the impact of beach face slope variation on SWI [26], subsurface dams to protect aquifers from SWI [27], the effect of an inclined boundary on SWI [28], and the impact of cutoff walls [29]. These experiments typically focus on a homogeneous geological environment, often represented by a sand-filled sandbox. However, the Earth consists of anisotropic behavior corresponding to various geological layers. Until simulation conditions closely resemble actual ground conditions, it becomes challenging to comprehend the interaction of salinity intrusion in complex geological environments. Therefore, accounting for the heterogeneity factor when conducting these laboratory experiments is crucial. There is always potential for research into the migration of the freshwater–saltwater interface during SWI experiments. This interface migrates when its natural state is disturbed by any external factor, and the extent of its intrusion depends on the rate at which saltwater flux intrudes. External factors such as groundwater overpumping result in deeper saltwater penetration into freshwater aquifers. Additionally, the migration of the interface depends on the geological heterogeneity of the media. The primary focus of the present study is to understand this migration behavior of the FSWI, particularly concerning the geological heterogeneity of the media.

The current study explores the boundary between freshwater and saltwater through a laboratory experiment using DC resistivity and numerical simulations. Initially, we conducted a sandbox experiment that involved two distinct scenarios: one characterized by a homogeneous setup with sand as the background material and the other featuring a heterogeneous environment incorporating sand, clay, and pebbles. Once a stable state was reached at the interface between freshwater and saltwater, DC resistivity measurements are obtained along the center of the profile (in a vertical cross-section). A forward model was generated to assess the sensitivity of the array utilized for resistivity data collection, followed by the creation of a one-dimensional inversion of the sounding model, which provides information about the depth of the interface. High-resolution imagery was captured throughout the experiment, and numerical solutions for both scenarios were obtained using the initial Henry parameters. Our main goal was to investigate the behavior of advancing saltwater wedges in various geological settings using a numerical model and the direct current resistivity method, focusing on the influence of clay layers in coastal environments. The present study emphasizes the effect of geological heterogeneity on the mechanisms of saltwater intrusion. In earlier research, uniform background materials like sand or artificial silica beads have typically been used to study intrusion behavior. For example, a homogeneous sandbox has been utilized in a study on subsurface dams to control SWI [30]. Few researchers have considered heterogeneity in their experiments by using artificial beads of different sizes [31,32]. However, the most significant aspect of this experimental study is the use of actual heterogeneous layers of sand, clay, and pebbles rather than artificial silica beads, making the study more realistic and applicable in terms of geological nature. Additionally, the study incorporates real-time geophysical data acquisition using DC resistivity to determine the depth of the freshwater–saltwater interface, providing a better understanding of subsurface contaminant flow. Considering the above parameters, this study suggests a more realistic observation and behavior of saltwater intrusion when it encounters different layers.

The present study holds significant utility as it offers valuable insights into the behavior of geological layers to SWI. The experimental findings were validated by an electrical resistivity tomography (ERT) investigation in the coastal area of West Bengal, India. The ERT study observed that anthropogenic activities, such as groundwater extraction for paddy crop cultivation and aquaculture practices, significantly impact groundwater quality. Consequently, it is essential to implement effective management strategies and policies, such as sustainable pond practices for aquaculture and crop rotation, to reduce dependency on water-intensive paddy crops, ensuring the sustainable use of groundwater in these regions.

Our study employs an integrated method that combines experimental data, numerical modelling, and geophysical approaches. This comprehensive approach provides valuable insights that can directly inform policymaking and the development of sustainable practices for managing coastal water. This research establishes a scientific foundation for developing targeted groundwater extraction regulations, implementing efficient land-use planning and designing physical barriers to protect freshwater supplies by demonstrating how various geological conditions influence saltwater intrusion dynamics.

2. Materials and Methods

2.1. Experimental Setup

The study conducted an experimental examination of saltwater intrusion, considering a conceptual model of a coastal scenario. This model accounts for the potential contribution of numerous human activities to the incursion of salinity in coastal aquifers (Figure 1). The experiment was carried out in a flow tank made up of acrylic glass with dimensions of 50 cm (length) × 30 cm (height) × 5 cm (width). The cross-section represents various coastal environments (Figure 2). The flow tank was divided into three chambers; the left chamber represents the freshwater reservoir, the middle chamber is filled with porous background material (as per two different scenarios), and the right chamber is the saltwater reservoir. Both freshwater and saltwater chambers are kept at a constant head for a continuous flow supply maintained by the outflow valves. Porous media are separated in these chambers using a US #18 fine mesh screen at a distance of 5.5 cm. Instead of artificial silica beads as a porous material, we used natural sand grains with a diameter ranging from 0.6 mm to 1.2 mm. We kept a head difference ($\Delta h = 1$ cm) between freshwater and saltwater reservoirs.

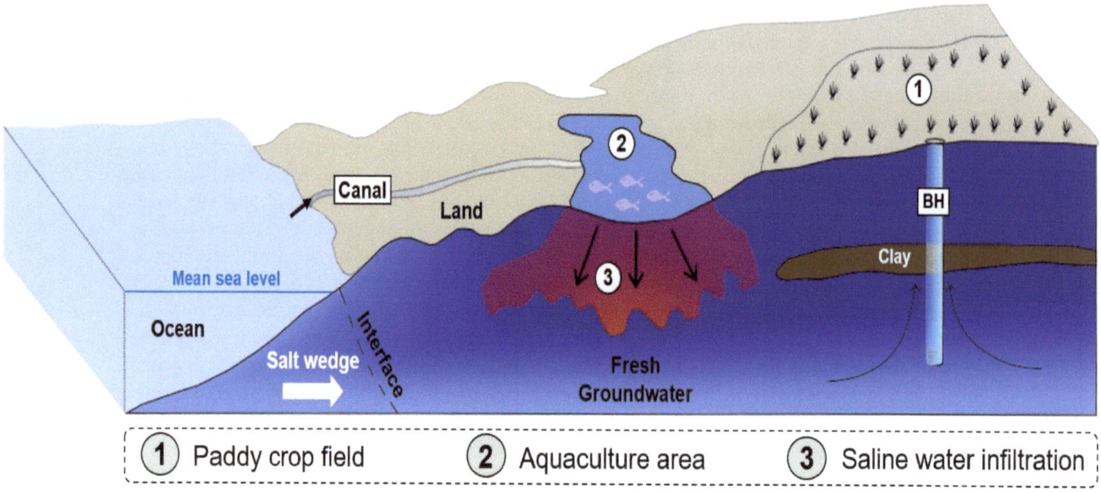

Figure 1. The conceptual model illustrating the saltwater intrusion scenario in the coastal area of West Bengal, India. Aquaculture activities lead to saltwater infiltration, where saltwater is collected through connected canals to the ocean. The saltwater intrusion is exacerbated by groundwater overpumping from boreholes (BHs) for paddy crop cultivation. This overextraction accelerates the rate of salinity incursion, resulting in increased contamination of groundwater quality with excessive salinity [33].

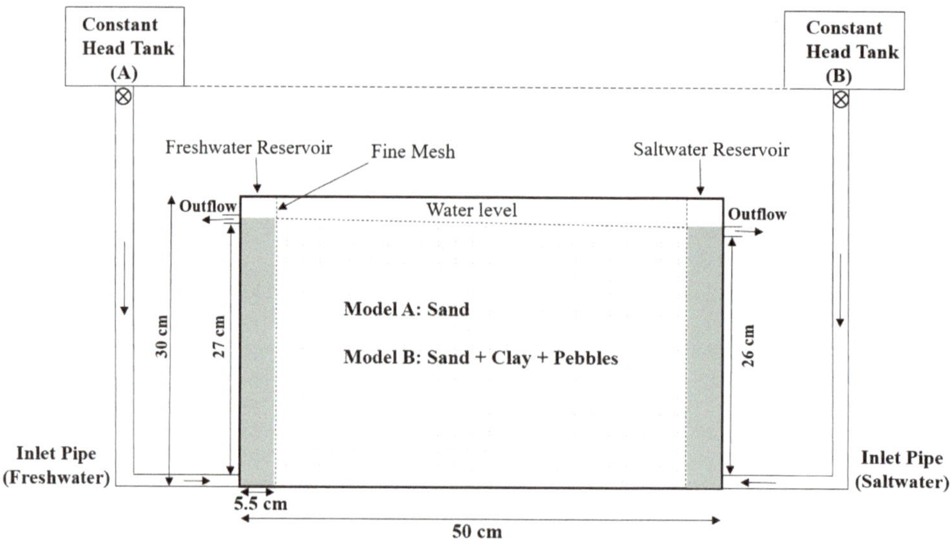

Figure 2. Schematic diagram of the experimental setup consisting of different lithologies, such as Model A (homogeneous) and Model B (heterogeneous).

The saltwater was prepared in a 50-litre barrel using commercial salt. To maintain the level of salinity (close to that of seawater), we dissolved 35 g of salt in 1 L of tap water. Saltwater was dyed with a carmine color in a 1 g/L solution concentration to distinguish it from freshwater. Instead of ordinary dye for mixing with saltwater, we used carmine color, as it does not show any adsorption effect on aquifer medium, and carmine can migrate at the same rate as cl⁻ ions [34]. The saltwater density was maintained at 1.025 kg/L and measured with a WKM hydrometer. The grid marking was performed at the base of the flow tank to monitor the saltwater flow. Two models were adopted for experimentation to simulate natural subsurface conditions. Model A is homogenous, in which sand (fine-to-medium grain) was used as a porous filled material, and Model B is heterogeneous, in which different layers of sand, clay, and sand are mixed with pebbles at various depths. Sand grains were delicately compressed to avoid air-filled voids and ensure the homogeneity of the media. For experimental purposes, natural sand was used instead of glass beads or silicon balls to improve the accuracy of replicating the geology of the real-field aquifer. A Terrascience instrument acquired the DC resistivity measurement using an external 90 V battery supply. Thin stainless-steel electrodes were used as two current electrodes (C1, C2) and two potential electrodes (P1, P2) in a dipole–dipole array. A high-resolution DSLR camera was used throughout the experiment to monitor the saltwater movement.

2.2. Experimental Procedures

The experimental procedure resembles that of prior research, although most previous studies only focused on homogeneous cases. This study conducted experiments for two scenarios to achieve optimal responses under controlled conditions. The middle chamber of the flow tank was filled with porous material (sand) (homogeneous case) and clay, sand, and pebbles (heterogeneous case) in multiple horizontal layers, as shown in Figure 3. Before the actual saltwater experiment started, the system was set to allow freshwater flow from left to right (to the saltwater chamber) under gradient conditions (Δh = 1 cm). Excess freshwater was allowed through the valves fixed at different heights.

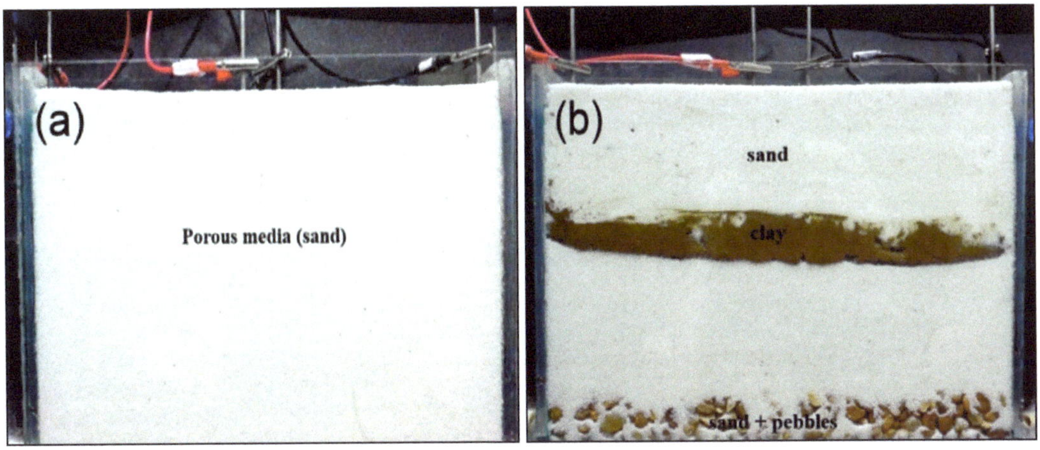

Figure 3. (**a**) Model A: Image of the experimental setup consisting of sand as homogeneous background material; (**b**) Model B: Image of the experimental setup consisting of heterogenous layers of sand, clay, and pebbles. On the upper side of the box, four stainless steel electrodes were used as the current pair (C1, C2) and the potential pair (P1, P2) for measuring the DC resistivity.

After achieving a steady equilibrium of freshwater flow from left to right, the intrusion experiment was initiated by opening the valve from constant saltwater head tank B. It was observed that saltwater rapidly flushed out the freshwater from the right chamber and began to invade the porous medium. Due to concentration differences, saltwater slowly migrated towards the freshwater chamber. As it was a density-dependent progression, the complete experiment was recorded, and high-resolution time-lapse images were captured to delineate freshwater and saltwater. During the investigation, considerable mixing of freshwater–saltwater flow was observed until the system achieved a steady-state condition.

After reaching a stable state, there were no further observational changes in the location of the saltwater wedge.

2.3. Numerical Modeling

In the realm of mathematical formulations regarding the dynamics of coastal salinity, the Henry problem stands out as a widely recognized and accepted tool by numerous researchers [8]. This problem aims to streamline the experimental behavior by focusing on a vertical SWI near coastal aquifers. In this context, a balance is maintained between the inland flow of freshwater and the intrusion of seawater from the coast, until it is disturbed by an external factor. The aquifer is presumed to exhibit homogeneity and isotropy. The SWI is influenced by various factors, including anisotropy [35], heterogeneity [36,37], hydraulic conductivity [38], dispersivity [39], and the impact of the inland boundary conditions on SWI [40]. The Henry problem (HP)'s formulation is based on the concept of density-dependent flow, which involves the integration of variable-density flow equations, an advection–dispersion equation, and mixture density as a function of saltwater concentration. The flow system is governed by Darcy's law in the following manner:

$$v = -K\left(\nabla H + \frac{\rho_{mix} - \rho_f}{\rho_f} g_z\right) \tag{1}$$

where v is Darcy's velocity (m/s), K is the hydraulic conductivity tensor (m/s), H is the hydraulic head (m), ρ_{mix} is the mixed fluid density (kg/m^3) which depends on concentration c, ρ_f is the freshwater density (kg/m^3), and g_z is the unit vector corresponding to the direction of gravity.

$$\rho_{mix} = \rho_f\left(1 + \alpha \frac{c}{c_s}\right) \tag{2}$$

where c_s is the concentration of seawater and $\alpha = \left(\rho_s - \rho_f\right)/\rho_f$

The medium's pressure gradient, fluid viscosity, and porosity structure influence the Darcy velocity field. The simulation modules provide the facility to solve Darcy's law, where diluted species are governed by diffusion and convection processes. The species are assumed to be diluted in such a way that density and viscosity are consistent in the mixture. The mass balance equation used for such a system is as follows:

$$V.\nabla c = \nabla.(D_F \nabla c) \tag{3}$$

where V is the flow velocity (m/s) obtained using Darcy's law, c is the species concentration (mol/m^3), and D_F is the diffusion-dispersion tensor (m^2/s).

The initial model parameters used in the present study for generating the SWI flow model are given in Table 1 and Equations (1)–(3) were implemented for numerical simulation are inspired by simple aquifer conditions [41]. The "subsurface flow" module environment of COMSOL version 4.4 was used, with boundary conditions set to no flow for both upper and lower boundary faces. The numerical solutions for SWI problems have previously been compared with semianalytical solutions and are well documented [42]. For a better approximation of the solution, the system was discretized and prepared to solve the problem using a finer mesh, as depicted in Figure 4. Finite element analysis (FEA) in COMSOL involves the use of the finite element method (FEM) to solve such fluid flow problems. FEM and appropriate mesh selection can effectively be used in simulations and analyses with high accuracy and computational efficiency. The numerical modeling approach offers a more comprehensive understanding of subsurface density-dependent flow conditions, particularly in addressing challenges such as salinity intrusion in coastal regions [43–53].

Table 1. Summary of initial simulation model parameters used for this study.

Input Parameters	Value
Porosity, ϕ	0.35
Freshwater density $(\mathrm{kg/m^3})$ ρ_f	1000
Saltwater density $(\mathrm{kg/m^3})$, ρ_s	1025
Saltwater concentration $(\mathrm{mol/m^3})$, c	1
Inflow velocity $(\mathrm{m/s})$, V_{Inflow}	3.3×10^{-5}
Pressure (Pa), P	0
Permeability $(\mathrm{m^2})$, k	1.02×10^{-9}
Fluid diffusion coefficient $(\mathrm{m^2/s})$, D	1.886×10^{-6}
Dynamic viscosity (kg/m.s), μ	0.001
Gravity $(\mathrm{m/s^2})$, g	9.8
Model Dimension	
Length, x (m)	0.39
Height, y (m)	0.27

Note: For each input parameter, symbols are included.

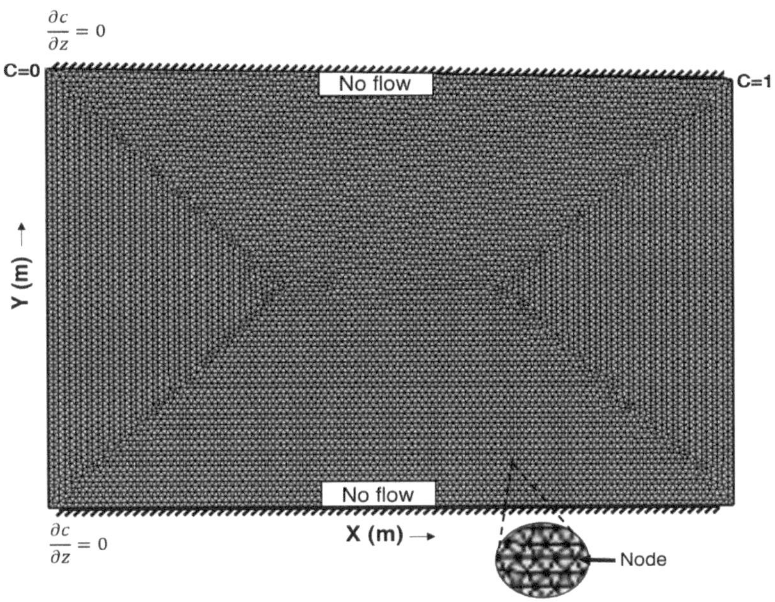

Figure 4. A finer mesh was used for the simulation, incorporating boundary conditions such as no flow (on the upper and lower faces), freshwater concentration (c = 0) at the left boundary, and saltwater concentration (c = 1) at the right boundary.

2.4. DC Resistivity Sounding

The resistivity data were collected once the freshwater–saltwater interface reached a steady state. The direct current (DC) resistivity method was used to determine the depth of the freshwater–saltwater interface. This method works on principle so that two current electrodes (C1, C2) are used to inject current (I) into the subsurface, and two potential electrodes (P1, P2) measure the potential difference (ΔV) generated due to the interaction of current lines with different geological layers. The ratio of potential difference and current provides the resistance value (R), which is multiplied by the geometric factor (G) to calculate the apparent resistivity (ρ_{a}) value.

The equation used to calculate apparent resistivity (ρ_a) is as follows:

$$\rho_a = G\left(\frac{\Delta V}{I}\right) \tag{4}$$

where the geometrical factor (G) is the linear arrangement of electrodes for the dipole–dipole array; G = πn (n + 1) (n + 2) a, where n is the dipole separation factor that varies from n = 1, 2, 3. . .; and a is electrode spacing.

Under the DC resistivity method, the Schlumberger array is conventionally used for vertical electrical sounding (VES) purposes to achieve a greater depth of investigation. However, it is less sensitive in identifying inclined subsurface bodies. Therefore, we have used a dipole–dipole array to investigate the saltwater wedge in this experiment (Figure 5). This array has greater sensitivity to detect the lateral resistivity variation and can detect vertical/inclined subsurface features with greater accuracy. The inclined features are progressive salinity contours associated with the freshwater–saltwater interface. The data were collected by readings in both forward and reverse modes to enhance the accuracy of resistivity measurements. After obtaining the average of these readings, the measured data were obtained. The current signal strength was also improved during the experiment by connecting the DC power supply batteries in series. This adjustment was necessary because the current was significantly attenuated due to the highly conductive nature of the host medium (water-saturated). A forward model response was initially generated using Res2dmod ver. 3.03 (Geotomosoft Solutions, Malaysia). For the dipole–dipole array, data points were measured along the center of the profile (vertical cross-section). The observed resistivity data were processed, and smooth models were obtained using Occam's inversion method [54,55].

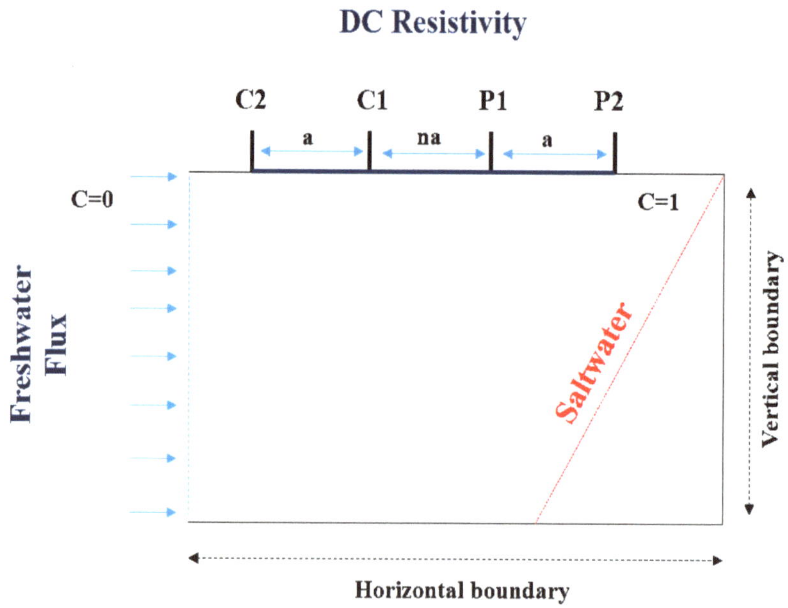

Figure 5. An illustrative arrangement of the dipole–dipole array used for the SWI experiment, where a is the distance between the electrodes and n = 1,2,3. . .. is the dipole separation factor. The freshwater flux concentration is kept at 0, while the saltwater flux boundary concentration is 1.

Some researchers have combined VES with other methods, such as time-domain electromagnetic (TDEM) techniques, to enhance our understanding of salinity intrusion in coastal aquifers [56,57]. In the present study, the multielectrode ERT field data collected for validation purposes using ABEM Terrameter with a Wenner array (Location 1) and Wenner–Schlumberger array (Location 2). After data acquisition, processing and interpretation were conducted utilizing Res2Dinv ver. 3.71 (Geotomo Software, Malaysia). The field data were inverted using a smoothness-constrained least-squares method. The choice of a suitable inversion scheme is crucial in obtaining high-resolution inverted subsurface images [58,59]. The forward model successfully detected a lateral change in resistivity distribution, enabling it to identify the inclined interface between freshwater and saltwater. The forward response generated is shown as apparent resistivity pseudosection (X vs. Ps.Z), which can later be used in an inversion engine to obtain an actual subsurface model (Figure 6).

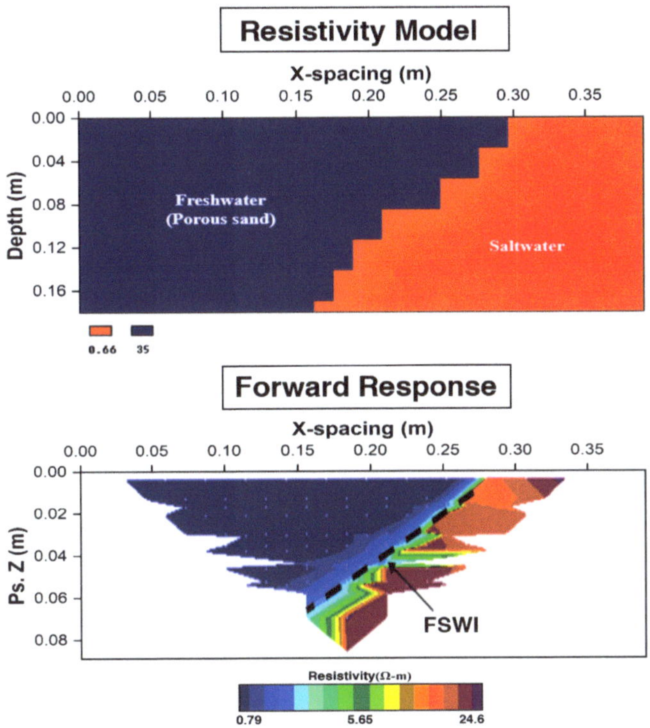

Figure 6. A forward model response was generated for the FSWI using the dipole–dipole array. The resistivity of salt water is 0.66 ohm-m, while the resistivity of freshwater sand is 35 ohm-m.

3. Results and Discussion

3.1. Qualitative Observations

The saltwater intrusion dynamic was examined for both Model A and B; the comparative movement of the saltwater wedge is shown in Figure 7. This approach was designed to enhance our understanding of the flow pattern and solute transport associated with the system. It was generated using a head difference of $\Delta h = 1$ cm between the left (freshwater) and right (saltwater) sides. When the experiment began, the time window was initiated and continuously recorded as the saltwater wedge (SW) advanced throughout the experiment. For Model A, the SW flow was smooth, and after 5 min, the initial height (y) of the SW was measured at 0.05 m, while its lateral extent (x) was found to be 0.4 m, as shown in Figure 7a. The FSWI SS-1 crossed the centre of the experimental box after 30 min, with a height of 0.23 m and a lateral extent of 0.14 m, as shown in Figure 7b. Model B began with a smooth progression of the SW, although minor shape alterations were attributed to the sand and pebble layer at the bottom of the container, as shown in Figure 7c. When the SW reached a height of 0.13 m and encountered a clay layer, the interface was disturbed, and the SW rate became slow compared to that of Model A. SS-2 developed after 40 min, showing a deviation from the expected smooth pattern.

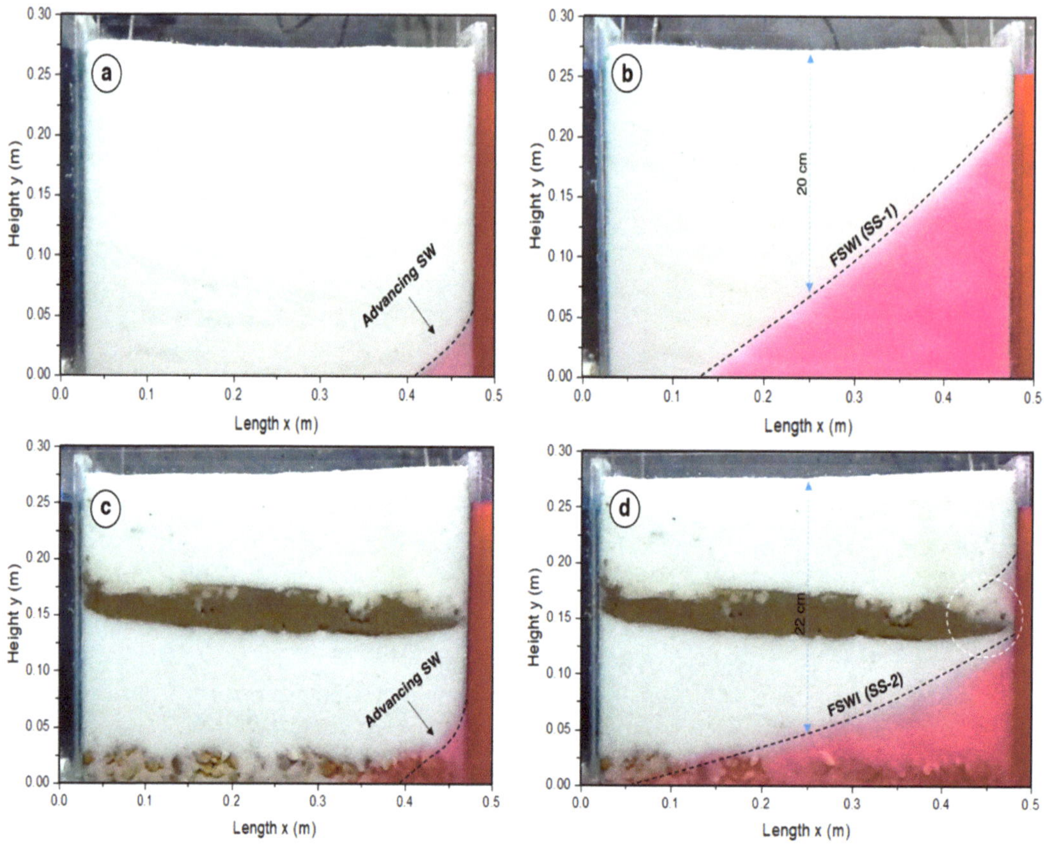

Figure 7. (**a**) Picture shows the advancement of the saltwater wedge (SW) for homogeneous Model A; (**b**) Picture shows the freshwater–saltwater interface (FSWI) SS-1 for homogeneous Model A after 30 min; (**c**) Picture shows the advancement of the SW for inhomogeneous Model B; (**d**) Picture shows the FSWI SS-2 for inhomogeneous Model B after 40 min. The disturbed FSWI, resulting from the presence of a tightly bound clay layer, is indicated by a dashed white circle. The depth of the FSWI is 20 cm for Model A and 22 cm for Model B.

The deviation was likely triggered by a tightly packed clay layer with a low porosity exhibiting a nearly impermeable behaviour. A small interface developed on the upper-right side of the box, as shown in Figure 7d, possibly due to the slightly porous clay layer that allowed its development. The comparative height and lateral positions of the SW were observed for both models, and a subsequent analysis revealed distinct placements once a stable interface was established. At a lateral distance (x) of 0.4 m, the SW for Model A reached a height of 0.17 m, while for Model B it was just 0.10 m. This was probably due to the existence of a clay layer that hindered the development of the contact. According to our observations, the clay layer behaves like a natural barrier, which does not allow the saline water easy passage into the freshwater zone, depending on the degree of saturation within the layer.

After reaching a steady condition, the SW was observed to have different heights for Models A and B, with δ representing the relative difference in their height. The development of saline contours with respect to length (m) and their respective heights (m) was predominantly influenced by subsurface heterogeneity, notably the presence of a clay layer (Figure 8). The advancement of the SW depends on the medium's geological composition, specifically whether it moves smoothly or encounters any obstacles. The key factors include the porosity (ϕ) and permeability (k) of the geological layer. A tightly packed layer with less pore space will not be easily penetrated by the saline water, whereas a loosely packed layer with more pore space will allow the saline water to penetrate it more easily.

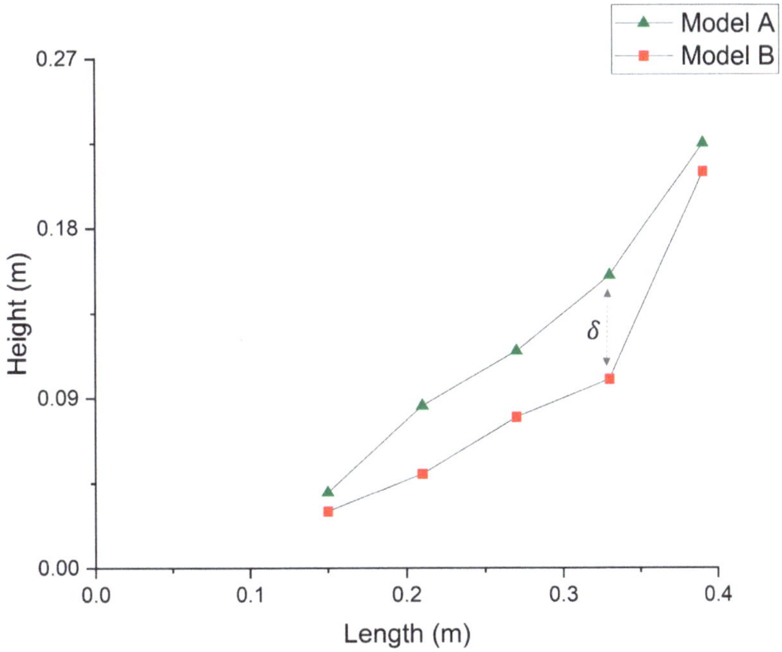

Figure 8. The plot represents the SW (height vs. length) for Models A and B when it reaches the steady state, and the δ is the relative difference in SW height.

The experimental behaviour exhibited a two-dimensional salinity incursion pattern across various geological conditions. A stable equilibrium was achieved between the freshwater and saltwater under steady-state conditions despite variations in the shape of the interface. However, it is important to understand that the occurrence of three-dimensional heterogeneity in coastal aquifers may not behave in the same way as two-dimensional heterogeneity. An effort was undertaken to investigate the response of various geological layers to SWI.

3.2. Numerical Model Based on Experimental Feedback

In this section, the experimental results obtained were compared to the numerical model to establish the relationship between the two. The numerical model was created for the homogeneous and heterogeneous scenarios using the initial model parameters (Table 1) shown in Figure 9. For Model A, the flow of the SW was smooth as it advanced in the homogenous media (Figure 9a). The 17% contour line reached its lateral position (x) at 0.14 m after 30 min (Figure 9c). For Model B, the 17% contour line crossed the lateral position

(x) of 0.10 m after 40 min but with a deviated shape (Figure 9d). For Model B, a numerical layer (L1) was created to have the similar properties of a geological clay layer, being both porous and impermeable (Figure 9b). To justify these properties, we assigned the upper and lower boundaries of the L1 layer as impermeable while keeping the left and right sides as open boundaries and allowing freshwater and saltwater flux to enter the layer from the left and right sides, respectively.

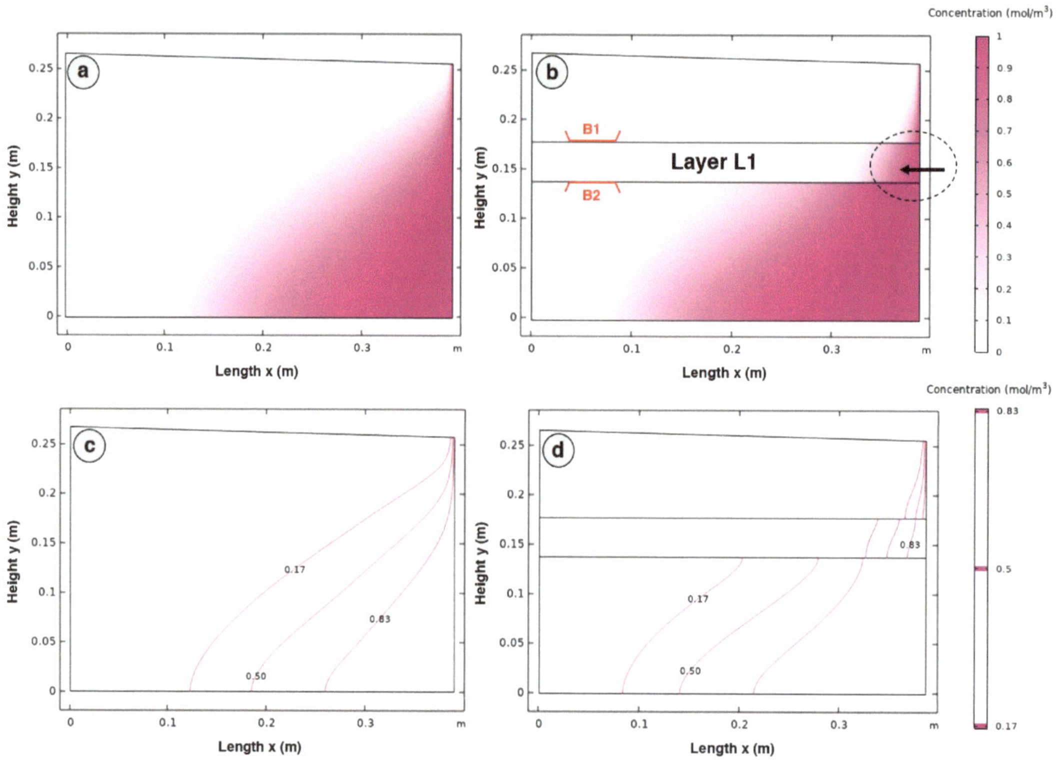

Figure 9. (**a**) The surface concentration (mol/m^3) plotted for homogeneous Model A showing the variation in salinity during the experiment; (**b**) The surface concentration (mol/m^3) plotted for homogeneous Model B showing the variation in salinity during the experiment. Layer L1 represents a numerically created layer that behaves like a porous yet impermeable stratum, bounded by upper and lower layers, denoted as B1 and B2, respectively. A dotted black circle with an arrow indicates the path through which saltwater flux enters this porous formation. The deviation of FSWI can also be observed above and below this layer. (**c**) Progressive isochlor contour lines (red color) of 83%, 50%, and 17% for Model A; (**d**) Progressive isochlor contour lines (red color) of 83%, 50%, and 17% for Model B.

However, as shown in Figure 3b, the clay layer used during the experiment is tightly packed (sandwiched between an upper sand and a lower sand mixed with pebbles), reducing its porosity. The surrounding compactness affects a layer's porosity (ϕ). Here, the clay layer behaves as an impermeable layer with a lower porosity. Meanwhile, there is a constraint with the numerical layer (L1) as its behaviour slightly differs from that of the experimental clay layer. This discrepancy arises from the boundary conditions (BCs) chosen for layer L1—impermeable on the upper and lower boundaries and open on the left and right sides. In contrast, an actual geological clay layer is porous and impermeable on all surface boundaries, making it difficult to assign the same boundary conditions in a numerical model. If we were to assign such boundary conditions to the numerical layer (L1), we would not be able to solve the 2D freshwater–saline water flow problem. Therefore, while a numerical layer cannot fully replicate the actual behaviour of a clay layer, an attempt can be made to observe the behaviour of the FSWI across impermeable boundaries.

The isochlor contour lines were plotted for both models with concentrations of 83%, 50%, and 17% (Figure 9c,d). These contour lines represent the decreasing concentration levels, expressed as a percentage, of the saline water flux (initially at concentration c = 1) as it permeates the medium against the freshwater flux.

The computational model exhibits a strong correlation with the experimental images in the case of Model A, where the FSWI crossed the centre of the experimental box after 30 min at a lateral extent (x) of around 0.14 m. In both the experimental and the model images, the SWs had relatively comparable heights at the centre of the vertical cross-section. For Model B, both the experimental and model images validate the deviation of the interface between freshwater and saltwater when it comes into contact with the clay layer (referred to as layer L1 in the model). However, there was a notable disparity in the height and the lateral extent of the SW between the experimental and model images, particularly evident at the centre of the profile. This significant variation in the SW can be attributed to an impermeable clay layer in the middle of the box. The behaviour of this layer differs between the experimental setup and the numerical model boundaries. For Model B, the numerical image depicts a greater SW height at the centre of the experimental box compared to that of the experimental image. The alignment between the experimental and numerical model results is illustrated in Figure 10. For Model A, E1, E2, and E3 denote experimental images taken at various x (m) positions, corresponding to numerical models N1, N2, and N3, respectively. Similarly, for Model B, E4, E5, and E6 represent experimental images captured at different x (m) positions, corresponding to numerical models N4, N5, and N6. These observations lead to two significant findings. First, it was observed that when advancing salt contours encounter impermeable layers like clay, the interface deviates from its initial trajectory. Additionally, the porous characteristics of the layer (such as clay) were observed to influence the height of the saltwater wedge. Specifically, when the layer is densely packed, the height of the saltwater wedge decreases at the same lateral distance. In contrast, a clay layer with a high level of porosity can lead to a greater height of the saltwater wedge, allowing for a smooth build-up of the interface.

3.3. Vertical Electrical Sounding (VES)

The DC resistivity data were obtained using the VES method once a steady-state equilibrium was achieved in both Model A and Model B. Data were collected along the center of the profile using a dipole–dipole array, with each reading noted in both the direct and reverse modes of the current (I) direction. The average value was plotted for each particular data point for better accuracy. Each observed data point for both Models A and B in Figure 11a,c represents the apparent resistivity value corresponding to the N-spacing (m) along the center of the experimental box. Figure 11b,d depict the one dimensional inverted response obtained from those data points, illustrating the variation in apparent resistivity values along the center of the profile with corresponding depth for experimental Models A and B.

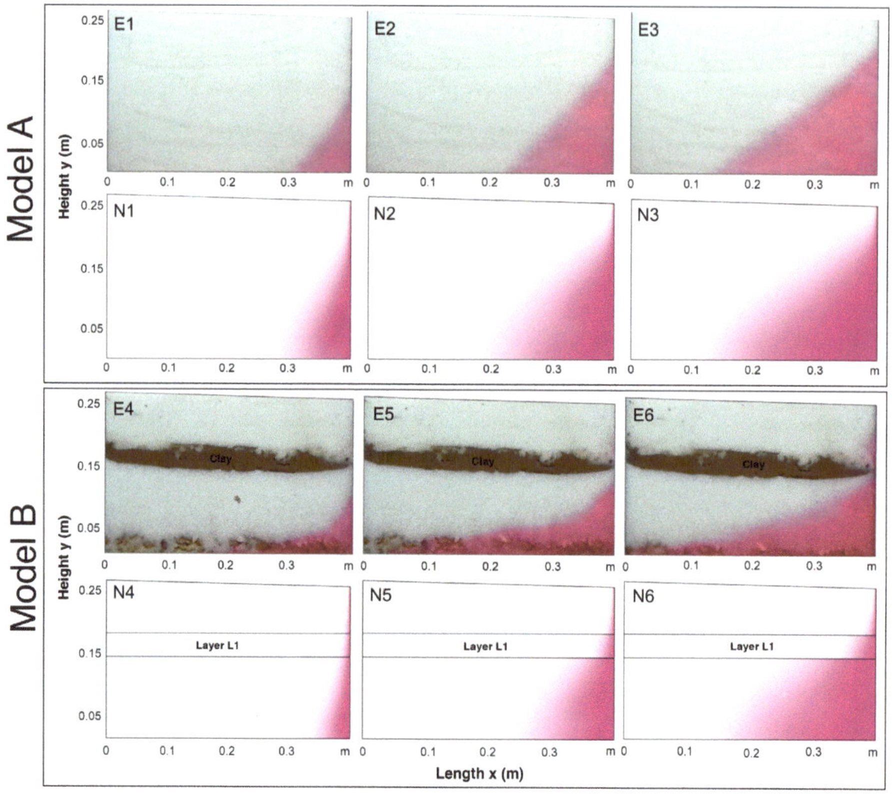

Figure 10. The comparative figures for the advancement of saltwater concentration with respect to different x (m) positions. For Model A, the experimental images are represented by E1, E2, and E3, corresponding to numerical model images N1, N2, and N3 respectively. For Model B, the experimental images are represented by E4, E5, and E6, while the numerical model images are represented by N4, N5, and N6.

For Models A and B, the inverted response is plotted using a smooth-layer model rather than a layered model. The noise-free data were acquired for model A due to having a consistent sand layer throughout its volume. In contrast, model B exhibited various layers, such as sand, clay, and sand mixed with pebbles, which can lead to an abrupt change in apparent resistivity values and thus may lead to noisy data points. If a layered model had been used, we would have encountered a high fitting error for the noisy points and would not have accurately represented the actual variation in resistivity values with respect to depth. Therefore, smooth layers were used for both models to obtain a better fit for such noisy points. Also, for a better comparison between models A and B, we maintained both models as smooth-layer models.

Model A shows a consistent and gradual change in the resistivity data due to the homogeneous geological medium (sand) within the box. The resistivity value obtained for freshwater-saturated sand was approximately 54 ohm-m at a depth of 3 cm below the measurement surface of the box. The depth of each layer was measured from the point at which the resistivity data were initially measured using the potential electrodes. It was observed that beyond the 17 cm mark, the resistivity begins to decrease upon encountering the FSWI. Beyond this depth, the measured resistivity was 0.72 ohm-m due to the saline sand, which indicates that the saline sand exhibited a higher level of conductivity (less resistivity) than that of the freshwater-saturated sand. For Model B, initially, the data points exhibited smooth characteristics. However, sparse data points were observed at a certain depth, indicating noisy data excluded from the one-dimensional inversion process. The first layer consisted of freshwater-saturated sand, with a measured resistivity of 26.40 ohm-m at a depth of 4 cm. In the second layer, the resistivity decreased to 6.90 ohm-m, corresponding to a clay layer situated at an 8 cm depth. A significant decrease in the resistivity data was observed at a depth of 20 cm, indicating the presence of the FSWI. Below this depth, the measured resistivity was 1.77 ohm-m, which was attributed to the presence of saline sand mixed with pebbles.

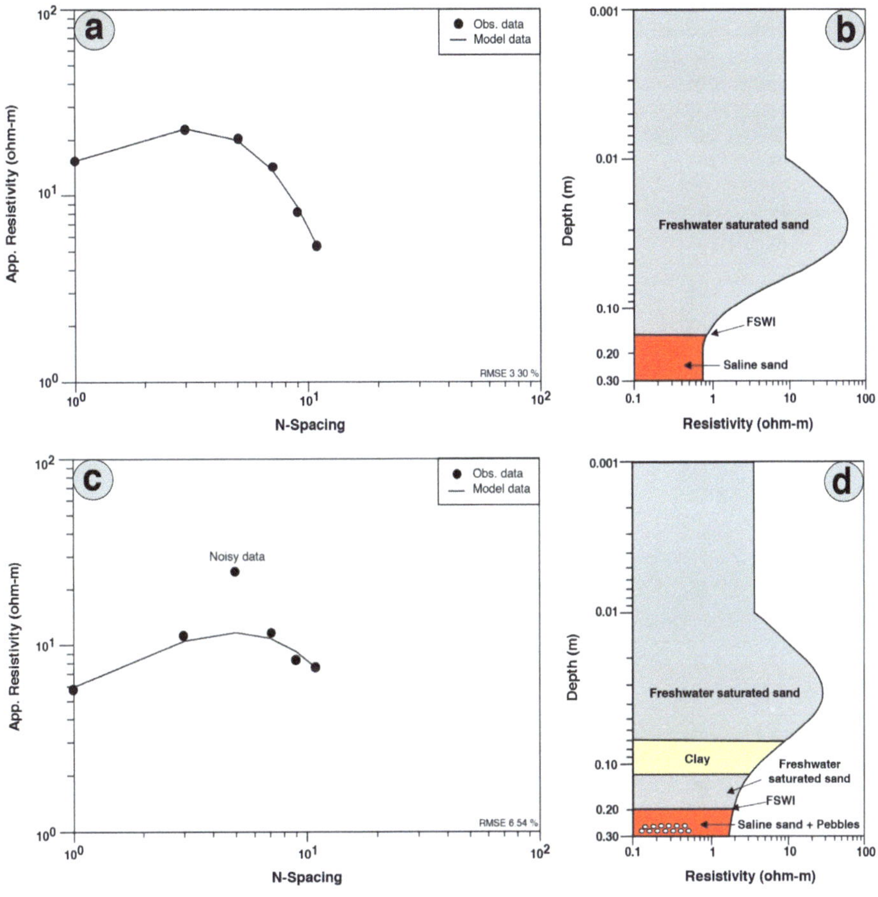

Figure 11. (**a**) Sounding data points acquired using the dipole–dipole array for Model A; (**b**) 1D inverted response obtained for homogeneous Model A; (**c**) Sounding data points acquired using the dipole–dipole array for Model B; (**d**) 1D inverted response obtained for heterogeneous Model B. The depth of the FSWI is approximately 17 cm in Model A and around 20 cm in Model B.

The observed resistivity data vary depending on the medium. In the case of Model A, the saline sand exhibited a resistivity value of 0.72 ohm-m. In contrast, for Model B, the resistivity was measured as 1.77 ohm-m, which is slightly higher due to the surrounding effect of the sand mixed with pebbles. The sounding results demonstrate a strong correlation with the experimental data. In the case of Model A, the interface depth at the center of the profile was approximately 20 cm according to the experimental image and 17 cm according to the depth derived from the one-dimensional inversion model. For Model B, the interface depth was approximately 22 cm according to the experimental image and 20 cm based on the one-dimensional inverted model. Slight discrepancies were noted with depth in the interface observations, as VES is an indirect geophysical measurement, and model error can arise from factors like data noise, signal attenuation, and errors in one-dimensional inversion, such as equivalence and suppression errors. However, our findings still fall within an acceptable range, and we successfully identified the FSWI utilizing the dipole–dipole array.

3.4. Validation with ERT Field Data for Location 1: Near Sea Coast

The ERT field data were used to validate the experimental findings. The correlation between the experimental and numerical results was evaluated by analysing the ERT subsurface image. The two-dimensional profile was acquired along the coast of the Mandarmani area of West Bengal in India, as illustrated in Figure 12a. This region is a tourist hotspot well known for its salt industries, aquaculture practices, and agricultural lands. Over the past few years, there has been a significant increase in water demand in this region.

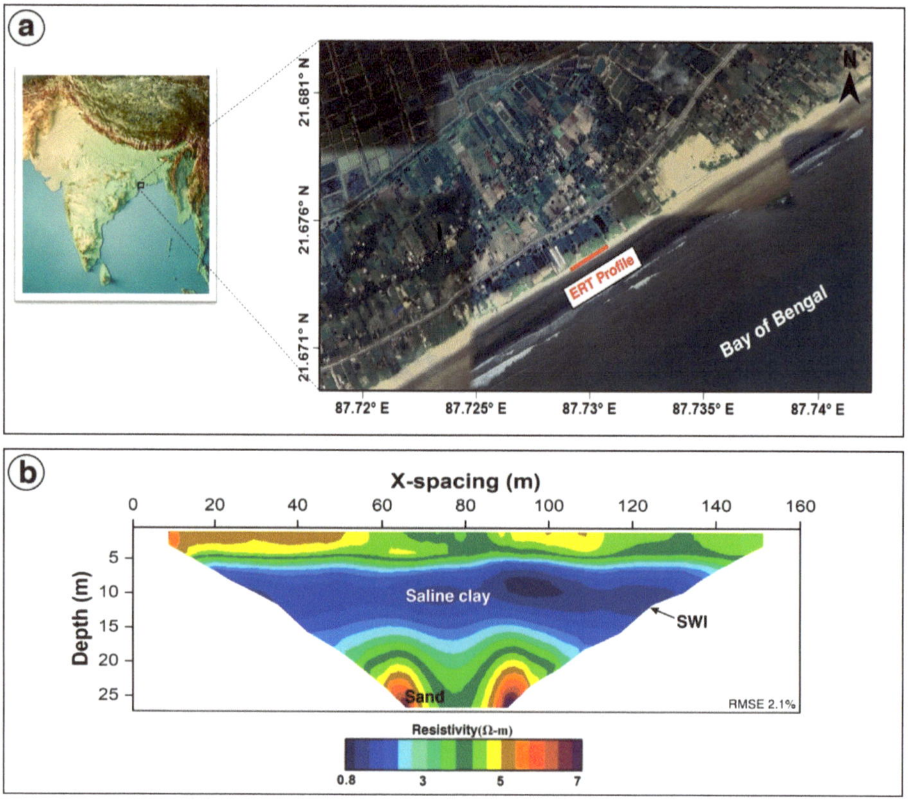

Figure 12. (**a**) Image showing the location of the ERT profile acquired near the coast of the Bay of Bengal, India; (**b**) The 2D inverted response of the ERT data shows the presence of a saline clay layer. RMSE stands for root mean square error.

Additionally, the overpumping of groundwater has been identified as a concern. Salinity intrusion presents a considerable challenge for the region, arising from natural phenomena and human interventions. The study area is characterized by alluvial deposits comprising sand, silt, and clay. Furthermore, an analysis of the borehole and lithology data in the area reveals that surface sediment deposits predominantly comprise medium-to-coarse sand interspersed with clay and small patches of dune sand [20]. The region is abundant in borewells utilized for agricultural practices, primarily for cultivating paddy crops and household water needs.

The ERT data were collected over a 160 m profile using a Wenner array, with electrodes positioned at 2 m intervals. This choice was made deliberately to maintain a fine electrode spacing to achieve high-resolution images. The inversion routine is based on the smoothness-constrained least-squares method with the L2 norm [60]. The two-dimensional inverted image reveals a prominent high-conductive zone with resistivity values ranging from 0.8 to 1.8 ohm-m, indicating the presence of a saline clay layer marked as a zone of SWI at a depth of 6 to 16 m below ground level. The sand layer at a depth of 20 to 26 m below ground level has a resistivity value of 6 to 7 ohm-m, as shown in Figure 12b. The area has resistivity values for a sandy freshwater formation that vary from 20 to 60 ohm-m. However, in ERT location 1, the shallow layers are completely saline (with resistivity values less than 1.8 ohm-m) for 6 to 16 m depths below ground level.

Saltwater intrusion naturally occurs near coastal areas. However, during the field survey, local people reported a high level of groundwater salinity in this coastal region. They have drilled borewells deeper than 20 m to access freshwater, as the shallow groundwater is completely saline. This issue arises from the extensive use of hand pumps and borewells in the confined coastal area. Overpumping lowers the water table, allowing saltwater from the sea to exert pressure and infiltrate deeper into freshwater aquifers. The ERT results confirm SWI in the coastal region, primarily caused by the disruption of the natural balance between saline and freshwater interfaces due to overpumping by residents.

Interestingly, it was observed that saltwater is trapped in unconfined shallow clay layers, preventing the further deterioration of groundwater in deeper sand layers. This study suggests that these clay layers can act as natural barriers, trapping saline water due to their porous but impermeable nature. However, the clay layer can only hold saline water up to a certain threshold. Excessive overpumping will increase salinity intrusion, causing the clay layer to fail to retain saline water and potentially impacting the surrounding layers. Similarly, the experimental results show that the tightly packed clay layer prevented the saltwater wedge from easily passing through, resulting in a deviated interface shape due to its impervious nature.

3.5. Validation with ERT Field Data for Location 2: Aquaculture Area

The ERT data were acquired in the aquacultural ponds with an area of 2.25 km^2 near the Mandarmani-Contai region of West Bengal in India, where they are artificially formed with seawater collected through a network of interconnected canals (Figure 13a). The area is committed to aquaculture practices focused on a unique breed of fish that thrives exclusively in saltwater environments. The geological area resembles fluvial deposits with clay, silt, and sand layers. The fish farming practices in this region serve as a crucial economic aspect of the local community's livelihood. Nonetheless, the extensive nature of such production methods also exerts a detrimental impact on environmental health.

The ERT data were collected over a 800 m profile using a Wenner–Schlumberger array with 10 m electrode spacing. The data were inverted using the smoothness-constrained least-squares method with the L2 norm. The two-dimensional inverted section reveals the presence of a saline zone (with levels of resistivity ranging from 0.9 to 2 ohm-m) extending to a depth of 40 m below ground level (Figure 13b), which correspond to silty clay layers. The deeper zone is indicated by a sand layer with resistivity values of more than 12 ohm-m. The resistivity values for the sandy freshwater formation in the area range from 20 to 60 ohm-m. It can be observed from the ERT result that deeper layers (more than 50 m below ground level) are not affected by saltwater intrusion as resistivity values are greater than 12 ohm-m. However, considering shallow layers, the likely cause of the heightened salinity in this area is attributed to aquacultural practices, particularly fish farming. In such practices, seawater is transported to artificial ponds through connected canals from coastal regions. A considerable portion of this saltwater infiltrates the subsurface. The two-dimensional section illustrates that saline water permeates silty clay layers. However, no further penetration of salt water into deeper sand layers was observed, likely due to the silty clay layer acting as an impermeable layer.

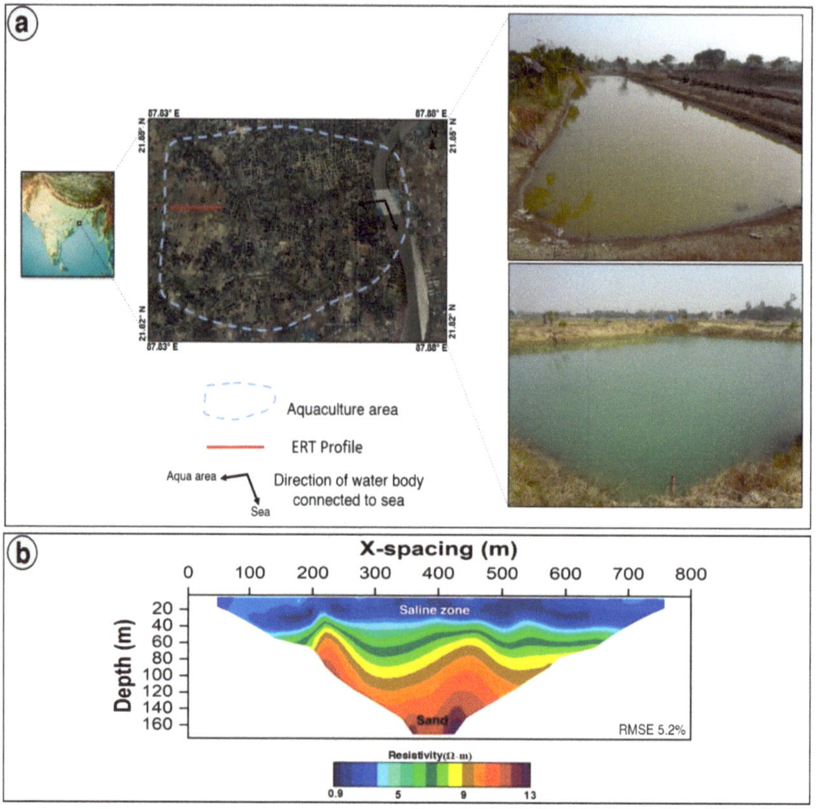

Figure 13. (**a**) The ERT profile of length of 800 m acquired along the aquaculture area of Mandarmani-Contai region, India, where seawater is collected through interconnected canals. The image also depicts many ponds filled with saline water specifically allocated for fish farming; (**b**) The inverted 2D response of the ERT data indicates a saline zone with silty clay layers and sand within the area. RMSE stands for root mean square error.

Researchers worldwide have reported that SWI has contributed to the shrinkage of lakes [61], degrading water quality and threatening freshwater coastal resources [62]. The present study can be useful as its originality lies in integrating experimental data with numerical modeling and geophysical techniques, offering a comprehensive analysis of SWI dynamics in homogeneous and heterogeneous models. This approach enhances our understanding SWI mechanisms, informing coastal groundwater management by predicting the impacts of groundwater extraction and aquaculture on SWI. The findings can be applied to other coastal regions with similar geological characteristics to the study area in West Bengal, India, such as the Nile Delta [63], the Gulf of Mexico [64], and the Mekong Delta [65], which face similar challenges of saltwater intrusion due to intensive groundwater extraction and agricultural practices. By adopting this integrated methodology, tailored groundwater extraction policies and land-use planning can be developed, improving the sustainability of coastal groundwater resources. This approach can inspire future research and the development of sustainable groundwater management practices in coastal areas globally.

3.6. Limitations of This Experimental Study and Future Directions

The present study introduces new insights by incorporating heterogeneous background materials in the laboratory simulation of SWI. In this study, the alignment of experimental and numerical model results was quite good, though some uncertainties remain due to the following factors: (1) The numerical models could better align with the experimental findings if the same type of filler material, such as clay, had been used in the numerical simulation. It can provide better results as, in our case, the layer (L1) with constrained boundary conditions was not able to fully justify the actual behavior of a clay layer; (2) Instead of using VES for the experimental study, which provides a one-dimensional inverted model, a two-dimensional ERT method could be adopted. This method offers a better resolution and accuracy in predicting the depth of the FSWI. Some researchers have used a multielectrode ERT setup for laboratory experiments [58], yielding promising results as they can observe the high-resolution two-dimensional resistivity distribution rather than the one-dimensional resistivity change. These one-dimensional inverted models can be affected by acquisition errors, cultural noise, and inversion errors, leading to false interpretations and inaccurate estimates of the actual depth of layers. This discrepancy is evident in our study, in which the FSWI depth is slightly mismatched between the experimental and one-dimensional results. These issues may need to be addressed to reduce uncertainties associated with such experimental studies.

4. The Significance of Research Outcomes in Informing Policies for Coastal Water Management and Sustainable Aquaculture Practices

The current research findings are crucial for shaping policies related to coastal water management and promoting sustainable practices in aquaculture ponds and lakes. In the study area of West Bengal, India, groundwater overpumping and aquaculture ponds have emerged as significant concerns due to elevated salinity levels in the region. By providing scientific evidence and insights into the environmental dynamics of coastal areas, research helps policymakers make informed decisions. Understanding factors such as saltwater intrusion, pollution levels, ecosystem health, and the impact of human activities allows for the development of effective management strategies. Moreover, research contributes to identifying best practices for sustainable aquaculture management, including aquaculture techniques, specifically fish farming, that minimize environmental degradation and maximize productivity. By integrating research outcomes into policy formulation, governments and organizations can work towards safeguarding coastal waters and preserving the ecological balance of ponds and lakes for future generations. The flowchart below illustrates the connection between sustainable water management (SWM) and sustainable aquaculture practices, which is crucial for the overall sustainability of the environment (Figure 14).

This study highlights the negative effects of artificial ponds and lakes used for aquaculture, especially in causing groundwater salinization. Effective governance of aquaculture practices is crucial in implementing measures that balance economic benefits with environmental preservation, ultimately contributing to sustainable water management efforts. This study employs an experimental approach to investigate salinity intrusion in diverse coastal environments. The ERT results validate the impact of SWI due to various anthropogenic activities. Additionally, such studies can aid in developing decision-making tools for policymakers to maintain coastal aquifer sustainability [66].

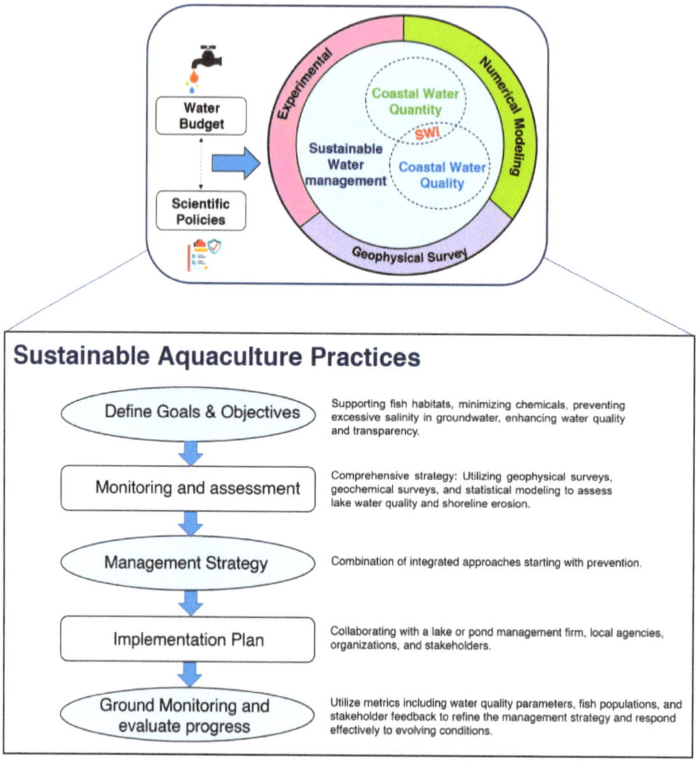

Figure 14. The flow chart illustrates sustainable aquaculture practices within the framework of sustainable water management (SWM).

The present study is also relevant to hydrogeology and environmental engineering, particularly for coastal groundwater management and mitigating saltwater intrusion. The integrated approach combines experimental data, numerical modeling, and geophysical techniques and offers valuable insights that can directly inform policymaking and sustainable coastal water management practices. By demonstrating how different geological conditions influence saltwater intrusion dynamics, this research provides a scientific basis for developing targeted groundwater extraction policies, implementing effective land-use planning, and designing physical barriers to protect freshwater resources. Emphasizing these potential applications can help policymakers and practitioners adopt more effective strategies for managing coastal groundwater resources and mitigating the adverse impacts of saltwater intrusion.

However, while the present study provides a better understanding of SWI behavior in a heterogeneous coastal environment, some mitigation measures still need to be physically implemented for the better prevention of SWI. Researchers have focused and aligned their experimental studies on physical barriers to minimize SWI [27,29]. Some physical and hydraulic management approaches can be used to mitigate SWI, such as abstraction barriers, cutoff walls, recharge wells, and tidal regulators. When exploring suitable locations for recharge wells, ERT combined with TDEM surveys can provide a better picture of the subsurface. Therefore, more geophysical studies are needed to complement laboratory experiments.

5. Conclusions

The present study employs an integrated approach, combining experimental, geophysical, and numerical modeling methodologies to gain insight into the dynamics of salinity intrusion in coastal aquifers. The experimental setup outlined in this study simulates various geological coastal scenarios by incorporating saltwater boundary conditions. A good agreement was found between the experimental results and models obtained from the numerical simulation and the DC resistivity method. The two critical findings indicate that the progression of the saline contours depends on the geological composition through which they propagate. First, homogeneous formations, such as sand, facilitate the unhindered infiltration of saline water into freshwater aquifers. Second, heterogeneous media featuring layers of sand, silt, and clay, especially impervious clay layers, act as natural barriers, impeding the advancement of saline water by trapping it within their structure.

This study offers valuable information to help us understand saltwater wedges' interaction and solute transport mechanisms under diverse geological conditions. In addition, this study serves as a valuable benchmark for validating density-coupled flow and transport models, particularly those that incorporate flux-type boundary conditions using different background materials. The ERT data also confirm SWI in coastal areas resulting from anthropogenic activities, such as groundwater overpumping (location 1) and aquacultural activities, leading to saline water infiltration into subsurface aquifers (location 2). This study also examines the behavior of clay layers, which intriguingly can function as natural barriers to salinity intrusion issues. However, it cannot be viewed as a complete solution to saltwater intrusion, as clay layers have saturation thresholds. Beyond these thresholds, they allow saline water infiltration into the surrounding layers. External physical barriers are necessary to prevent saline water intrusion into freshwater aquifers. Furthermore, to ensure groundwater sustainability, a continuous subsurface investigation is needed for monitoring the health of coastal aquifers. A comprehensive approach that can be adopted involves integrating a three-dimensional time-lapse ERT survey with a geochemical analysis.

The experimental findings hold validity for geological environments characterized by clay, silt, and sand layers and are applicable to coastal and river depositional settings, such as the coastline margin area of India, where the biggest challenge is saltwater intrusion. The present experimental model incorporating a heterogeneous geology can serve as input for future investigations. It can be enhanced by introducing physical barriers, pumping wells, recharge wells, and inclined slopes, facilitating multiple attributes for detailed experimental studies of SWI. The findings can be applied to other coastal regions with similar geological characteristics, such as delta regions worldwide, which face similar challenges of saltwater intrusion due to intensive groundwater extraction and agricultural practices; further remedial measures can be planned accordingly.

Author Contributions: Conceptualization, P.T. and R.R.; methodology, P.T. and R.R.; software, P.T.; validation, P.T., R.R. and S.P.S.; formal analysis, P.T.; investigation, P.T. and R.R.; resources, P.T., R.R., J.C. and S.P.S.; data curation, P.T. and R.R.; writing—P.T.; writing—review and editing, P.T. and J.C.; visualization, P.T., R.R. and J.C.; supervision, S.P.S. All authors have read and agreed to the published version of the manuscript.

Funding: This research received no external funding.

Data Availability Statement: The data presented in this study are available on request from the corresponding author. The data are not publicly accessible due to privacy concerns.

Acknowledgments: The authors would like to express their gratitude to the Department of Geology and Geophysics at the Indian Institute of Technology (IIT) Kharagpur, West Bengal, India, for generously providing the research facilities. We want to convey our sincere appreciation to Arkoprovo Biswas, at the Department of Geology, BHU, for providing insightful comments during the experimental study, and Rudra Mohan Pradhan, NPDF at IIT Kharagpur, for his invaluable guidance and support throughout the preparation of the manuscript.

Conflicts of Interest: The authors declare no conflicts of interest.

References

1. Neumann, B.; Vafeidis, A.T.; Zimmermann, J.; Nicholls, R.J. Future Coastal Population Growth and Exposure to Sea-Level Rise and Coastal Flooding—A Global Assessment. *PLoS ONE* **2015**, *10*, e0118571. [CrossRef]
2. Costall, A.; Harris, B.; Pigois, J.P. Electrical Resistivity Imaging and the Saline Water Interface in High-Quality Coastal Aquifers. *Surv. Geophys.* **2018**, *39*, 753–816. [CrossRef]
3. Hermans, T.; Paepen, M. Combined Inversion of Land and Marine Electrical Resistivity Tomography for Submarine Groundwater Discharge and Saltwater Intrusion Characterization. *Geophys. Res. Lett.* **2020**, *47*, e2019GL085877. [CrossRef]
4. Muzzillo, R.; Zuffianò, L.E.; Rizzo, E.; Canora, F.; Capozzoli, L.; Giampaolo, V.; De Giorgio, G.; Sdao, F.; Polemio, M. Seawater Intrusion Proneness and Geophysical Investigations in the Metaponto Coastal Plain (Basilicata, Italy). *Water* **2021**, *13*, 53. [CrossRef]
5. Oiro, S.; Comte, J.C. Drivers, Patterns and Velocity of Saltwater Intrusion in a Stressed Aquifer of the East African Coast: Joint Analysis of Groundwater and Geophysical Data in Southern Kenya. *J. Afr. Earth Sci.* **2019**, *149*, 334–347. [CrossRef]
6. Rasmussen, P.; Sonnenborg, T.O.; Goncear, G.; Hinsby, K. Assessing Impacts of Climate Change, Sea Level Rise, and Drainage Canals on Saltwater Intrusion to Coastal Aquifer. *Hydrol. Earth Syst. Sci.* **2013**, *17*, 421–443. [CrossRef]
7. Siemon, B.; Costabe, S.; Voß, W.; Meyer, U.; Deus, N.; Elbracht, J.; Günther, T.; Wiederhold, H. Airborne and Ground Geophysical Mapping of Coastal Clays in Eastern Friesland, Germany. *Geophysics* **2015**, *80*, WB21–WB34. [CrossRef]
8. Henry, H.R. Salt Intrusion into Fresh-Water Aquifers. *J. Geophys. Res.* **1959**, *64*, 1911–1919. [CrossRef]
9. Goswami, R.R.; Clement, T.P. Laboratory-Scale Investigation of Saltwater Intrusion Dynamics. *Water Resour. Res.* **2007**, *43*, W04418. [CrossRef]
10. Luyun, R.; Momii, K.; Nakagawa, K. Laboratory-Scale Saltwater Behavior Due to Subsurface Cutoff Wall. *J. Hydrol.* **2009**, *377*, 227–236. [CrossRef]
11. Thorenz, C.; Kosakowski, G.; Kolditz, O.; Berkowitz, B. An Experimental and Numerical Investigation of Saltwater Movement in Coupled Saturated-Partially Saturated Systems. *Water Resour. Res.* **2002**, *38*, 5-1–5-11. [CrossRef]
12. Zhang, Q.; Volker, R.E.; Lockington, D.A. Influence of Seaward Boundary Condition on Contaminant Transport in Unconfined Coastal Aquifers. *J. Contam. Hydrol.* **2001**, *49*, 201–215. [CrossRef] [PubMed]
13. De Franco, R.; Biella, G.; Tosi, L.; Teatini, P.; Lozej, A.; Chiozzotto, B.; Giada, M.; Rizzetto, F.; Claude, C.; Mayer, A.; et al. Monitoring the Saltwater Intrusion by Time Lapse Electrical Resistivity Tomography: The Chioggia Test Site (Venice Lagoon, Italy). *J. Appl. Geophys.* **2009**, *69*, 117–130. [CrossRef]
14. Gopinath, S.; Srinivasamoorthy, K.; Saravanan, K.; Suma, C.S.; Prakash, R.; Senthinathan, D.; Sarma, V.S. Vertical Electrical Sounding for Mapping Saline Water Intrusion in Coastal Aquifers of Nagapattinam and Karaikal, South India. *Sustain. Water Resour. Manag.* **2018**, *4*, 833–841. [CrossRef]
15. Hamzah, U.; Samsudin, A.R.; Malim, E.P. Groundwater Investigation in Kuala Selangor Using Vertical Electrical Sounding (VES) Surveys. *Environ. Geol.* **2007**, *51*, 1349–1359. [CrossRef]
16. Niculescu, B.M.; Andrei, G. Using Vertical Electrical Soundings to Characterize Seawater Intrusions in the Southern Area of Romanian Black Sea Coastline. *Acta Geophys.* **2019**, *67*, 1845–1863. [CrossRef]
17. Zarroca, M.; Bach, J.; Linares, R.; Pellicer, X.M. Electrical Methods (VES and ERT) for Identifying, Mapping and Monitoring Different Saline Domains in a Coastal Plain Region (Alt Empordà, Northern Spain). *J. Hydrol.* **2011**, *409*, 407–422. [CrossRef]
18. Goebel, M.; Pidlisecky, A.; Knight, R. Resistivity Imaging Reveals Complex Pattern of Saltwater Intrusion along Monterey Coast. *J. Hydrol.* **2017**, *551*, 746–755. [CrossRef]
19. Inim, I.J.; Udosen, N.I.; Tijani, M.N.; Affiah, U.E.; George, N.J. Time-Lapse Electrical Resistivity Investigation of Seawater Intrusion in Coastal Aquifer of Ibeno, Southeastern Nigeria. *Appl. Water Sci.* **2020**, *10*, 232. [CrossRef]
20. Kumar, P.; Tiwari, P.; Singh, A.; Biswas, A.; Acharya, T. Electrical Resistivity and Induced Polarization Signatures to Delineate the Near-Surface Aquifers Contaminated with Seawater Invasion in Digha, West-Bengal, India. *Catena* **2021**, *207*, 105596. [CrossRef]
21. Walther, M.; Delfs, J.O.; Grundmann, J.; Kolditz, O.; Liedl, R. Saltwater Intrusion Modeling: Verification and Application to an Agricultural Coastal Arid Region in Oman. *J. Comput. Appl. Math.* **2012**, *236*, 4798–4809. [CrossRef]
22. Koohbor, B.; Fahs, M.; Ataie-Ashtiani, B.; Belfort, B.; Simmons, C.T.; Younes, A. Uncertainty Analysis for Seawater Intrusion in Fractured Coastal Aquifers: Effects of Fracture Location, Aperture, Density and Hydrodynamic Parameters. *J. Hydrol.* **2019**, *571*, 159–177. [CrossRef]
23. Jakovovic, D.; Werner, A.D.; Simmons, C.T. Numerical Modelling of Saltwater Up-Coning: Comparison with Experimental Laboratory Observations. *J. Hydrol.* **2011**, *402*, 261–273. [CrossRef]
24. Mehdizadeh, S.S.; Vafaie, F.; Abolghasemi, H. Assessment of Sharp-Interface Approach for Saltwater Intrusion Prediction in an Unconfined Coastal Aquifer Exposed to Pumping. *Environ. Earth Sci.* **2015**, *73*, 8345–8355. [CrossRef]
25. Zhu, S.; Zhou, Z.; Werner, A.D.; Chen, Y. Experimental Analysis of Intermittent Pumping Effects on Seawater Intrusion. *Water Resour. Res.* **2023**, *59*, e2022WR032269. [CrossRef]
26. Dalai, C.; Dhar, A. Impact of Beach Face Slope Variation on Saltwater Intrusion Dynamics in Unconfined Aquifer under Tidal Boundary Condition. *Flow Meas. Instrum.* **2023**, *89*, 102298. [CrossRef]
27. Abd-Elaty, I.; Kuriqi, A.; Pugliese, L.; Ahmed, A. Shoreline Subsurface Dams to Protect Coastal Aquifers from Sea Level Rise and Saltwater Intrusion. *Appl. Water Sci.* **2024**, *14*, 49. [CrossRef]

28. Sharma, V.; Chakma, S. Experimental Investigation to Check the Feasibility of an Under-Surface Barrier for Stratified Layers to Remediate Seawater Intrusion Considering an Inclined Boundary. *J. Hydrol.* **2024**, *636*, 131283. [CrossRef]
29. Abdoulhalik, A.; Ahmed, A.A. The Effectiveness of Cutoff Walls to Control Saltwater Intrusion in Multi-Layered Coastal Aquifers: Experimental and Numerical Study. *J. Environ. Manag.* **2017**, *199*, 62–73. [CrossRef]
30. Chang, Y.; Chen, X.; Liu, D.; Tian, C.; Xu, D.; Wang, L. Study on the Control of Saltwater Intrusion Using Subsurface Dams. *Water* **2023**, *15*, 3938. [CrossRef]
31. Abdoulhalik, A.; Ahmed, A.A.; Abd-Elaty, I. Effects of Layered Heterogeneity on Mixed Physical Barrier Performance to Prevent Seawater Intrusion in Coastal Aquifers. *J. Hydrol.* **2024**, *637*, 131343. [CrossRef]
32. Ahmed, A.; Robinson, G.; Hamill, G.; Etsias, G. Seawater Intrusion in Extremely Heterogeneous Laboratory-Scale Aquifer: Steady-State Results. *Water* **2022**, *14*, 1069. [CrossRef]
33. Cantelon, J.A.; Guimond, J.A.; Robinson, C.E.; Michael, H.A.; Kurylyk, B.L. Vertical Saltwater Intrusion in Coastal Aquifers Driven by Episodic Flooding: A Review. *Water Resour Res.* **2022**, *58*.
34. Li, F.-L.; Chen, X.-Q.; Liu, C.-H.; Lian, Y.-Q.; He, L. Laboratory Tests and Numerical Simulations on the Impact of Subsurface Barriers to Saltwater Intrusion. *Nat. Hazards* **2018**, *91*, 1223–1235. [CrossRef]
35. Qu, W.; Li, H.; Wan, L.; Wang, X.; Jiang, X. Numerical Simulations of Steady-State Salinity Distribution and Submarine Groundwater Discharges in Homogeneous Anisotropic Coastal Aquifers. *Adv. Water Resour.* **2014**, *74*, 318–328. [CrossRef]
36. Held, R.; Attinger, S.; Kinzelbach, W. Homogenization and Effective Parameters for the Henry Problem in Heterogeneous Formations. *Water Resour. Res.* **2005**, *41*, W11420. [CrossRef]
37. Kerrou, J.; Renard, P. Analyse Numérique Des Effets Dimensionnels et Des Hétérogénéités Sur Les Intrusions d'eau Marine en Milieu Dispersif. *Hydrogeol. J.* **2010**, *18*, 55–72. [CrossRef]
38. Mehnert, E.; Jennings, A.A. The effect of salinity-dependent Hydraulic Conductivity on saltwater intrusion episodes. *J. Hydrol.* **1985**, *80*, 283–297. [CrossRef]
39. Abarca, E.; Carrera, J.; Sánchez-Vila, X.; Dentz, M. Anisotropic Dispersive Henry Problem. *Adv. Water Resour.* **2007**, *30*, 913–926. [CrossRef]
40. Sun, D.-M.; Niu, S.-X.; Zang, Y.-G. Impacts of Inland Boundary Conditions on Modeling Seawater Intrusion in Coastal Aquifers Due to Sea-Level Rise. *Nat. Hazards* **2017**, *88*, 145–163. [CrossRef]
41. Mozafari, B.; Fahs, M.; Ataie-Ashtiani, B.; Simmons, C.T.; Younes, R. On the Use of COMSOL Multiphysics for Seawater Intrusion in Fractured Coastal Aquifers. *E3S Web Conf.* **2018**, *54*, 20. [CrossRef]
42. Fahs, M.; Ataie-Ashtiani, B.; Younes, A.; Simmons, C.T.; Ackerer, P. The Henry Problem: New Semianalytical Solution for Velocity-Dependent Dispersion. *Water Resour. Res.* **2016**, *52*, 7382–7407. [CrossRef]
43. Abarca, E.; Vázquez-Suñé, E.; Carrera, J.; Capino, B.; Gámez, D.; Batlle, F. Optimal Design of Measures to Correct Seawater Intrusion. *Water Resour. Res.* **2006**, *42*, W09415. [CrossRef]
44. Chang, S.W.; Clement, T.P. Experimental and Numerical Investigation of Saltwater Intrusion Dynamics in Flux-Controlled Groundwater Systems. *Water Resour. Res.* **2012**, *48*, W09527. [CrossRef]
45. Dalai, C.; Munusamy, S.B.; Dhar, A. Experimental and Numerical Investigation of Saltwater Intrusion Dynamics on Sloping Sandy Beach under Static Seaside Boundary Condition. *Flow Meas. Instrum.* **2020**, *75*, 101794. [CrossRef]
46. Etsias, G.; Hamill, G.A.; Águila, J.F.; Benner, E.M.; McDonnell, M.C.; Ahmed, A.A.; Flynn, R. The Impact of Aquifer Stratification on Saltwater Intrusion Characteristics. Comprehensive Laboratory and Numerical Study. *Hydrol. Process* **2021**, *35*, e14120. [CrossRef]
47. Etsias, G.; Hamill, G.A.; Campbell, D.; Straney, R.; Benner, E.M.; Águila, J.F.; McDonnell, M.C.; Ahmed, A.A.; Flynn, R. Laboratory and Numerical Investigation of Saline Intrusion in Fractured Coastal Aquifers. *Adv. Water Resour.* **2021**, *149*, 103866. [CrossRef]
48. Guo, Q.; Huang, J.; Zhou, Z.; Wang, J. Experiment and Numerical Simulation of Seawater Intrusion under the Influences of Tidal Fluctuation and Groundwater Exploitation in Coastal Multilayered Aquifers. *Geofluids* **2019**, *2019*, 2316271. [CrossRef]
49. Memari, S.S.; Bedekar, V.S.; Clement, T.P. Laboratory and Numerical Investigation of Saltwater Intrusion Processes in a Circular Island Aquifer. *Water Resour. Res.* **2020**, *56*, e2019WR025325. [CrossRef]
50. Oswald, S.E.; Kinzelbach, W. Three-Dimensional Physical Benchmark Experiments to Test Variable-Density Flow Models. *J. Hydrol.* **2004**, *290*, 22–42. [CrossRef]
51. Robinson, G.; Ahmed, A.A.; Hamill, G.A. Experimental Saltwater Intrusion in Coastal Aquifers Using Automated Image Analysis: Applications to Homogeneous Aquifers. *J. Hydrol.* **2016**, *538*, 304–313. [CrossRef]
52. Sebben, M.L.; Werner, A.D.; Graf, T. Seawater Intrusion in Fractured Coastal Aquifers: A Preliminary Numerical Investigation Using a Fractured Henry Problem. *Adv. Water Resour.* **2015**, *85*, 93–108. [CrossRef]
53. Vats, O.P.; Sharma, B.; Stamm, J.; Bhattacharjya, R.K. Groundwater Circulation Well for Controlling Saltwater Intrusion in Coastal Aquifers: Numerical Study with Experimental Validation. *Water Resour. Manag.* **2020**, *34*, 3551–3563. [CrossRef]
54. Constable, S.C.; Parker, R.L.; Constable, C.G. Occam' s Inversion: A Practical Algorithm for Generatlng Smooth Models from Electromagnetic Sounding Data. *Geophysics* **1987**, *52*, 289–300. [CrossRef]
55. De-Groot Hedlin, C.; Constable, S. Occam's inversion to generate smooth, two-dimensional models from magnetotelluric data. *Geophysics* **1990**, *55*, 1613–1624. [CrossRef]
56. Ammar, A.I.; Gomaa, M.; Kamal, K.A. Applying of SP, DC-Resistivity, DC-TDIP and TDEM Soundings in High Saline Coastal Aquifer. *Heliyon* **2021**, *7*, e07617. [CrossRef] [PubMed]

57. Khalil, M.A.; Abbas, A.M.; Santos, F.M.; Masoud, U.; Salah, H. Application of VES and TDEM Techniques to Investigate Sea Water Intrusion in Sidi Abdel Rahman Area, Northwestern Coast of Egypt. *Arab. J. Geosci.* **2013**, *6*, 3093–3101. [CrossRef]
58. Goebel, M.; Knight, R.; Kang, S. Enhancing the Resolving Ability of Electrical Resistivity Tomography for Imaging Saltwater Intrusion through Improvements in Inversion Methods: A Laboratory and Numerical Study. *Geophysics* **2021**, *86*, WB101–WB115. [CrossRef]
59. Pradhan, R.M.; Singh, A.; Ojha, A.K.; Biswal, T.K. Structural Controls on Bedrock Weathering in Crystalline Basement Terranes and Its Implications on Groundwater Resources. *Sci. Rep.* **2022**, *12*, 11815. [CrossRef] [PubMed]
60. Loke, M.H.; Acworth, I.; Dahlin, T. A Comparison of Smooth and Blocky Inversion Methods in 2D Electrical Imaging Surveys. *Explor. Geophys.* **2003**, *34*, 182–187. [CrossRef]
61. Vahidipour, M.; Raeisi, E.; van der Zee, S.E.A.T.M. Active Saltwater Intrusion of Shrinking Bakhtegan -Thask Lakes in South Iran Threatens the Freshwater Resources of Coastal Aquifers. *J. Hydrol. Reg. Stud.* **2021**, *34*, 100790. [CrossRef]
62. Chithra, S.; Joseph, S.; Kannan, N. A Study of Saltwater Intrusion in the Kallada River, Southwest Coast of Kerala, India. *Water Supply* **2022**, *22*, 2194–2211. [CrossRef]
63. Sherif, M.M.; Singh, V.P. Effect of Climate Change on Sea Water Intrusion in Coastal Aquifers. *Hydrol. Process* **1999**, *13*, 1277–1287. [CrossRef]
64. Barlow, P.M.; Reichard, E.G. L'intrusion d'eau Salée Dans Les Régions Côtières d'Amérique Du Nord. *Hydrogeol. J.* **2010**, *18*, 247–260. [CrossRef]
65. Syvitski, J.P.M.; Saito, Y. Morphodynamics of Deltas under the Influence of Humans. *Glob. Planet. Chang.* **2007**, *57*, 261–282. [CrossRef]
66. Miky, Y.; Issa, U.H.; Mahmod, W.E. Developing Functional Recharge Systems to Control Saltwater Intrusion via Integrating Physical, Numerical, and Decision-Making Models for Coastal Aquifer Sustainability. *J. Mar. Sci. Eng.* **2023**, *11*, 2136. [CrossRef]

water

MDPI

Article

Characterization of a Contaminated Site Using Hydro-Geophysical Methods: From Large-Scale ERT Surface Investigations to Detailed ERT and GPR Cross-Hole Monitoring

Mirko Pavoni [1,*], Jacopo Boaga [1], Luca Peruzzo [1], Ilaria Barone [1], Benjamin Mary [2] and Giorgio Cassiani [1]

1 Department of Geosciences, University of Padova, 35100 Padova, Italy; luca.peruzzo.1@unipd.it (L.P.); ilaria.barone@unipd.it (I.B.); giorgio.cassiani@unipd.it (G.C.)
2 Tec4Agro Group, Institute of Agricultural Sciences—CSIC, 28006 Madrid, Spain; benjamin.mary@ica.csic.es
* Correspondence: mirko.pavoni@unipd.it

Abstract: This work presents the results of an advanced geophysical characterization of a contaminated site, where a correct understanding of the dynamics in the unsaturated zone is fundamental to evaluate the effective management of the remediation strategies. Large-scale surface electrical resistivity tomography (ERT) was used to perform a preliminary assessment of the structure in a thick unsaturated zone and to detect the presence of a thin layer of clay supporting an overlying thin perched aquifer. Discontinuities in this clay layer have an enormous impact on the infiltration processes of both water and solutes, including contaminants. In the case here presented, the technical strategy is to interrupt the continuity of the clay layer upstream of the investigated site in order to prevent most of the subsurface water flow from reaching the contaminated area. Therefore, a deep trench was dug upstream of the site and, in order to evaluate the effectiveness of this approach in facilitating water infiltration into the underlying aquifer, a forced infiltration experiment was carried out and monitored using ERT and ground-penetrating radar (GPR) measurements in a cross-hole time-lapse configuration. The results of the forced infiltration experiment are presented here, with a particular emphasis on the contribution of hydro-geophysical methods to the general understanding of the subsurface water dynamics at this complex site.

Keywords: hydrogeophysics; contaminated site; infiltration experiment; time-lapse geophysics; ERT; GPR

1. Introduction

In the characterization of contaminated sites, direct drilling methods and geochemical analysis of groundwater may not be completely representative of the entire investigated area, leaving some uncertainties in the management of remediation and secure strategies [1]. Therefore, utilizing near-surface geophysical techniques is a relatively fast and economical approach to retrieve further information about the structure of the subsurface and the hydrological dynamics [2], particularly in terms of the extensive spatial coverage and refined sampling that geophysics can provide. Among the different geophysical methods, electrical resistivity tomography (ERT) and ground-penetrating radar (GPR) are widely used for the characterization of contaminated sites [3,4], since electrical and dielectric properties are closely related to the lithology, soil texture and particularly to the water content, water quality and the presence of non-aqueous-phase contaminant liquids [5,6].

In this work, we present the main results of an extensive geophysical survey campaign conducted at a contaminated site in the Friuli High Plain, north-eastern Italy. The presence of contaminants in groundwater was detected around a former industrial site. The geological conditions of the site can be described as a relatively simple stratigraphic sequence composed of an upper unsaturated gravel layer, housing also a perched aquifer, discontinuously interrupted by a thin layer of clay (a paleo-soil) at a 6–7 m depth. Below the clay layer, another thick unsaturated gravelly zone reaches a depth of about 100 m, where the water table of the regional main aquifer is located. Thus, the development and consistency of the clay layer are critical from a hydraulic point of view in order to prevent deeper infiltration of pollution from the surface [7]. In the investigated area, to prevent the subsurface flow from reaching the contaminated site, it was planned to force the water infiltration into the deeper underlying thick unsaturated zone and the deep aquifer by interrupting the continuity of the clay layer with a draining trench upstream of the industrial site.

Firstly, to define the most suitable position for the trench, large-scale surface ERT surveys were performed to map the extension and continuity of the clay layer around and within the industrial site. Once the optimal position was defined, ERT cross-borehole investigations were performed before and after the realization of a pilot stretch of the draining trench to evaluate the effects of excavation. Finally, to verify the effectiveness of the trench in draining the shallow subsurface water flow into the deeper aquifer, two forced infiltration experiments were carried out and monitored using both ERT and GPR cross-borehole measurements in time-lapse configuration [8–10].

Preliminary synthetic models, based on a priori geological information, were tested in order to optimize the data acquisition strategies. In particular, we tested different ERT survey strategies (with different electrode array lengths, spacings and configurations) against different possible geological models, varying the thickness and depth of the conductive clay layer.

The goal of this work is to demonstrate the effectiveness of the use of geophysical investigations in the overall strategy of characterization and remediation of contaminated sites. In order to fully benefit from its information content, geophysics must be carefully planned, both for characterization and monitoring (time-lapse) modes, often with the support of synthetic experiments in the planning stage.

2. Site Description

2.1. Geological Framework

The investigation site is located in a High Plain of north-eastern Italy. This is the easternmost portion of the foreland area of the alpine chain, and the plain is a consequence of the progressive accumulation of fluvio-glacial sediments. During the last glacial maximum (LGM), the glacial and periglacial conditions in the mountain basins promoted a considerable production of sediments, and the development of the glaciers down to the valley outlets guaranteed efficient transport, feeding large fluvio-glacial systems. In this Eastern High Plain, this high sedimentation rate promoted the development of large river systems (e.g., the Isonzo, Torre, Tagliamento, Cellina, and Meduna rivers), and consequently the formation of extensive alluvial megafans, i.e., a fan-shaped depositional system with an extension that can exceed thousands of km^2, characterized by an apical portion consisting of gravelly deposits, and a distal portion essentially composed of fine sediments [11]. In the High Plain, the subsoil is mainly composed of alluvial deposits of braided systems that continuously migrated due to periodical avulsions. Therefore, in the evolution of the megafans, coarse sedimentation phases alternated with destructive and steady-state moments, even prolonged, which ensured the formation of soils that may have been subsequently buried (paleosoils) [12]. Finally, during the Holocene, glacier retreat led to a decrease in the sedimentation rate, the confinement of rivers within incised channels, and the transition of large areas of megafans into bypass surfaces without deposition. In these areas, the gravels have been weathered, generating soils with clay layers [13].

2.2. Experimental Area

At the contaminated site, the subsoil is composed of the Vivaro Unit, i.e., gravels with a sub-horizontal coarse stratification, sometimes with a slightly silty–sandy matrix, in which buried paleo-soils are interspersed [14]. This stratigraphic structure is confirmed by the boreholes realized for ERT and GPR investigations. As shown in Figure 1a, a 20 m deep perforation highlights the presence of a thick layer of gravel, which is cut by a thin level of clay at about a 7 m depth. Four such boreholes have been drilled and equipped with 24 electrodes each from a depth of 1.6 m to the bottom of the boreholes (0.8 m spacing). Figure 1b presents the geometry of the boreholes (red circles) and the pilot-scale draining trench. The latter has a length of approximately 17 m, a depth of 13 m (as shown in Figure 1c) and is about 1 m wide. After excavation it was completely filled with permeable coarse gravel. This stretch of trench was intended as a pilot test to confirm the effectiveness of the proposed intervention; the full trench was planned to have a length exceeding 100 m. This pilot trench was realized in correspondence with an ERT surface transect collected prior to the excavation. Four ERT-equipped boreholes were drilled on both sides of the trench, allowing for meaningful cross-hole acquisitions. Note that the boreholes also allow for cross-hole GPR acquisitions given the relatively small borehole distance (8 m).

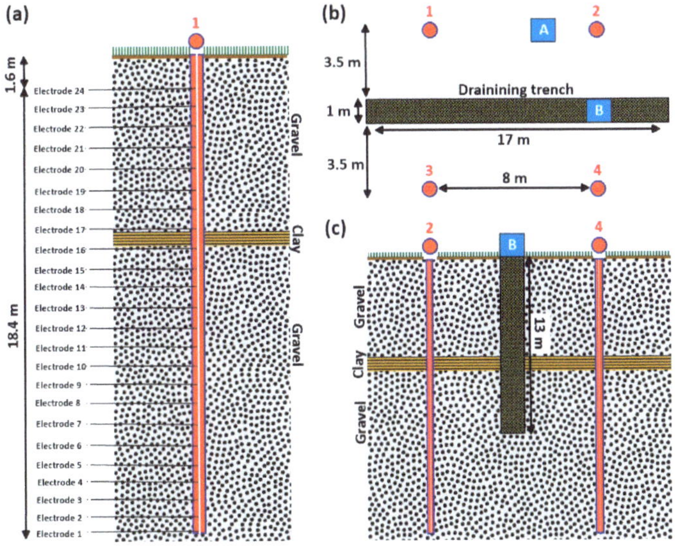

Figure 1. (**a**) Schematization of borehole 1 and the stratigraphy found during the drilling: 24 electrodes placed from the bottom of the borehole (20 m depth) to a 1.6 m depth; (**b**) planned geometry of the area with four boreholes (red dots), the draining trench (parallel to the ERT surface line), and the position of the water injections (blue square A–B); (**c**) vertical section of the area between borehole 2 and borehole 4, and the draining trench in between. Note that the thickness of the clay layer is in the 0.5 m range and less (considering all four drilled boreholes).

Two infiltration experiments were conducted in the pilot zone, with the purpose of assessing to what extent the dug trench, interrupting the continuity of the clay layer, was capable of conveying water into the deeper unsaturated zone, as expected. The first infiltration experiment (indicated by the blue square "A" in Figure 1b) was realized on the side of the trench, with the injection point located between borehole 1 and borehole 2. The second infiltration experiment (blue square "B" in Figure 1b,c) was realized by placing the water injection point right above the trench, between borehole 2 and borehole 4. In both cases, a square meter at the surface was isolated using wooden walls and a constant water flow rate into the box was ensured by a water tanker for a total injected volume of nearly 17 cubic meters.

3. Methods

3.1. Electrical Resistivity Tomography (ERT)

ERT surveys are performed with multi-electrode instruments to retrieve the electrical resistivity distribution of the subsurface. The measurements were carried out with an array of several dozens of electrodes, either placed at the ground surface (the most common arrangement) or in dedicated boreholes, and are galvanically coupled with the soil. The measurement consists of injecting an electrical current with two electrodes and recording the voltage difference that arise at other pairs of electrodes [5]. Based on the target of the survey, different types of acquisition schemes can be adopted; Wenner-alpha and Wenner–Schlumberger schemes (injection dipole outside the potentiometric one) guarantee a higher vertical resolution, while a dipole–dipole scheme (injection dipole is adjacent to the potentiometric one) allows for a higher lateral sensitivity (for details see [15]). The penetration depth of a measured point is linked to the length of the quadrupole; consequently, the maximum depth of investigation is defined by the total length of the electrode array. For the same reason, in cross-hole configuration, the distance between neighbouring boreholes cannot be too large, generally no larger than the depth of the boreholes themselves. On the other hand, the resolution of the investigation is linked to the spacing between the electrodes, and it is higher close to the electrodes [15]. Therefore, the resolution of ERT surface measurements decreases in depth. Consequently, when a high resolution at depth is required, cross-hole ERT with borehole electrodes can be adopted. Nevertheless, since the number of the electrodes controlled by a multi-electrode device is limited, usually ranging between 48 and 120 channels, it is necessary to find a compromise between the spacing and the length of the array, considering the requested resolution of the survey (e.g., the spacing should not be larger than the thickness of the layers) and its penetration (about ¼ of the total length of the array) [5].

The quality of ERT datasets can be evaluated by a combination of two approaches: (i) stacking errors, where each quadrupole is measured several times and a standard deviation is calculated for each one and (ii) reciprocal errors, where for each quadrupole, the measurement is performed by exchanging the injection dipole and the potentiometric dipole, and the difference between direct and reciprocal measurements is calculated for each quadrupole [16]. Finally, since the measurements are influenced by the contribution of different materials that compose the subsurface, the acquired ERT datasets need to be inverted to identify the most suitable subsurface resistivity model that reproduces the measurements [17].

3.1.1. ERT Forward Modelling

Starting from a known subsoil structure with a defined resistivity distribution, and an array of electrodes at the surface, we can discretize the investigated domain with a mesh and numerically calculate, by applying Poisson's equation (for details, see [15]), the voltages that arise at the electrodes of the array if we inject the electrical current i with a dipole A-B. Therefore, we can find the potential difference ΔV for any pair of potentiometric dipoles M-N of the array and obtain a synthetic dataset of apparent resistivities ρ_a by applying Equation (1), which applies to surface electrode configurations only. This process is called the "forward problem" or "forward modelling".

$$\rho_a = \frac{\Delta V}{i} 2\pi \left(\frac{1}{AM} - \frac{1}{BM} - \frac{1}{AN} + \frac{1}{BN}\right)^{-1} = \frac{\Delta V}{i} K \quad (1)$$

Here, K is the geometric factor of the measured quadrupole. In borehole measurements, the electric current is flowing in all directions. Consequently, K needs to be recalculated, as shown in Equation (2) (where z is the depth of the electrodes):

$$K = 4\pi \left[\left(\frac{1}{|z_A - z_M|} + \frac{1}{z_A + z_M} \right) - \left(\frac{1}{|z_B - z_M|} + \frac{1}{z_B + z_M} \right) - \left(\frac{1}{|z_A - z_N|} + \frac{1}{z_A + z_N} \right) + \left(\frac{1}{|z_B - z_N|} + \frac{1}{z_A + z_N} \right) \right]^{-1} \quad (2)$$

Once a synthetic dataset is computed, an inversion framework can be applied to the calculated apparent resistivities to obtain an inverted synthetic resistivity model.

In the presented work, the Python-based software ResIPy (v3.4.2) [16] was used to perform the forward modelling process and define the optimal acquisition parameters for the preliminary ERT surface measurements. Based on geological information, we created different subsurface models, as shown in Figure 2. In each model, an upper (relatively) high resistive layer of 300 Ωm represents a weathered gravel with a slightly silty–sandy matrix, which is cut by a low resistive layer (i.e., clay) of 10 Ωm at a 5 m depth in Figure 2a,b and at a 10 m depth in Figure 2c,d. In the models of Figure 2a,c, the low resistive layer has a thickness of 0.50 m, while in the models of Figure 2b,d, the thickness is 2 m. In all models, the bottom layer of 450 Ωm represents the deeper, unsaturated silty gravel layer.

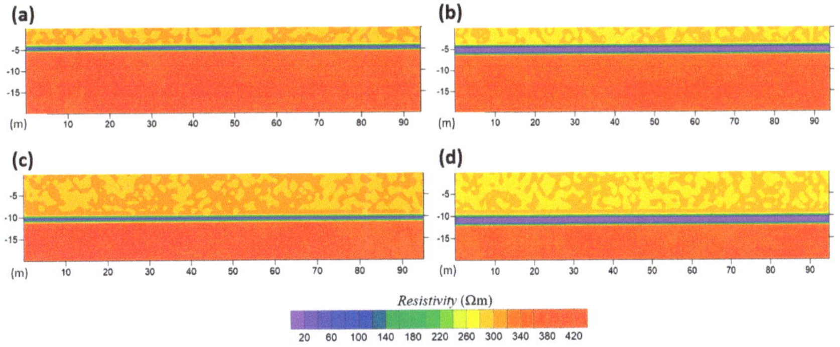

Figure 2. The subsurface models used for ERT surface forward modelling. In each model, an upper high resistive layer of 300 Ωm is cut by a low resistive layer of 10 Ωm at 5 m of depth in (**a**,**b**), at 10 m of depth in (**c**,**d**). In the models of (**a**,**c**), the low resistive layer has a thickness of 0.50 m, while in the models of (**b**,**d**) the thickness is 2 m.

Firstly, forward modelling was performed considering a surface ERT investigation with different electrode spacings: a dipole–dipole acquisition scheme and testing different values of skip (i.e., the number of electrodes skipped to create a dipole).

Afterwards, the same subsurface structure was used to perform a cross-borehole ERT survey. As shown in Figure 3, we considered a low-resistive layer with both a thickness of 0.50 m (Figure 3a–c) and 2 m (Figure 3d–f) at a depth of 5 m. We tested boreholes with different electrode spacings (0.5 m in Figure 3a,b,d,e, and 1 m in Figure 3c,f), a skip 4 AB-MN configuration (injection dipole electrodes in the same borehole, and potentiometric dipole electrodes in the other borehole), and different borehole separations (5 m in Figure 3a,d, and 8 m Figure 3b,c,e,f).

3.1.2. ERT Inverse Modelling

Inverse modelling is used to calculate the most reliable resistivity model that can reproduce the measured apparent resistivities within data error bounds. This is an iterative process where forward models are calculated repeatedly in order to minimize the misfit between the predicted and observed data [17]. However, different resistivity models can lead to practically the same response—within data error levels—and a unique solution can be found by formulating the inversion process as a regularized optimization problem, applying Occam's approach [18], which searches for the smoothest model that can fit the measured data within their error level (for details, see [15]).

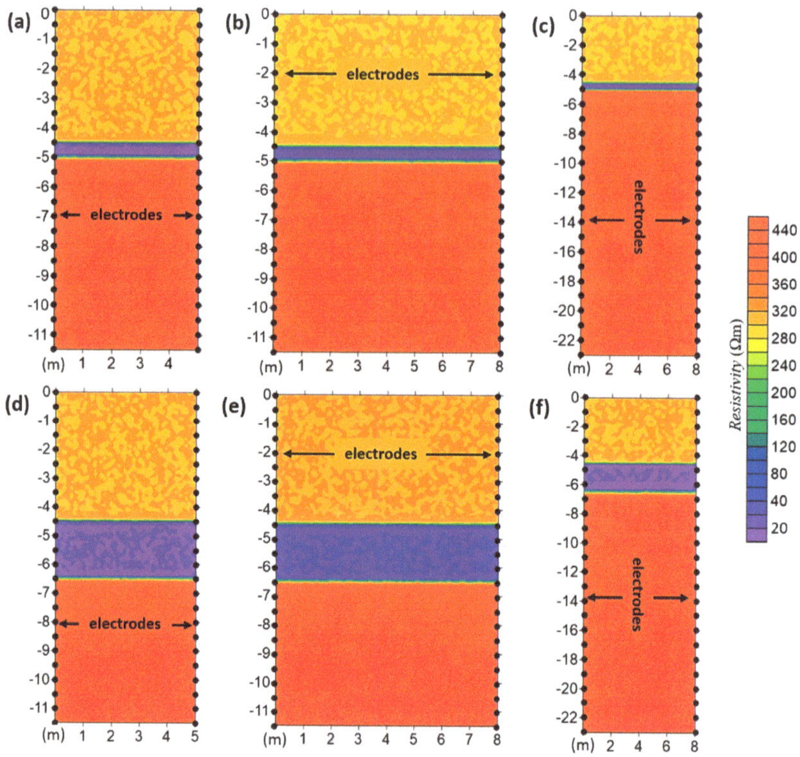

Figure 3. The subsurface models used to perform the ERT cross-borehole forward modelling process. In each model, an upper high resistive layer of 300 Ωm is cut by a low resistive layer of 10 Ωm at 5 m of depth in (**a**,**b**), at 10 m of depth in (**c**,**d**). In the models of (**a**–**c**), the low resistive layer has a thickness of 0.50 m, while in the models of (**d**–**f**) the thickness is 2 m.

Defining an optimal error degree for the inversion process is a key factor to avoid unrealistic artifacts in the final inverted resistivity model [5]. The expected data error can be estimated by using the stacking error, but usually this approach overestimates the data quality and a better assessment of the error can be achieved by considering reciprocal measurements [19]. Once a reciprocal error threshold value is defined, all the quadrupoles with higher values are filtered out and not considered during the inversion process. High-quality ERT datasets have reciprocal errors of lower than 5%, while low-quality datasets have values as high as 20%, typically when the galvanic contact between electrodes and the ground surface is poor [20].

In this work, inverse modelling of the acquired datasets was performed using the Python-based software ResIPy [16] and, for each dataset, we defined a boundary threshold for the reciprocal error that allowed for a reliable quality of the measured apparent resistivities but at the same time a homogeneous distribution of measured points in the pseudo-section [21,22]. For the surface ERT measurements, an expected data error of 10% was defined, while for the ERT cross-borehole datasets, we selected a more comfortable 5% error level. All inverted models presented here have a final RMS (Root Mean Square—evaluation of the normalized misfit between calculated and measured data) close to 1.

3.1.3. ERT Time-Lapse Survey

ERT surveys can also be used to monitor time-dependent subsurface processes through changes in resistivity over time. This kind of investigation is particularly useful to monitor natural processes in the subsoil (e.g., moisture variation during the year [23,24] or as a consequence of forced irrigation experiments, with several hundreds or thousands of litres of water injected into the subsurface [10,25]). In order to enhance the changes from one time frame to the next, ratio or difference inversion approaches are usually applied [5]. For each quadrupole in the dataset, the parameter to be inverted is the ratio or the difference between the resistance ($R = \Delta V/I$) measured in the considered time step and the initial resistance measured before the water injection.

Time-lapse inversion is a powerful approach to highlight small variations in resistivity that would otherwise be overwhelmed by error differences in subsequent absolute resistivity images [26]. For each time step, the results are consequently given in terms of resistivity variation with respect to the initial model obtained before the water injection. Therefore, after the inversion process, only the initial reference model will be plotted in terms of absolute resistivities, while the following time step results will be plotted in terms of variations in resistivity with respect to such an initial model.

In the presented work, two forced infiltration experiments were carried out and monitored with ERT cross-borehole time-lapse measurements, as well as repeated GPR measurements—as described below. In each survey, a representative value of reciprocal error was defined for the acquired ERT datasets, and only the common quadrupoles in all filtered datasets were used to perform the time-lapse ratio inversion.

3.1.4. ERT Data Acquisition

All ERT surveys (surface, cross-borehole, and time-lapse) were carried out using a Syscal Pro resistivimeter (Iris Instruments, Orléans, France), using a stacking range between 3 and 6 (5% standard deviation threshold), and with direct reciprocal measurements.

As previously discussed, several ERT surface investigation lines were collected around and within the contaminated area to evaluate the depth and continuity of the clay layer. In this work, we only focused on the survey line acquired upstream of the industrial site, in correspondence with the area chosen for the realization of the draining trench and the boreholes. Based on the results of the previous forward modelling tests, surface ERT measurements were performed using an array of 120 electrodes spaced at 0.80 m and a dipole–dipole skip 8 acquisition scheme.

To evaluate the effectiveness of the trench, ERT cross-borehole measurements were collected before and after the pilot trench excavation and backfilling with coarse permeable material. The datasets were acquired using 48 electrodes, with 24 in each borehole (see Figure 1), using an AB-MN skip 4 acquisition scheme. To better evaluate the resistivity variation linked to the excavation, a ratio inversion process was also applied to the data acquired before and after excavation in correspondence with the section between boreholes 1 and 3 (see Figure 1).

Two forced infiltration experiments were also performed using ERT cross-borehole time-lapse measurements (also using four electrodes in the surface between the boreholes) using an AB-MN skip 4 acquisition scheme. As shown in Figure 1b, the first experiment was performed between borehole 1 and borehole 2, while the second experiment was performed between borehole 2 and 4 (one week after the first experiment). In each experiment, a time-zero dataset was acquired in natural dry conditions; afterwards, about 16,600 litres of salt water were injected into the subsurface over 9 h, and twelve ERT measurements were acquired during the following 28 h (as shown in Table 1 for the first infiltration experiment—the same acquisition scheme was used for the second infiltration experiment).

Table 1. Time steps of the first forced infiltration experiment (between borehole 1 and 2—see blue square A in Figure 1b) monitored using ERT and GPR time-lapse measurements. The second infiltration experiment was realized with the same acquisition parameters and time steps (one week after the first experiment).

Time	ERT Time Step	Water Released
11.00	t0	-
12:00	-	Start
12:35	t1	1000 L
13:08	t2	2000 L
13:45	t3	3000 L
14:20	t4	4000 L
14:55	t5	5000 L
15:30	t6	6000 L
16:00	t7	7000 L
21:30	-	16,600 L (End)
07:50 (+1 day)	t8	-
10:00 (+1 day)	t9	-
12:00 (+1 day)	t10	-
14:00 (+1 day)	t11	-
16:00 (+1 day)	t12	-

3.2. Ground-Penetrating Radar (GPR)

Spatiotemporal variations in water content in shallow soil layers can be efficiently estimated using techniques that measure the electrical permittivity ε or the relative electrical permittivity (also called the dielectric constant), ε_r, of porous media [27–30]. ε_r is strongly affected by the presence of water in soil pores, as the corresponding dielectric constant of water is so high that it overcomes any other solid or fluid component in the porous medium. GPR is particularly suitable to measure ε_r, as the propagation of electromagnetic waves, to a first approximation, can be directly related to the value of the bulk ε_r of the material:

$$\sqrt{\varepsilon_r} = \frac{c}{v} \quad (3)$$

where c is the electromagnetic wave speed in the vacuum (0.3 m/ns) and v is the measured velocity of propagation of the radar waves. In order to measure GPR velocities, the transmitter probe must be separated from the receiver one (using a bi-static GPR), and the time of arrival of the GPR waves must be defined at a known distance between the two. This can be achieved on the surface or in the borehole. In the infiltration test zone of the study site, thanks to the drilled boreholes, which all have a plastic casing that does not impede the propagation of electromagnetic waves outside the boreholes, we could exploit the same boreholes and equip then with ERT electrodes to acquire cross-hole GPR data. For this purpose, we used a Pulse-Ekko 100 MHz antenna system. Both the ZOP (zero offset profile, e.g., [31]) and MOG (multi-offset gather, e.g., [32,33]) were tested. In this case, VRP (vertical radar profiling [34]) had an unfavourable geometry and was not tested. In the ZOP configuration, the receiver and transmitter antennas go down at the same time along the boreholes, thus keeping the distance between the two probes, i.e., the distance between the two boreholes, fixed. In the MOG configuration, the transmitter antenna is moved in different positions independently from the receiver antenna, collecting a much more detailed survey between the boreholes and allowing for a 2D tomographic reconstruction. Cross-hole GPR data were collected for 2 days before, during and after the second infiltration test using boreholes 2 and 4 for a total of 14 different datasets.

4. Results

4.1. ERT Forward Modelling Results

Figure 4 shows the inverted synthetic models derived from the forward modelling process applied to the subsurface structure of Figure 2a using surface ERT configurations.

The low-resistive layer of clay is in this case very thin (0.5 m) and is smeared into a much thicker, less-conductive layer in the results of Figure 4a, obtained by simulating an array of 96 electrodes with a 0.5 m electrode spacing and a dipole–dipole skip 0 acquisition scheme. This is completely consistent with the very well-known equivalence problem in classical geoelectrical acquisitions. In fact, albeit with different final results configurations, the same phenomenon manifests itself in all other configurations in Figure 4 and is the result of the thin, very conductive layer being able to short circuit the current coming from the surface through itself, effectively shielding any other structure lying below this layer [35]. Note that this phenomenon is linked to the continuity of the clay layer—should the layer have any sizeable discontinuity, the discontinuity would become visible and part of the underlying structure would also be revealed.

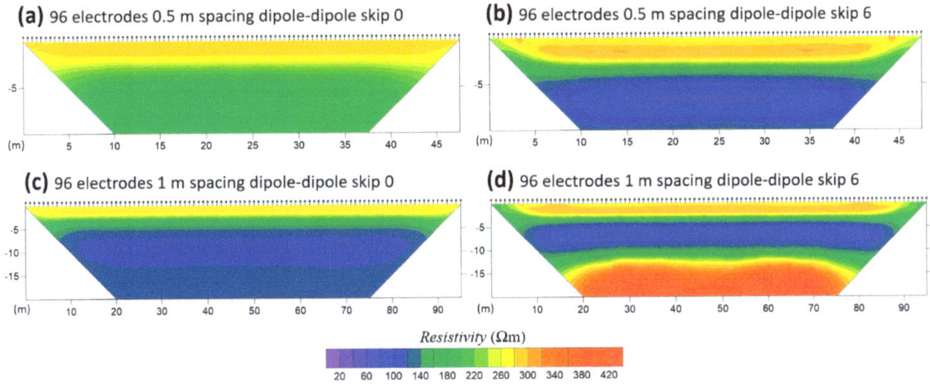

Figure 4. Inverted synthetic models obtained considering the subsurface model in Figure 2a. (**a**,**b**) show, respectively, the inverted synthetic models found with 96 electrodes, a 0.5 m spacing, and dipole–dipole skip 0 (**a**) and skip 6 (**b**) configurations; (**c**,**d**) show, respectively, the inverted synthetic models found with 96 electrodes, a 1 m spacing, and dipole–dipole skip 0 (**c**) and skip 6 (**d**) configurations.

Figure 5 presents the inverted synthetic models obtained from the forward modelling process applied to the subsurface structure of Figure 2b, with a 2 m-thick clay layer at a 5 m depth. The results are very similar to the ones in Figure 4, as the short-circuiting effect of the clay layer is the same, if not more pronounced.

Considering the clay layer at a 10 m depth (Figure 2c,d), and testing the same electrode array and acquisition schemes, Figure 6 shows again the limitation of surface ERT acquisitions in the face of a continuous conductive layer: its thickness cannot be ascertained and the structures below this layer remain unknown. Yet, this phenomenon is strictly linked to the continuity of this electrically conductive, hydraulically impeding layer. Thus, even though the overall deep structure of the subsoil cannot be imaged entirely, the continuity of the layer can be proven using surface ERT.

Figure 7 shows the inverted synthetic models considering an ERT cross-borehole configuration. Figure 7a–c present, respectively, the results obtained considering the subsurface structures and the electrode geometries in Figure 3a–c. In Figure 7a, with boreholes placed 5 m apart and an electrode spacing of 0.5 m, the inverted synthetic model correctly defines the subsurface structure, i.e., the depth, thickness and resistivity value of the clay layer. Given the geometry of the current injection, the short-circuiting phenomenon observed in surface acquisitions does not take place. Increasing the distance between the boreholes to 8 m, as shown in Figure 7b, leads to some uncertainties; i.e., the clay layer seems to have a larger thickness in the central area of the model, and a low resistivity area (~65 Ωm) is found close to the surface. These are effects of the loss in resolution away from the electrode areas, with the corresponding prevalence of the smoothing effect of Occam's inversion. As shown in Figure 7c, modifying the electrode spacing to 1 m (considering a borehole separation of 8 m) allows for a reduction in these artifacts; i.e., the clay layer thickness is more homogeneously retrieved in the central area of the model. Considering a 2 m-thick clay layer, the configuration with a 5 m borehole separation and a 0.5 m electrode spacing, as shown in Figure 7d, correctly reproduces the subsurface structure in Figure 3d. Increasing the borehole separation to 8 m (considering a 0.5 m electrode spacing), as shown in Figure 7e, does not produce artifacts, and allows for the correct definition of the model of Figure 3e. A reliable inverted synthetic model is also found by increasing the electrode spacing to 1 m, as shown in Figure 7d.

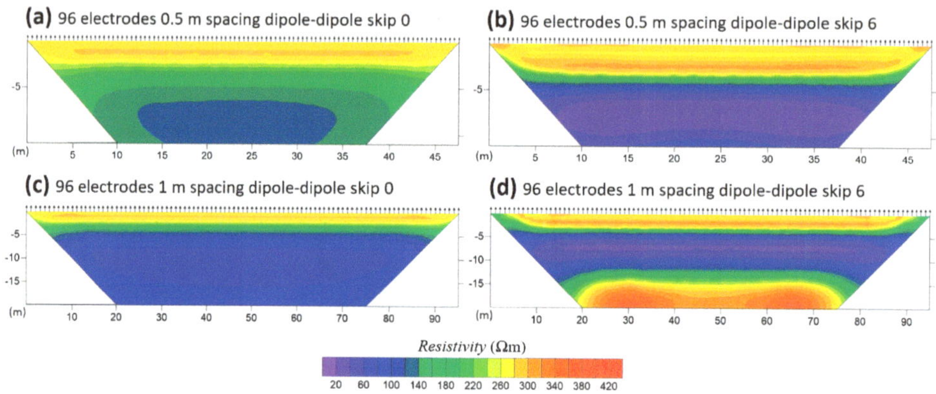

Figure 5. Inverted synthetic models obtained considering the subsurface model in Figure 2b. (**a**,**b**) show, respectively, the inverted synthetic models found with 96 electrodes, a 0.5 m spacing, and dipole–dipole skip 0 (**a**) and skip 6 (**b**) configurations; (**c**,**d**) show, respectively, the inverted synthetic models found with 96 electrodes, a 1 m spacing, and dipole–dipole skip 0 (**c**) and skip 6 (**d**) configurations.

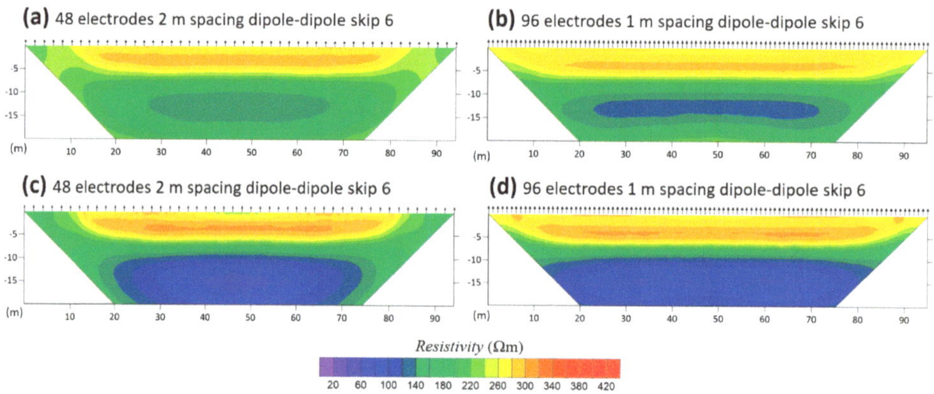

Figure 6. (**a**) shows the inverted synthetic model of Figure 2c found with 48 electrodes and a 2 m spacing; (**b**) shows the inverted synthetic model of Figure 2c found with 96 electrodes and a 1 m spacing; (**c**) shows the inverted synthetic model of Figure 2d found with 48 electrodes and a 2 m spacing; (**d**) shows the inverted synthetic model of Figure 2d found with 96 electrodes and a 1 m spacing. All the models were calculated using a dipole–dipole skip 6 acquisition scheme.

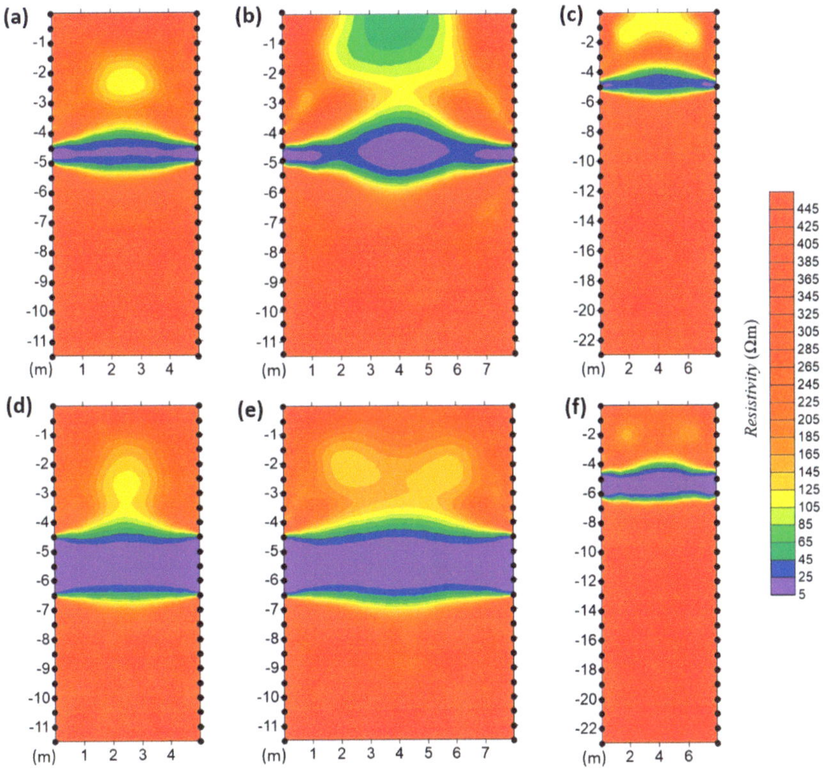

Figure 7. (**a**) shows the inverted synthetic model of Figure 3a; (**b**) shows the inverted synthetic model of Figure 3b; (**c**) shows the inverted synthetic model of Figure 3c; (**d**) shows the inverted synthetic model of Figure 3d; (**e**) shows the inverted synthetic model of Figure 3e; (**f**) shows the inverted synthetic model of Figure 3f.

4.2. ERT Field Data Inversion

Figure 8 shows the inverted resistivity model obtained from the ERT surface measurements performed upstream of the contaminated area and in correspondence with the line where the trench was subsequently excavated. In the shallow portion of the image, a high-resistivity layer (ρ > 2000 Ωm) corresponds to an unsaturated gravel layer. At about a 7 m depth, a sharp resistivity variation is found (ρ < 200 Ωm), corresponding to the contact with the clay layer. In the right part (x > 50 m), this boundary is deeper, and the high-resistivity layer reaches a depth of about 10 m. Finally, the clay layer seems to extend to the bottom of the section along the entire measured transect. However, as discussed above for synthetic modelling, this is only an indication of the continuity of the clay layer, and is not to any extent an evaluation of its thickness. In fact, 200 Ωm is a value far too high for the resistivity of clay: this testifies to the emergence of an example of the equivalence problem. In fact, the clay layer is much thinner and much more conductive.

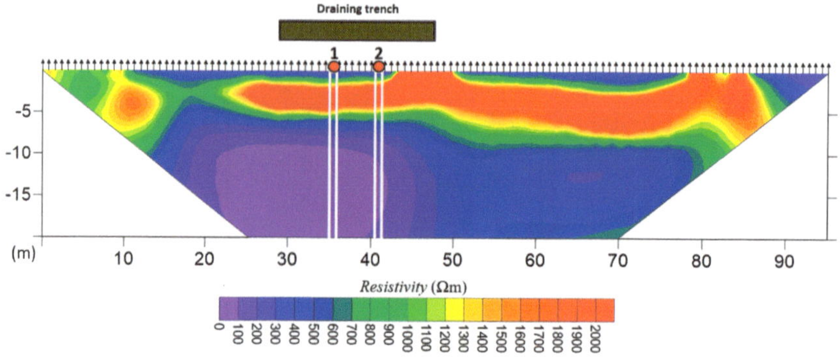

Figure 8. Inverted resistivity model of the surface ERT survey line collected upstream of the contaminated area. Based on this result, the draining trench (brownish rectangle) and the boreholes (red dots and white polygons) were realized approximately in the middle of the transect.

Figure 9 presents the inverted resistivity models obtained with the measurements performed using the ERT cross-borehole configuration before the trench excavation. The measurements were collected across boreholes 1–2 (Figure 9a), boreholes 1–3 (Figure 9b), and boreholes 2–4 (Figure 9c). All models show practically the same subsurface structure: an upper high-resistivity layer ($\rho \approx 1000\ \Omega$m), i.e., a gravel layer, a low-resistivity layer ($\rho < 10\ \Omega$m), i.e., a clay layer at a 6–7 m depth with a thickness of about 0.50 m, and a deeper high-resistivity layer of gravel.

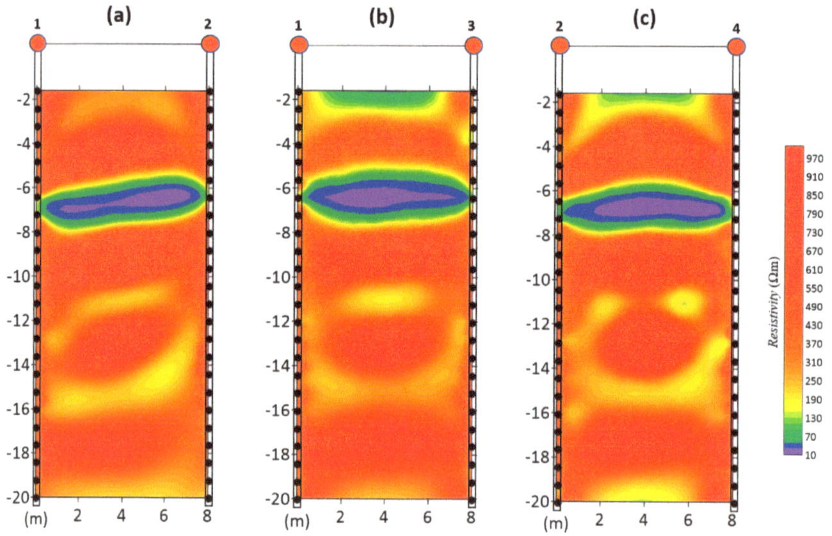

Figure 9. Inverted resistivity models of the cross-borehole (red dots) ERT surveys: (**a**) model between borehole 1 and 2; (**b**) model between borehole 1 and 3; (**c**) model between borehole 2 and 4.

Figure 10a shows the result of the time-lapse ratio inversion applied to the ERT cross-borehole measurements in the boreholes 1–3 section, acquired before and after the excavation of the trench. On the other hand, Figure 10b presents the result of the time-lapse ratio inversion applied to the same ERT cross-borehole datasets (boreholes 1–3), both collected after the realization of the trench (one month apart).

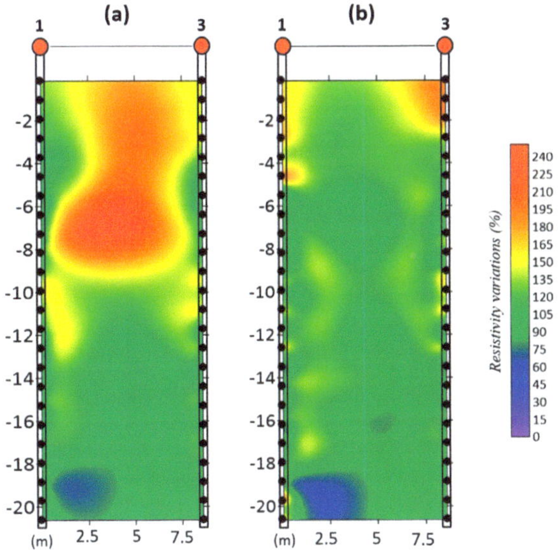

Figure 10. (**a**) ERT cross-borehole (boreholes 1–3) ratio inversion result: first dataset was collected before the realization of the draining trench and the second one collected after the realization; (**b**) ERT cross-borehole (boreholes 1–3) ratio inversion result of two datasets, both collected after the realization of the draining trench.

From the results shown in Figure 10a, the effect of the excavation is clear, with an increase in the resistivity of the central area of the section down to a 9–10 m depth, corresponding to the depth of the trench, interrupting the continuity of the clay layer. The material used to fill the trench (permeable coarse gravel) most likely has a very similar electrical resistivity to the lowermost gravel layer; for this reason, no great resistivity variations are detected below 9 m. This result is confirmed by the model in Figure 10b, which does not show high apparent variations in resistivity in the post-excavation inverted models.

4.3. ERT Time-Lapse Inversions during Water Infiltration Experiments

Figure 11 shows the results of the time-lapse cross-hole ERT monitoring of the first infiltration experiment monitored with ERT time-lapse cross-borehole measurements (along the profile of boreholes 1–2).

It is clear that, during the entire experiment, resistivity variations do not affect the at 6–7 m-deep clay layer (white dashed line in the panels of Figure 11, derived from the evidence from Figure 9a), and it is therefore not directly crossed by the water flow along this profile. This is consistent with the fact that along the borehole 1–2 profile, the clay layer is not cut by the trench. The shallow part of the subsurface above the clay layer is clearly affected by the water infiltration (Figure 11a–e). Afterward, in Figure 11f, evidence suggests that the water flow propagates deeper, possibly as an effect of water getting into the nearby, off-section trench and finding its way to the deeper gravels. These negative resistivity variations increase until water is injected (Figure 11g), but once the irrigation is stopped, a progressive decrease in this effect is observed in the subsequent time steps (Figure 11h–k), and the overall infiltration effect slowly fades away until the last measurement (Figure 11l).

Figure 12 shows the result of the second infiltration experiment monitored via ERT time-lapse cross-borehole measurements using the cross-hole configuration of boreholes 2–4.

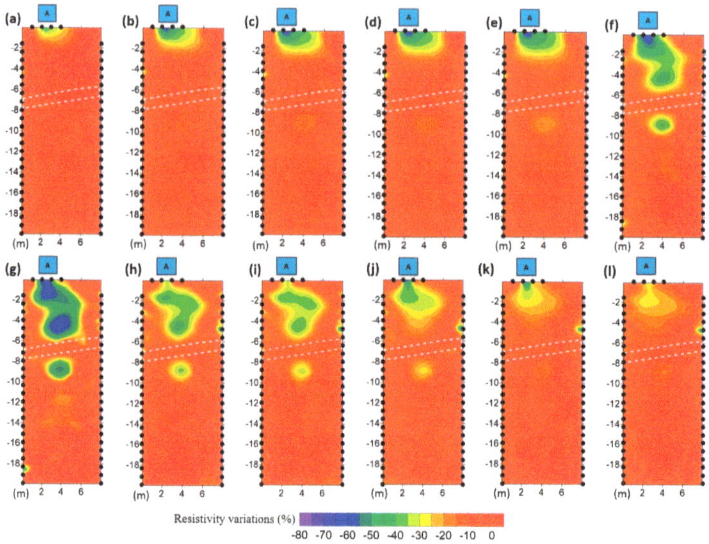

Figure 11. Cross-hole time-lapse ERT results corresponding to infiltration experiment 1. The injection point (blue square A) is located between boreholes 1 and 2; the same used for the ERT cross-hole acquisition. The white dashed line represents the clay layer detected in the inverted resistivity model of Figure 9a—note that along this profile, the clay layer is NOT interrupted by the trench (see Figure 2b). (**a**–**l**) represent, respectively, the results of the time-lapse ratio inversion applied at time steps t1–t12 presented in Table 1 using dataset t0 as the reference data.

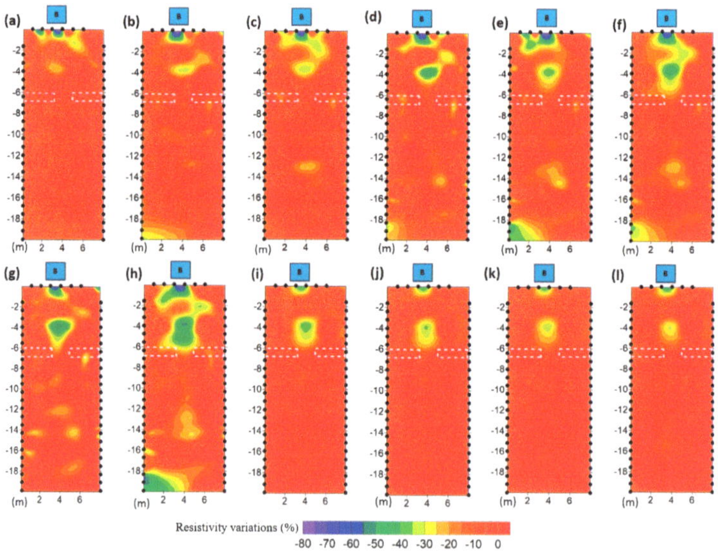

Figure 12. Cross-hole time-lapse ERT results corresponding to infiltration experiment 2. The injection point (blue square B) is located between boreholes 2 and 4; the same used for ERT acquisition. The white dashed line represents the clay layer detected in the inverted resistivity model of Figure 9c, which is cut in the middle with a discontinuity of 1 m representing the draining trench. (**a**–**l**) represent the results of the ERT time-lapse ratio inversion applied at time steps t1–t12 presented in Table 1 using dataset t0 as the reference data.

In this case, the clay layer at 6–7 m depth is discontinuous as an effect of trench digging along the monitoring section (boreholes 2–4) and the water infiltration point is right on the vertical of the trench in the middle of the section itself. Once the water injection begins, clear negative resistivity variations propagate vertically (Figure 12a–e) and cross the discontinuity of the clay layer (Figure 12f–h), albeit losing intensity possibly because of lateral spread, as detected in the companion results of the first infiltration experiment (Figure 11). Note that the colour scale in Figure 12 highlights considerable negative variations in resistivity, and the changes in resistivity across the trench (e.g., in Figure 12h) are substantial, reaching a depth of about 16 m, where infiltration seems to stop. As in the first infiltration experiment, once infiltration ceases from the surface, a progressive decrease in the negative resistivity variation signal takes place (Figure 12i,j), and is practically null in the latest measurements (Figure 12k,l).

4.4. GPR Monitoring during Water Infiltration Experiments

We also conducted cross-hole GPR monitoring during the second infiltration test. In order to efficiently follow the infiltration process, we planned to use the same step of the time-lapse ERT. Thus, we collected both ZOP surveys and more detailed MOG data. In the following, we will only present the more effective MOG results. These were acquired with a vertical spacing between the antenna positions equal to 0.5 m for the entire length of both boreholes 2 and 4.

The results are shown in Figure 13, where both the initial and final (AM and PM in Figure 13) instants are shown. The data were inverted using the PRONTO inversion code [36]. Travel-time data (Figure 13a,d) were thus inverted into GPR velocity distributions (Figure 13b,e), and these in turn into estimations of moisture content (Figure 13c,f) using the relationship presented by Topp et al. (1980) [37]. While subtle differences in moisture content are visible in the 2D images, more readable results are obtained by averaging the results along horizontal lines, thus producing 1D vertical profiles (Figure 13g). It is clear that there is a substantial overall increase in moisture content across the profile—with a strong peak corresponding to the wet clay layer—extending to a maximum depth of 16 m, consistent with the time-lapse ERT evidence (Figure 12).

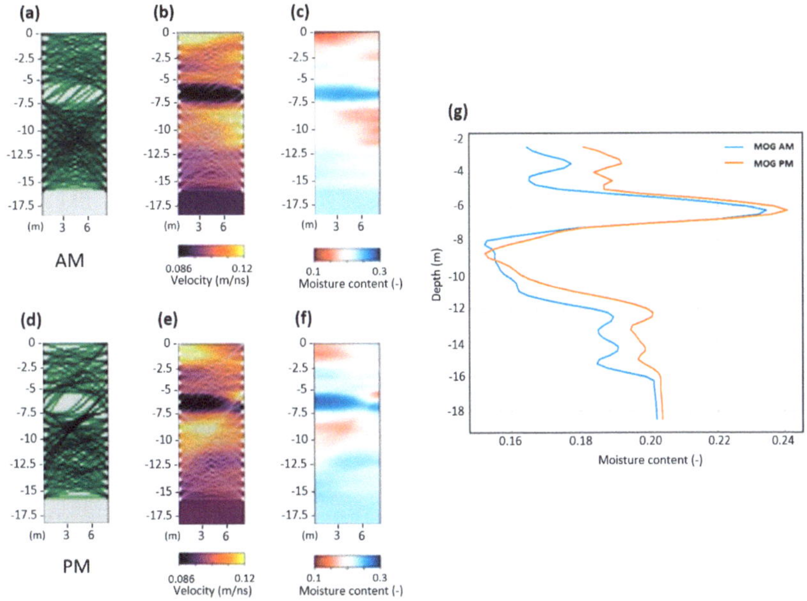

Figure 13. Cross-hole time-lapse GPR results corresponding to infiltration experiment 2.

5. Discussion

The obtained results demonstrate that advanced geophysical applications can be fundamental tools to improve the characterization of a contaminated site, both in terms of its structure and fluid dynamics. In particular, ERT and GPR cross-borehole surveys in time-lapse configuration can be efficiently integrated to monitor and understand local infiltration dynamics.

In this specific context, the existence of a highly conductive layer of clay at a 5 m depth poses a severe challenge to surface ERT investigations. Forward modelling shows that short transects and low dipole skips do not allow us to define the correct structure of the subsoil (Figures 4a–c and 5a–c). Increasing the length of the electrode array and the dipole skip leads to slightly better inverted models, but the thickness of the clay layer is still much larger than expected (Figures 4d and 5d). Inverted synthetic models with clay layers with thicknesses of 0.5 m (Figure 4d) and 2 m (Figure 5d) are quite similar and it is difficult to discriminate between them. This problem is further emphasized by increasing the depth of the clay layer to 10 m (Figure 6), which would require longer transects to be correctly detected. The inversion results of the field datasets confirm the limits of the surface ERT surveys to define a subsoil structure with a clay level between high-resistive gravel layers, as expected from the well-known equivalence problem. Although, in reality, the dataset was acquired with an electrode array of 96 m and with a dipole–dipole skip 8 configuration, in the inverted resistivity model obtained upstream of the contaminated site (Figure 8), it is possible to correctly define the depth of the clay layer (6–7 m) but not its thickness, which is clearly amplified till the bottom. This is probably due to a short-circuiting problem in the subsoil, as the electric current tends to flow in the less resistive layer of clay instead of propagating into the underlying resistive gravel layer.

On the contrary, forward modelling using cross-borehole configurations demonstrates that this method can define the real subsurface structure, i.e., the depth, thicknesses and resistivity values of each layer. However, the geometry of the boreholes (distance and depth), and the spacing of the electrodes must be adequately designed. The distance between boreholes should not be larger than half the length of the electrode array in the borehole [15]. Larger separations lead to a significant decrease in sensitivity at the centre of the model and the probable development of unrealistic artefacts, e.g., Figure 7b,c. The spacing of electrodes also affects the quality of the inverted models; i.e., with a spacing lower than the thickness of the layers, it is possible to precisely define their depth, thickness, and resistivity values (e.g., Figure 7a,d). The results of forward modelling are confirmed by the results obtained with the field datasets, and the corresponding inverted models (Figure 9) show the same subsurface structure found in borehole cores.

The results obtained from the cross-borehole ERT measurements, performed to verify the effect of draining trench excavation, demonstrate the reliability of the ratio inversion approach. The excavation, filled with coarse and permeable granular sediment, is clearly apparent in the inverted ERT results in Figure 10a (obtained by comparing pre-trench and post-trench ERT datasets), with an increase in resistivity developing from the surface down to a 8–9 m depth, thus interrupting the continuity of the clay layer.

Finally, the infiltration experiments with cross-borehole ERT measurements in time-lapse configuration verify the effectiveness of the draining trench. Even when injecting the water in a lateral position with respect to the trench, such as in the first experiment (Figure 11), the water is drained below the clay layer and 18 h after the end of the injection (t12), the subsoil has practically returned to the initial natural conditions (t0). In the second infiltration experiment (Figure 12), carried out right above the trench, after the beginning of the experiment, the water infiltrated more quickly depth-wise compared to the first experiment (compare the negative resistivity variations in Figures 11a–e and 12a–e), and the flow crosses the discontinuity of the clay layer, as shown in Figure 12f–h. Once the water injection is completed, the conditions return more quickly to the initial natural conditions (t0) compared to the first experiment (compare the negative resistivity variations in Figures 11i–l and 12i–l), as expected considering that the injection position was exactly above the trench. Therefore, the results of these two experiments confirm that the trench is correctly draining the sub-surface water flow into the deeper gravel layer, thus preventing it from reaching the critical contaminated area.

The GPR data, and particularly the high-resolution MOG data (Figure 13), fully confirm the evidence from ERT, in particular the fast infiltration of the injected water above the trench and its resting on top of a further discontinuity at about a depth of 16 m.

6. Conclusions

The obtained results demonstrate that geophysical investigations are a valuable tool for the characterization of contaminated sites and for the management of remediation and secure strategies. In relatively short times and with limited costs, it is possible to obtain additional and more extensive information compared to direct investigations, e.g., variations in depths and thicknesses and discontinuities in the layers in the subsoil.

The use of preliminary information from the investigation site to perform a forward modelling process is an excellent strategy to properly define the acquisition parameters of ERT field measurements, i.e., the array length, electrode spacing, acquisition scheme, and borehole separation.

In environments with very conductive layers interspersed between more resistive layers, it is necessary to pay particular attention to the interpretation of results derived from surface ERT measurements, particularly in the evaluation of the thicknesses and depths of the layers. As apparent from the obtained results in this case study, both from synthetic datasets and field datasets, it is not possible to correctly define the real thickness of the clay layer, which seems to extend deeper in the inverted models. However, this is an important indication of the continuity of such a layer that has key implications for the water subsurface circulation at the site. However, to define the correct structure of the subsurface, it is necessary to perform ERT survey in cross-hole configuration, using appropriate borehole and electrode geometries, i.e., borehole length and separation and electrode spacing, and acquisition schemes.

The ratio inversion applied to ERT time-lapse surveys is a reliable approach for verifying natural or induced variations in the subsoil, which would be difficult to identify by comparing the resistivity models obtained by individual inversions. At the investigated site, the resistivity variation results (Figures 10–12) allow for an easy evaluation of the effect of the excavation of the trench and its effectiveness in draining the sub-surface water flow in the deeper gravel layer.

Cross-hole GPR is also a powerful technique, and in this case fully corroborates the ERT results, pointing more directly towards a quantitative estimation of moisture content and its space–time changes.

Author Contributions: Conceptualization, G.C.; methodology, G.C.; validation, G.C.; formal analysis, M.P. and B.M.; investigation, G.C., J.B., L.P., I.B., B.M. and G.C.; resources, G.C.; data curation, M.P. and B.M.; writing—original draft preparation, M.P.; writing—review and editing, G.C., J.B., L.P., I.B. and B.M.; visualization, M.P., J.B. and G.C.; supervision, G.C.; project administration, G.C.; funding acquisition, G.C. All authors have read and agreed to the published version of the manuscript.

Funding: This study was carried out within the RETURN Extended Partnership and received funding from the European Union Next-Generation EU (National Recovery and Resilience Plan—NRRP, Mission 4, Component 2, Investment 1.3—D.D. 1243 2/8/2022, PE0000005).

Data Availability Statement: Datasets used in this research will be sent to interested researchers upon request.

Acknowledgments: The authors acknowledge that this work is based on the thesis of Mirko Pavoni—Department of Geosciences—University of Padova.

Conflicts of Interest: The authors declare no conflicts of interest.

References

1. Cassiani, G.; Binley, A.; Kemna, A.; Wehrer, M.; Orozco, A.F.; Deiana, R.; Boaga, J.; Rossi, M.; Dietrich, P.; Werban, U.; et al. Noninvasive Characterization of the Trecate (Italy) Crude-Oil Contaminated Site: Links between Contamination and Geophysical Signals. *Environ. Sci. Pollut. Res.* **2014**, *21*, 8914–8931. [CrossRef] [PubMed]
2. Binley, A.; Hubbard, S.S.; Huisman, J.A.; Revil, A.; Robinson, D.A.; Singha, K.; Slater, L.D. The Emergence of Hydrogeophysics for Improved Understanding of Subsurface Processes over Multiple Scales. *Water Resour. Res.* **2015**, *51*, 3837–3866. [CrossRef]
3. Wang, T.P.; Chen, C.C.; Tong, L.T.; Chang, P.Y.; Chen, Y.C.; Dong, T.H.; Cheng, S.N. Applying FDEM, ERT and GPR at a site with soil contamination: A case study. *J. Appl. Geophys.* **2015**, *121*, 21–30. [CrossRef]
4. Godio, A.; Arato, A.; Stocco, S. Geophysical Characterization of a Nonaqueous-Phase Liquid–Contaminated Site. *Environ. Geosci.* **2010**, *17*, 141–161. [CrossRef]
5. Binley, A. Tools and Techniques: Electrical Methods. In *Treatise on Geophysics*; Elsevier: Amsterdam, The Netherlands, 2015; pp. 233–259.
6. Doetsch, J.; Linde, N.; Pessognelli, M.; Green, A.G.; Günther, T. Constraining 3-D Electrical Resistance Tomography with GPR Reflection Data for Improved Aquifer Characterization. *J. Appl. Geophys.* **2012**, *78*, 68–76. [CrossRef]
7. Santi, P.M.; McCray, J.E.; Martens, J.L. Investigating Cross-Contamination of Aquifers. *Hydrogeol. J.* **2006**, *14*, 51–68. [CrossRef]
8. Doetsch, J.; Linde, N.; Binley, A. Structural Joint Inversion of Time-lapse Crosshole ERT and GPR Traveltime Data. *Geophys. Res. Lett.* **2010**, *37*. [CrossRef]
9. Deiana, R.; Cassiani, G.; Kemna, A.; Villa, A.; Bruno, V.; Bagliani, A. An Experiment of Non-invasive Characterization of the Vadose Zone via Water Injection and Cross-hole Time-lapse Geophysical Monitoring. *Near Surf. Geophys.* **2007**, *5*, 183–194. [CrossRef]
10. Steelman, C.M.; Klazinga, D.R.; Cahill, A.G.; Endres, A.L.; Parker, B.L. Monitoring the evolution and migration of a methane gas plume in an unconfined sandy aquifer using time-lapse GPR and ERT. *J. Contam. Hydrol.* **2017**, *205*, 12–24. [CrossRef]
11. Fontana, A.; Mozzi, P.; Marchetti, M. Alluvial Fans and Megafans along the Southern Side of the Alps. *Sediment. Geol.* **2014**, *301*, 150–171. [CrossRef]
12. Fontana, A.; Mozzi, P.; Bondesan, A. Alluvial Megafans in the Venetian–Friulian Plain (North-Eastern Italy): Evidence of Sedimentary and Erosive Phases during Late Pleistocene and Holocene. *Quat. Int.* **2008**, *189*, 71–90. [CrossRef]
13. Hippe, K.; Fontana, A.; Hajdas, I.; Ivy-Ochs, S. A High-Resolution ^{14} C Chronology Tracks Pulses of Aggradation of Glaciofluvial Sediment on the Cormor Megafan between 45 and 20 Ka BP. *Radiocarbon* **2018**, *60*, 857–874. [CrossRef]
14. Fontana, A.; Rossato, S.; Poli, M.E.; Furlani, S.; Stefani, C. *Carta delle Unità Geologiche della Pianura del Friuli Venezia Giulia: Scala 1:150.000*; Università degli Studi di Trieste: Trieste, Italy, 2019; ISBN 9788894039429.
15. Binley, A.; Kemna, A. DC Resistivity and Induced Polarization Methods. In *Hydrogeophysics*; Rubin, Y., Hubbard, S.S., Eds.; Springer: Dordrecht, The Netherlands, 2005; pp. 129–156.
16. Blanchy, G.; Saneiyan, S.; Boyd, J.; McLachlan, P.; Binley, A. ResIPy, an Intuitive Open Source Software for Complex Geoelectrical Inversion/Modeling. *Comput. Geosci.* **2020**, *137*, 104423. [CrossRef]
17. Wagner, F.M.; Uhlemann, S. An Overview of Multimethod Imaging Approaches in Environmental Geophysics. In *Advances in Geophysics*; Elsevier: Amsterdam, The Netherlands, 2021; pp. 1–72.
18. Constable, S.C.; Parker, R.L.; Constable, C.G. Occam's Inversion: A Practical Algorithm for Generating Smooth Models from Electromagnetic Sounding Data. *Geophysics* **1987**, *52*, 289–300. [CrossRef]
19. Slater, L.; Binley, A. Synthetic and Field-Based Electrical Imaging of a Zerovalent Iron Barrier: Implications for Monitoring Long-Term Barrier Performance. *Geophysics* **2006**, *71*, B129–B137. [CrossRef]
20. Pavoni, M.; Carrera, A.; Boaga, J. Improving the Galvanic Contact Resistance for Geoelectrical Measurements in Debris Areas: A Case Study. *Near Surf. Geophys.* **2022**, *20*, 178–191. [CrossRef]
21. Pavoni, M.; Boaga, J.; Carrera, A.; Zuecco, G.; Carturan, L.; Zumiani, M. Brief Communication: Mountain Permafrost Acts as an Aquitard during an Infiltration Experiment Monitored with Electrical Resistivity Tomography Time-Lapse Measurements. *Cryosphere* **2023**, *17*, 1601–1607. [CrossRef]
22. Pavoni, M.; Boaga, J.; Wagner, F.M.; Bast, A.; Phillips, M. Characterization of Rock Glaciers Environments Combining Structurally-Coupled and Petrophysically-Coupled Joint Inversions of Electrical Resistivity and Seismic Refraction Datasets. *J. Appl. Geophys.* **2023**, *215*, 105097. [CrossRef]
23. Nielson, T.; Bradford, J.; Pierce, J.; Seyfried, M. Soil Structure and Soil Moisture Dynamics Inferred from Time-Lapse Electrical Resistivity Tomography. *Catena* **2021**, *207*, 105553. [CrossRef]
24. Lapenna, V.; Perrone, A. Time-Lapse Electrical Resistivity Tomography (TL-ERT) for Landslide Monitoring: Recent Advances and Future Directions. *Appl. Sci.* **2022**, *12*, 1425. [CrossRef]
25. Doetsch, J.; Linde, N.; Vogt, T.; Binley, A.; Green, A.G. Imaging and quantifying salt-tracer transport in a riparian groundwater system by means of 3D ERT monitoring. *Geophysics* **2012**, *77*, B207–B218. [CrossRef]
26. Daily, W.; Ramirez, A.; LaBrecque, D.; Nitao, J. Electrical Resistivity Tomography of Vadose Water Movement. *Water Resour. Res.* **1992**, *28*, 1429–1442. [CrossRef]
27. Topp, G.C.; Davis, J.L. Measurement of Soil Water Content Using Time-domain Reflectrometry (TDR): A Field Evaluation. *Soil Sci. Soc. Am. J.* **1985**, *49*, 19–24. [CrossRef]

28. Roth, K.; Schulin, R.; Flühler, H.; Attinger, W. Calibration of Time Domain Reflectometry for Water Content Measurement Using a Composite Dielectric Approach. *Water Resour. Res.* **1990**, *26*, 2267–2273. [CrossRef]
29. van Overmeeren, R.A.; Sariowan, S.V.; Gehrels, J.C. Ground Penetrating Radar for Determining Volumetric Soil Water Content; Results of Comparative Measurements at Two Test Sites. *J. Hydrol.* **1997**, *197*, 316–338. [CrossRef]
30. Robinson, D.A.; Friedman, S.P. A Method for Measuring the Solid Particle Permittivity or Electrical Conductivity of Rocks, Sediments, and Granular Materials. *J. Geophys. Res. Solid Earth* **2003**, *108*. [CrossRef]
31. Rossi, M.; Cassiani, G.; Binley, A. A Stochastic Analysis of Cross-Hole Ground-Penetrating Radar Zero-Offset Profiles for Subsurface Characterization. *Vadose Zone J.* **2012**, *11*, vzj2011-0078. [CrossRef]
32. Parkin, G.; Redman, D.; Von Bertoldi, P.; Zhang, Z. Measurement of soil water content below a wastewater trench using ground-penetrating radar. *Water Resour. Res.* **2000**, *36*, 2147–2154. [CrossRef]
33. Deiana, R.; Cassiani, G.; Villa, A.; Bagliani, A.; Bruno, V. Calibration of a Vadose Zone Model Using Water Injection Monitored by GPR and Electrical Resistance Tomography. *Vadose Zone J.* **2008**, *7*, 215–226. [CrossRef]
34. Pringle, J.K.; Westerman, A.R.; Clark, J.D.; Guest, J.A.; Ferguson, R.J.; Gardiner, A.R. The use of vertical radar profiling (VRP) in GPR surveys of ancient sedimentary strata. In *Ground Penetrating Radar in Sediments*; Geological Society: London, UK, 2003; Volume 211, pp. 225–246. [CrossRef]
35. Doetsch, J.A.; Coscia, I.; Greenhalgh, S.; Linde, N.; Green, A.; Günther, T. The Borehole-Fluid Effect in Electrical Resistivity Imaging. *Geophysics* **2010**, *75*, F107–F114. [CrossRef]
36. LaBrecque, D.; Alumbaugh, D.L.; Yang, X.; Paprocki, L.; Brainard, J. Chapter 15 Three-Dimensional Monitoring of Vadose Zone Infiltration Using Electrical Resistivity Tomography and Cross-Borehole Ground-Penetrating Radar. In *Methods in Geochemistry and Geophysics*; Elsevier: Amsterdam, The Netherlands, 2002; pp. 259–272.
37. Topp, G.C.; Davis, J.L.; Annan, A.P. Electromagnetic Determination of Soil Water Content: Measurements in Coaxial Transmission Lines. *Water Resour. Res.* **1980**, *16*, 574–582. [CrossRef]

water

MDPI

Article

Assessing and Improving the Robustness of Bayesian Evidential Learning in One Dimension for Inverting Time-Domain Electromagnetic Data: Introducing a New Threshold Procedure

Arsalan Ahmed [1,*], Lukas Aigner [2], Hadrien Michel [3], Wouter Deleersnyder [1,4], David Dudal [4,5], Adrian Flores Orozco [2] and Thomas Hermans [1,*]

1 Department of Geology, Ghent University, Krijgslaan 281-S8, 9000 Gent, Belgium; wouter.deleersnyder@kuleuven.be
2 Research Unit Geophysics, Department of Geodesy and Geoinformation, TU Wien, 1040 Vienna, Austria; lukas.aigner@geo.tuwien.ac.at (L.A.); adrian.flores-orozco@geo.tuwien.ac.at (A.F.O.)
3 Urban and Environmental Engineering, Faculty of Applied Sciences, University of Liege, 4000 Liege, Belgium; hadrien.michel@uliege.be
4 Department of Physics, KU Leuven Campus Kortrijk—KULAK, Etienne Sabbelaan 53, 8500 Kortrijk, Belgium; david.dudal@kuleuven.be
5 Department of Physics and Astronomy, Ghent University, Krijgslaan 281-S9, 9000 Gent, Belgium
* Correspondence: arsalan.ahmed@ugent.be (A.A.); thomas.hermans@ugent.be (T.H.)

Abstract: Understanding the subsurface is of prime importance for many geological and hydrogeological applications. Geophysical methods offer an economical alternative for investigating the subsurface compared to costly borehole investigations. However, geophysical results are commonly obtained through deterministic inversion of data whose solution is non-unique. Alternatively, stochastic inversions investigate the full uncertainty range of the obtained models, yet are computationally more expensive. In this research, we investigate the robustness of the recently introduced Bayesian evidential learning in one dimension (BEL1D) for the stochastic inversion of time-domain electromagnetic data (TDEM). First, we analyse the impact of the accuracy of the numerical forward solver on the posterior distribution, and derive a compromise between accuracy and computational time. We also introduce a threshold-rejection method based on the data misfit after the first iteration, circumventing the need for further BEL1D iterations. Moreover, we analyse the impact of the prior-model space on the results. We apply the new BEL1D with a threshold approach on field data collected in the Luy River catchment (Vietnam) to delineate saltwater intrusions. Our results show that the proper selection of time and space discretization is essential for limiting the computational cost while maintaining the accuracy of the posterior estimation. The selection of the prior distribution has a direct impact on fitting the observed data and is crucial for a realistic uncertainty quantification. The application of BEL1D for stochastic TDEM inversion is an efficient approach, as it allows us to estimate the uncertainty at a limited cost.

Keywords: uncertainty; saltwater intrusion; TDEM; BEL1D; SimPEG

1. Introduction

Geophysical methods offer an economical alternative for investigating the subsurface compared to the use of direct methods. Most geophysical methods rely on a forward model to link the underlying physical properties (e.g., density, seismic velocity, or electrical conductivity) to the measured data and by solving an inverse problem. Deterministic inversions typically use a regularization approach to stabilize the inversion and resolve the non-unicity of the solution, yielding a single solution. However, uncertainty quantification is generally limited to linear noise propagation [1–4]. In contrast, stochastic inversion methods based on a Bayesian framework compute an ensemble of models fitting the data, based on the exploration of the prior model space [5]. Bayesian inversion is rooted in the fundamental principle that a posterior distribution can be derived from the product of the likelihood function and the prior distribution. Various strategies have been developed in this regard, as evidenced by the literature across several disciplines, including, but not limited to, hydrology and hydrogeology [6–9] or geophysics [10,11]. Although the increase in computer performance has advanced the use of stochastic approaches, long computational time remains an important issue for their broader adoption [12–15]. Indeed, most stochastic approaches rely on Markov chain Monte Carlo (McMC) methods for sampling the posterior model space [5], which require a large number of iterations and forward model computations.

Alternatives have been developed to estimate the posterior distribution at a limited cost such as Kalman ensemble generators [16,17] or Bayesian Evidential learning (BEL) [18,19]. BEL is a simulation-based prediction approach that has been initially proposed to by-pass the difficult calibration of subsurface reservoir models and to directly forecast targets from the data [20,21], with recent applications in geothermal energy [22–24], reservoir modelling [25–28], experimental design [29] and geotechnics [30]. It has also been quickly adopted by geophysicists to integrate geophysical data into model or properties prediction [24,31,32]. BEL has also been recently proposed as an efficient alternative for the 1D inversion of geophysical data (BEL1D) [18,19]. BEL1D circumvents the inversion process by using a machine learning approach derived from Monte Carlo sampling of the prior distribution. It has been proven efficient for the estimation of the posterior distribution of water content and relaxation time from nuclear magnetic resonance data [18], and the derivation of seismic velocity models from the analysis of the dispersion curve [19]. The main advantage of BEL1D is to rely on a smaller number of forward model runs than McMC approaches to derive the posterior distribution, leading to a reduced computational effort. Earlier work has shown that BEL1D converges towards the solution obtained from an McMC procedure but it slightly overestimates the uncertainty, especially in the case of large prior uncertainty [18]. The use of iterative prior resampling followed by a filtering of models based on their likelihood has been recently proposed to avoid uncertainty overestimation [19]. Although this increases the computational cost of BEL1D, it remains about one order of magnitude faster than McMC [19]. In this contribution, we propose to apply a threshold on the data misfit after the first BEL1D first iteration to circumvent the need for multiple iterations when prior uncertainty is large.

So far, BEL1D has only been applied to a limited number of geophysical methods. In this contribution, we apply the algorithm to the inversion of time-domain electromagnetic (TDEM) data. We combine BEL1D with the TDEM forward modelling capabilities of the open-source Python package SimPEG (version 0.14.1) [33,34] for the stochastic inversion of TDEM data. Electromagnetic surveys have proven to be efficient for delineating groundwater reservoir structure and water quality (e.g., [35–37]). In the last decades, the popularity of TDEM has largely increased with the adoption of airborne TDEM surveys for mineral but also hydrogeological applications (e.g., [38–40]). More recently, towed transient electromagnetic (tTEM) systems [41] and waterborne TDEM systems [42,43] were designed for continuous measurements of TDEM data, thus allowing to the coverage of large areas in relatively short times.

To date, the inversion of such extensive surveys relies on deterministic quasi-2D or -3D inversion [44], i.e., using a 1D forward model with lateral constraints. In the process of resolving inverse problems, which entails fitting observational data, the forward model representing the underlying physical processes is pivotal. However, this model is susceptible to errors inherent in the modelling process. The employment of an accurate numerical forward model imposes substantial computational demands, consequently constraining the feasible quantity of forward simulations [45].

In light of these computational constraints, it is a prevalent practice to resort to rapid approximation strategies for the forward solver [46,47], to work with a coarser discretization [48,49] or to deploy surrogate models to replace the expensive simulations [50–52].

Modelling errors may introduce significant biases in the posterior statistical analyses and may result in overly confident parameter estimations if these errors are not accounted for [52,53]. Hansen et al. [53] studied the effect of using approximate forward models on the inversion of GPR cross-hole travel time data and demonstrated that the modelling error could be more than one order of magnitude larger than the measurement error, leading to unwanted artifacts in the realizations from the posterior probability. For EM methods, studies have demonstrated the negative effect of using fast approximations of the forward model on the accuracy of the inversion [54–56]. In particular, for TDEM methods, using accurate models is computationally too expensive to be attractive for stochastic inversion of large data sets, as are those obtained from surveys with airborne or tTEM methods. Therefore, in this study we investigate a possible approach that balances accuracy in the modelling and reduced computational costs.

Stochastic approaches for the inversion of TDEM are therefore still uncommon (e.g., [13,15,56,57], yet these are computationally demanding for large data sets. Typically, the whole inversion needs to be re-run for every sounding, independently. Hence, developing a fast alternative is highly relevant for to-date hydro-geophysical investigations.

In this paper, we focus on the robustness of BEL1D to retrieve the posterior distributions of electrical subsurface model parameters from the inversion of TDEM data. The novelties of our contribution lie in the following:

1. Demonstrating that BEL1D is an efficient approach for the stochastic inversion of TDEM data.
2. Exploring the impact of the accuracy of the forward solver to estimate the posterior distribution, and finding a compromise between accuracy and computational cost.
3. Proposing and validating a new thresholding approach to circumvent the need for iterations when the prior uncertainty is large.
4. Applying the new approach to field TDEM data collected in the Luy River catchment in the Binh Thuan province (Vietnam) for saltwater intrusion characterization. This data set was selected because electrical resistivity tomography (ERT) data are available for comparison, but lack sensitivity at greater depth. The case study is also used to illustrate the impact of the selection of the prior on the posterior estimation.

The computational undertakings in this study are performed using the pyBEL1D package (version 1.1.0) [58] which serves as the computational backbone for our analyses. This integration of theoretical insights and practical applications is intended to advance the understanding and uncertainty quantification of TDEM surveys.

2. Materials and Methods

2.1. BEL1D

In contrast to deterministic approaches, BEL1D does not rely on the stabilization of the ill-posed inverse problem through regularization. Instead, BEL1D learns a statistical relationship between the target (the set of parameters of interest, in this case a subsurface layered model of the electrical conductivity) and the predictor (the geophysical data). This statistical relationship is derived from a combination of models and data (typically a few thousand) drawn from the prior distribution which reflects the prior geological knowledge. For each sampled model, the forward model is then run to generate the corresponding data set [18]. Next, a statistical relationship is learned in a lower dimensional space and used to calculate the posterior distribution corresponding to any data set consistent with the prior, without the need to run any new forward model. We refer to [18,19] for details about the algorithm. Here, we only provide a short overview. BEL1D consists of seven steps:

Step 1: Prior sampling and forward modelling

As in any stochastic inversion, the first step is to assign the range of prior uncertainty based on earlier field knowledge. For TDEM 1D inversion, we need to define the number of layers, their thickness and electrical conductivity. A set of n prior models is sampled. For each sampled model, the corresponding TDEM data are simulated using the forward model. In this step, it is important to state the size of the transmitting and receiving loop and the waveform and magnetic momentum of the primary field, as well as the acquisition time and sampling of the decay-curve.

More specifically, this first step entails defining the prior model using a finite set of N_L layers, with the final layer simulating the half-space. Except for this layer, which is defined by its conductivity only, the other layers are defined by their conductivity and thickness. Thus, the total number of model parameters or unknowns is $q = 2 \times N_L - 1$. For each of those q parameters, a prior distribution is described, which must reflect the prior understanding of the survey site. Such information can be based on either previous experiments or more general geological and geophysical considerations. Random models are sampled within the prior range, and the forward model is run for each one to calculate the corresponding noise-free data set $\boldsymbol{d}$ (Figure 1, boxes 1 and 2):

$$\boldsymbol{d} = f(\boldsymbol{m}) \tag{1}$$

where $\boldsymbol{m}$ is the set of q model parameters and f is the forward model solving the physics (see Section 2.2).

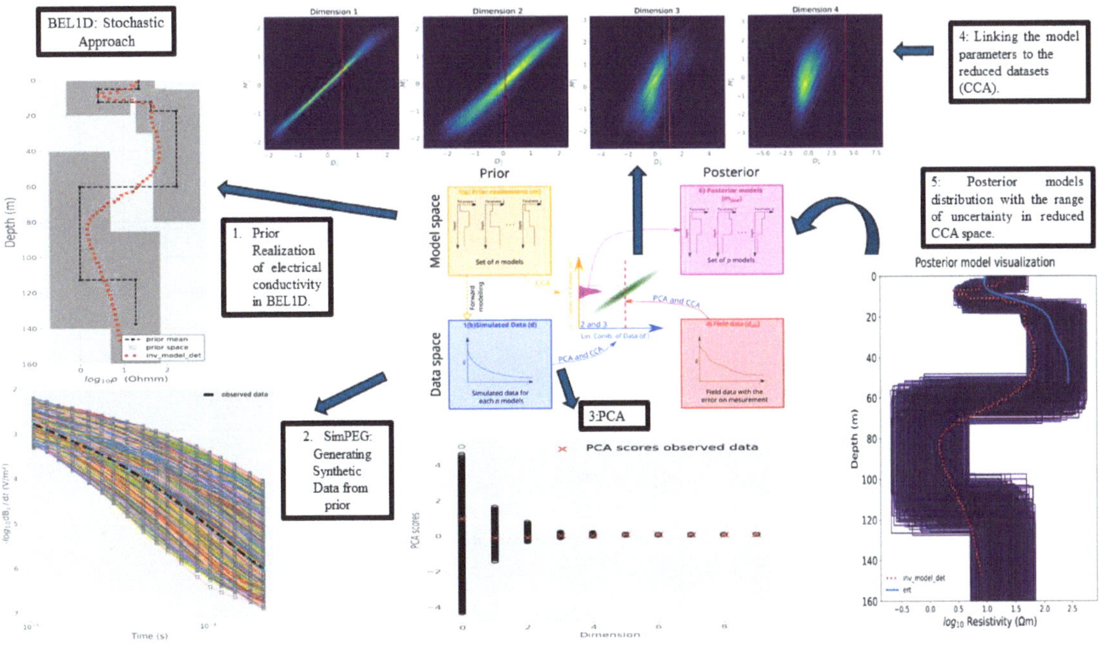

Figure 1. The schematic diagram of BEL1D applied to TDEM data (modified from [18]).

Step 2: Reducing the dimensionality of data.

Lowering the dimensionality of the data is required to determine a statistical connection between the target and the predictor. Dimension reduction also helps to limit the impact of noise on the inversion [31]. Principal component analysis (PCA) identifies linear combinations of variables that explain most of the variability by using the eigenvalue decomposition [59]. Higher dimensions typically exhibit less variability and can be disregarded. Noise is propagated using Monte Carlo simulation [18,31] to estimate the uncertainties of the PCA scores caused by data noise (Figure 1 box 3). Similarly, the dimensions of model parameters q can be reduced if necessary.

Step 3: Statistical relationship between target (model parameters) and predictor (the reduced dataset)

Canonical correlation analysis (CCA) is used to determine a direct correlation between the target and predictor [18]. CCA essentially calculates the linear combinations of (reduced) predictor variables and target variables that maximize their correlation, producing a set of orthogonal bivariate relationships [59]. The correlation typically decreases with the dimensions, the first dimension being the most correlated (Figure 1 box 4).

Note that CCA is not the only approach for deriving a statistical relationship. Due to the expected non-linearity in the statistical relationship between seismic data and reservoir properties, [32] have used summary statistics extracted from unsupervised- and supervised-learning approaches including discrete wavelet transform and a deep neural network combined with approximated Bayesian computation to derive a relationship. Similarly, [60] used a probabilistic Bayesian neural network to derive the relationship.

Step 4: Generation of the posterior distributions in CCA space

In the CCA reduced space, kernel density estimation (KDE) with a Gaussian kernel [61] is used to map the joint distribution $f_H(\boldsymbol{m}_c, \boldsymbol{d}_c)$, where the suffix c refers to the canonical space and m and d stand for model and data. We employ a multi-Gaussian kernel with bandwidths selected in accordance with the point density [18]. The resulting distributions are not restricted to any specific distribution with a predetermined shape. As a result, a simple and useful statistical description of the bivariate distribution can be generated (Figure 1 box 4).

Using KDE, results are partly dependent on the choice of the kernel, especially the bandwidth, which can result in posterior samples falling out of the prior space [18]. The py-BEL1D code allows for the filtering of erroneous posterior samples resulting from KDE [58]. This limitation partly explains why BEL1D tends to overestimate the posterior distribution [18,19], as the derived joint distribution is an approximation in a lower dimensional space, not relying on the calculation of a likelihood function, such as in McMC, which would ensure convergence to the actual posterior distribution. However, [19] has empirically shown that the approach was efficient and yielded similar results to McMC. An alternative to KDE is to use transport maps [24].

Step 5: Sampling of the constituted distributions

The KDE maps are then used to extract from the joint distribution the posterior distribution $f_H(\boldsymbol{m}_c|\boldsymbol{d}_{obs,c})$ for any observed data set projected into the canonical space $\boldsymbol{d}_{obs,c}$. Using the inverse transform sampling method [62], we can now easily generate a set of samples from the posterior distributions in the reduced sample (Figure 1 box 5).

Step 6: Back transformation into the original space

The set of the posterior samples in CCA space are back-transformed into the original model space. The only restriction is that more dimensions must be kept in the predictor than the target in order to support this back transformation. The forward model is then run for all sampled models to compute the root-mean-squared error (RMSE) between observed and simulated data.

Step 7: Refining the posterior distribution by IPR or a threshold

In case of large prior uncertainty, [19] recommend applying iterative prior resampling (BEL1D-IPR). The idea is to enhance the statistical relationship by sampling more models in the vicinity of the solution. In short, models of the posterior distribution are added to the prior distributions, and steps 2 to 6 are repeated. This iterative procedure is followed by a filtering of the posterior models based on their likelihood, using a Metropolis sampler. This allows for the sampling of the posterior distribution more accurately, but at a larger computational cost. BEL1D-IPR is used as the reference solution in this study as it has been benchmarked against McMC [19].

We propose to reduce the computational effort of BEL1D-IPR by applying a filtering procedure after the first iteration. The threshold criterion is defined based on the expected relative RMSE (rRMSE) estimated from the data noise. The rRMSE is calculated in log space to account for the large range of variations in the amplitude of the measured TDEM signal, so that a systematic relative error expressed in % corresponds to a predictable value of the rRMSE calculated in log space. For each time window, we assume the systematic error can be expressed as a percentage of the expected signal d_i

$$e_i = a\, d_i$$

where a is the expected relative error. The measured data could then be expressed as

$$d_{i,m} = (1 + a)d_i$$

Expressing the error on a log scale, we have

$$e_{i,log} = \log d_{i,m} - \log d_i = \log \frac{d_{i,m}}{d_i} = \log(1 + a)$$

which is independent from the absolute value of the data. It is then possible to predict the rRMSE value if a systematic relative error a was contaminating the data set. This value is 0.18, 0.135, and 0.05 for a systematic error on the data of 20, 15 and 5%, respectively.

Since the actual error on field data is not systematic but has a random component, and since the estimation of the level of error from stacking might underestimate the error level, the choice of the threshold is somehow subjective. In the field case, for example, a stacking error of about 5% was estimated, which we found to underestimate the actual noise level so that we chose a threshold corresponding to 3 times that value (15%, or a threshold of 0.135 on the rRMSE).

With such an approach deviating from the Bayesian framework, the posterior solution is only an approximation of the true posterior distribution. The main advantage is to eliminate the need to run new forward models and to ensure that the same prior distribution can be used for several similar data sets, making the prediction of the posterior very fast in surveys with multiple soundings. We refer to this new approach as BEL1D-T.

2.2. *SimPEG: Forward Solver*

We use the open-source python package SimPEG to obtain the TDEM response for a given set of model parameters and the acquisition set-up [33,34]. The main advantage of SimPEG is that it provides an open source and modular framework for simulating and inverting many types of geophysical data. We opted for a numerical implementation instead of the more classical semi-analytical solution such as the one provided in empymod [63] to assess the impact of a modelling error in the forward model on the estimation of the posterior. This step is crucial to assess how an error in the forward model propagates into the posterior distribution. Indeed, for the field data inversion, we initially experienced some inconsistencies between the prior and the data, and we wanted to rule out the forward solver being responsible for it. We nevertheless limit ourselves to a strictly 1D context, yet the approach could be extended to assess the error introduced by multi-dimensional effects (through a 2D or 3D model), and is therefore flexible. However, the use of a 3D model increases the computational cost, and it is beyond the scope of this study to compare numerical and semi-analytical forward solvers [55].

The SimPEG implementation uses a staggered-grid discretization [64] for the finite volume approach [34], which calls for the definition of the physical properties, fields, fluxes, and sources on a mesh [65–67]. The details of the implementation can be found in [33,34]. For the 1D problem, SimPEG makes use of a cylindrical mesh. The accuracy and computational cost of the forward solver depend on the time and space discretization.

2.2.1. Temporal Discretization

For the temporal discretization, it is a good practice to start with short time steps at the early times when the electromagnetic fields change rapidly [65]. At later stages, the time steps can be increased as the variations in the EM fields are more gradual and the signal-to-noise ratio (S/N) decreases. Shorter time steps increase the accuracy of the forward model but also the calculation time. Hence, it is important to find an adequate trade-off between accuracy and computational cost. In this paper, we tested three sets of temporal discretization with increased minimum and average size for the time steps (Table 1 and Figure 2).

Table 1. Description of the different temporal discretization. F (fine), I (intermediate) and C (coarse) are the corresponding acronyms.

Temporal Discretization	Total Number of Time Steps	Maximum Size of Time Steps (s)	Weighted Average Length of Time Steps (s)
Fine (F)	1710	10^{-5}	0.581×10^{-6}
Intermediate (I)	510	10^{-5}	1.95×10^{-6}
Coarser (C)	185	10^{-4}	5.38×10^{-6}

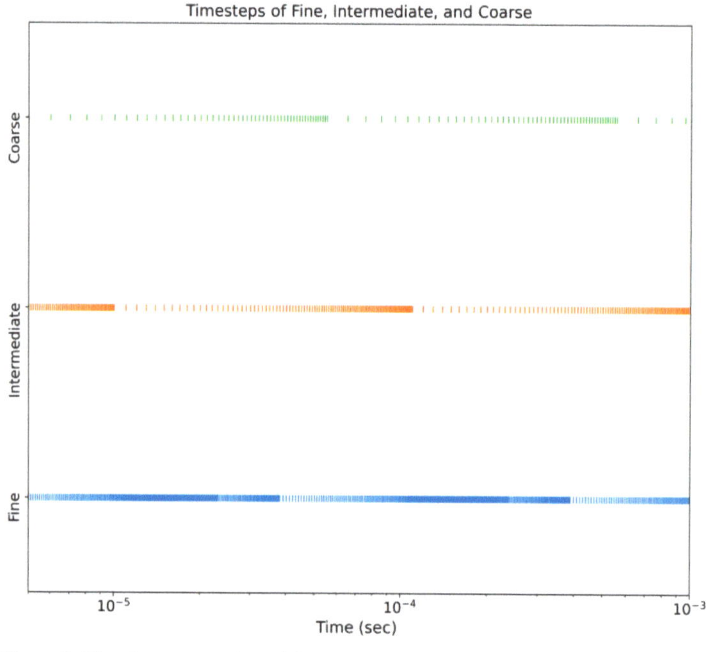

Figure 2. Visual representation of the time discretization. The Y-axis shows the time discretization and the X-axis shows the logarithmic scale of the time-step size.

2.2.2. Spatial Discretization

Spatial discretization also has a direct impact on the accuracy of the forward solver [65]. When creating the mesh, as shown in Figure 3, the discretization in the vertical direction is controlled by the cell size in the z-direction, whereas the horizontal discretization is controlled by the cell size in the x-direction. A finer discretization results in a more accurate solution but is also more computationally demanding. Note that a coarse discretization might also prevent an accurate representation of the layer boundaries as defined in the prior. If the layer boundary does not correspond to the edge of the mesh, a linear interpolation is used. In this paper, we selected five values for the vertical discretization to test the impact of the spatial discretization on the estimated posterior (Table 2).

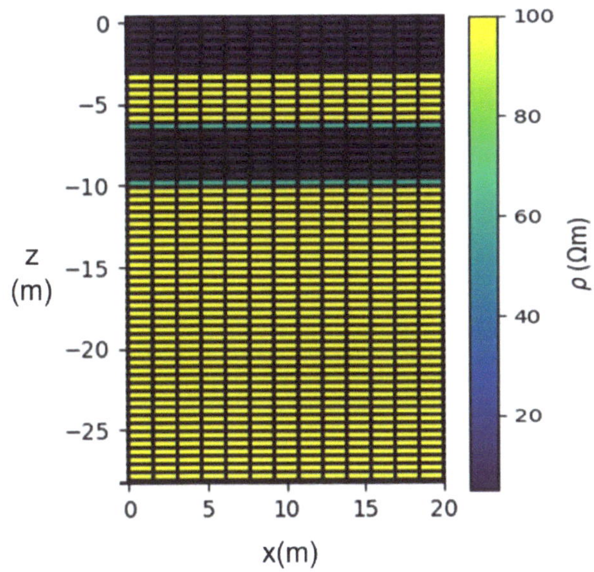

Figure 3. Example of the cylindrical mesh used for the forward model with a vertical discretization of 0.5 m and a horizontal discretization of 1.5 m. The cells with positive z represent the air, and are modelled with a very high resistivity and logarithmically increasing cell size.

Table 2. Cell size in z-direction for the different spatial discretization. The letters in brackets, VF (very fine), F (fine), M (medium), C (coarse) and VC (very coarse) are used as acronyms in the remainder of this paper.

Spatial Discretization	Thickness of Grid Cells (in m)
Very Fine (VF)	0.25
Fine (F)	0.5
Medium (M)	1
Coarse (C)	1.5
Very Coarse (VC)	2

2.3. Synthetic Benchmark

We analysed the impact of both temporal and spatial discretization on the accuracy of the posterior distribution for all fifteen combinations of the temporal and spatial discretization (see Tables 1 and 2), using synthetic data. A single combination is referred to by its acronyms, starting with the time discretization. The combination F-C, for example, corresponds to the fine time discretization combined with the coarse spatial discretization.

The synthetic data set is created with the finest discretization using the benchmark model parameters in brackets (see Table 3) defined by a five-layer model, with the last layer having an infinite thickness. The posterior distribution obtained with that same discretization and BEL1D-IPR is used as a reference. The prior is also the same for all tests and consists of uniform distributions for the nine model parameters (Table 3). The acquisition settings mimic the field set-up; see the following subsection.

Table 3. Prior range of values for all parameters of the model. Benchmark model parameters for the synthetic model are shown in brackets.

Layers	Thickness (m)	Resistivities (ohmm)
Layer 1	0.5–6.5 (5)	10–55 (20)
Layer 2	5–15 (10)	1–15 (4.5)
Layer 3	0.5–10 (5)	20–100 (50)
Layer 4	35–50 (42)	50–115 (75)
Layer 5	∞ (∞)	5–20 (10)

3. Field Site

Understanding the interactions between salt and freshwater dynamics is crucial for managing coastal aquifers, yet it is difficult due to the required subsurface information, with high spatial and temporal resolution not always accessible from borehole data. The study area for the field tests is located in the Luy River catchment in the Binh Thuan province (Vietnam, Figure 4), which has been facing saltwater intrusions problems for many years [68–70].

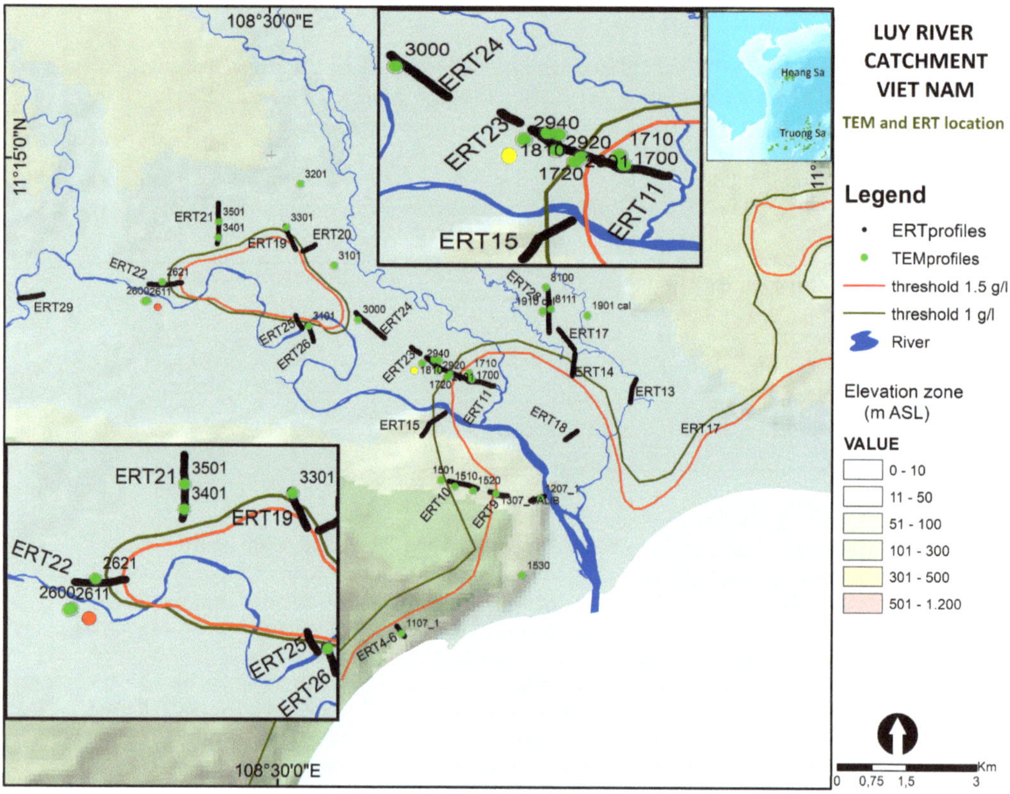

Figure 4. The Luy River catchment in Vietnam with location of TDEM soundings (green points) and ERT profile (black line). The red and yellow dots represent the location of the soundings (2611 and 1307) used in this paper [68,69].

The data were collected using the TEM-FAST 48 equipment (Applied Electromagnetic Research, Utrecht, The Netherlands), with a 25 m square loop with a single turn acting as both transmitter and receiver. The injected current was set to 3.3 A with a dead-time of 5 μs. The data were collected using 42 semi-logarithmic time windows ranging from 4 μs to 4 ms. The signal was stacked allowing for noise estimation. A 50 Hz filter was applied to remove noise from the electricity network. For the inversion, the early time and late time were manually removed (see Sections 4.3 and 4.4). The recorded signals at an early time steps, i.e., below 10^{-5} μs, were impacted by the current switch-off phenomena, while above 1 ms the signal-to-noise ratio was too low. We therefore filtered the TDEM data to a time range from 8 μs to 500 μs. In the forward model, we implemented the current shut-off ramp from the TEM-FAST48 system following the approach proposed by [71].

4. Results

We subdivide the results into four subsections. In the first subsection, we analyse the impact of the accuracy of the forward solver on the accuracy of the posterior in BEL1D-IPR. In the second section, we test the impact of a threshold on the rRMSE applied after the first BEL1D iteration (BEL1D-T). The third subsection is dedicated to the selection of the prior. Finally, the last section corresponds to the application of BEL1D-T to the field data.

4.1. Impact of Discretization

In this section, we tested in total 15 combinations of temporal and spatial discretization to study their behaviour on both the computation time and the accuracy of the posterior distribution computed with BEL1D-IPR (four iterations). The reference used the finest time and spatial discretization (F-VF). Since the computational costs of BEL1D are directly related to the number of prior samples and the computational cost of running one forward model [19], computing the solution for the F-VF combination is more than 150 times more expensive that running it with the C-VC combination (Table 4). An initial set of 1000 models is used in the prior. All calculations and simulation were carried out on a desktop computer with the following specifications: Processor intel ® CORE ™ i7-9700 CPU @ 3.00 GHz, RAM 16.0 GB.

Table 4. Time (in seconds) to compute one forward model in SimPEG for the 15 combinations of time and space discretization. The red colour corresponds to posterior distributions whose mean is biased, whereas the blue colour represents an under- or overestimation of the uncertainty for the two shallowest layers.

	Spatial Discretization				
Time	**VF**	**F**	**M**	**C**	**VC**
F	389.02	73.88	33.4	25.92	17.7
I	114.79	22.38	6.3	3.55	2.73
C	44.98	11.48	3.90	2.46	2.02

We first analyse the impact of the forward solver in BEL1D-IPR. A very similar behaviour is noted for all combinations using the VF spatial discretization, in combination with the three temporal discretization for all parameters (Figure 5). The parameters (thickness and resistivity) of the two first layers are recovered with relatively low uncertainty, while the uncertainty remains quite large for deeper layers, showing the intrinsic uncertainty of the methods related to the non-unicity of the solution. The results look globally similar, but a detailed analysis of the posterior distribution focusing on the resolved parameters (two first layers, see Figure 6) shows a slight bias of the mean value in C-VF and I-VF for the thickness of the second layer. This bias is small (less than 0.5 m) and could be the result of the sampling. A slightly larger uncertainty range can also be observed for the I and C time discretization.

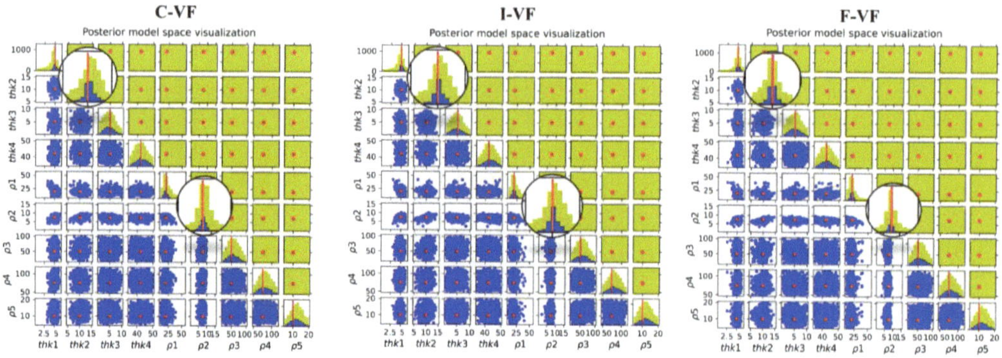

Figure 5. Posterior model space visualization of fine-, intermediate-, and coarse-time discretization with very fine spatial discretization symbolized as C-VF, I-VF and F-VF. Thickness in meters and resistivity in ohm.m. Yellow dots correspond to prior models, blue dots to posterior models.

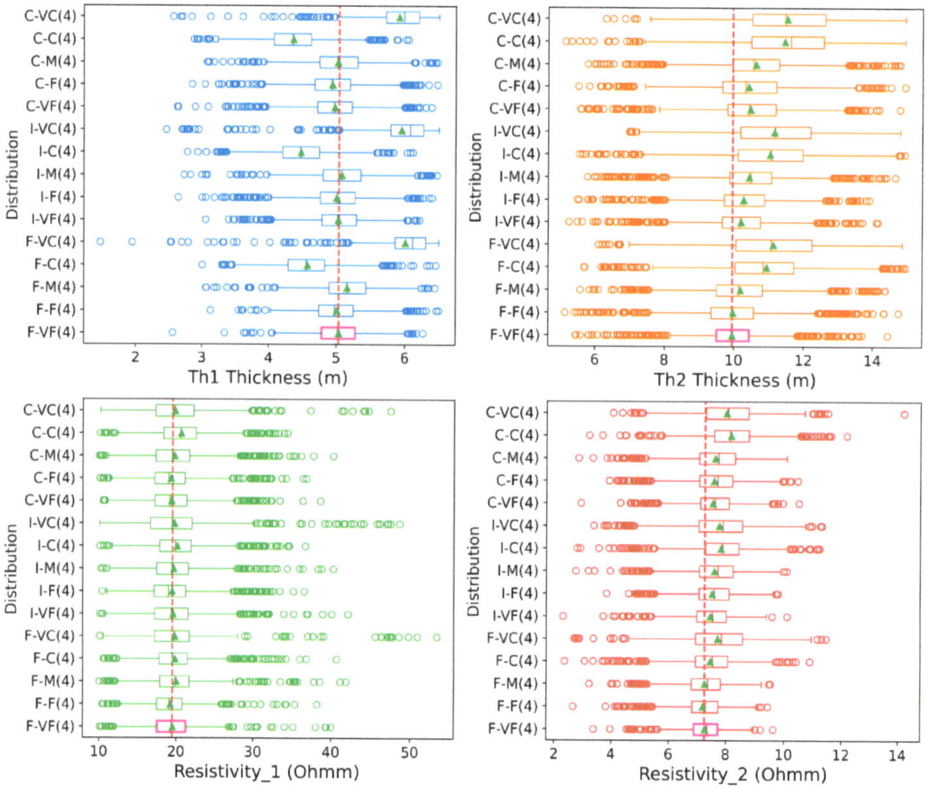

Figure 6. Box plot of thickness and resistivity for the two first layers for BEL-IPR (four iterations). The red line shows the benchmark value and the F-VF(4) is the reference solution.

Globally, a systematic bias is observed for the largest spatial discretization (VC and C) for the thickness of layers 1 and 2 (Figure 6), which can likely be attributed to the difficulty in properly representing thin layers with a coarse discretization. A bias in the thickness of layer 2 is also noted for all coarse-time discretization, and to a lesser extent for the intermediate-time discretization, although this is limited when combined with F and VF spatial discretization. There is no significant bias visible in the estimation of the resistivity of layer 1, while most combinations have a small but not significant bias for layer 2, and the uncertainty range tends to be overestimated or underestimated for most combinations with large spatial discretization. Eventually, combinations with a VF or F spatial discretization combined with all time discretization, as well as the F-M combination, provide relatively similar results to the reference F-VF.

The time and spatial discretization for simulating the forward response of TDEM have therefore a strong impact, not only on the accuracy of the model response, but also on the estimation of the parameters of the shallow layers after inversion. In particular, the coarser spatial discretization biases the estimation of the thicknesses of the shallow layer. The same is also observed for the combination of a coarse- or intermediate-time discretization with a medium spatial discretization. As shallow layers correspond to the early times, this bias is likely related to an inaccurate simulation of the early TDEM response by the forward solver due to the chosen discretization. Although it comes with a high computation cost, we recommend keeping a relatively fine time and space discretization to guarantee the accuracy of the inversion. The cheapest option in terms of computational time with a minimum impact on the posterior distribution corresponds in this case to the C-F combination.

4.2. *Impact of the Threshold*

Because of the additional costs associated with the iterations, we compare the posterior distributions obtained with BEL1D-IPR to our new BEL1D-T approach, applying a threshold after the first iteration. The selected threshold based on the rRMSE calculated on the logarithm of the data are 0.18, 0.135, 0.05, corresponding, respectively, to a systematic error on the data of 20%, 15% and 5%. Various values of the threshold are tested for the reference solution (F-VF discretization) (Figure 7) and the analysis of the discretization is repeated (Figure 8). The threshold is applied after the first iteration to avoid additional computational time. The corresponding posterior distribution retains only the models that fit the data to an acceptable level. Note that the corresponding posterior distribution has a lower number of models than the IPR on BEL1D, as the latter enriches the posterior with iterations.

For solutions without a threshold, the colour scale is based on the quantiles of the RMSE in the posterior distribution. The threshold thus removes the models with the largest RMSE (yellow-green). Without the threshold (Figure 7A,B,G), some models not fitting the data are present in the posterior. The threshold approach after one iteration succeeds in obtaining a posterior closer to the reference solution (Figure 7D–F,J–L). The benchmark model, which is the true model, lies in the middle of the posterior.

The impact of the selected threshold value on the posterior distribution is illustrated in Figure 7D–F,J–L. Since the threshold is based on the rRMSE, decreasing its value is equivalent to rejecting the models with the largest data misfit from the posterior, while only models fitting the data with minimal variations are kept in the posterior. This rejection efficiently removes poor models from the posterior. If a low value is selected, only the very few best-fitting models are kept, and these are very similar to the reference model, hence reducing the posterior uncertainty range in the selected models (overfitting), while a high value of the threshold might retain models that do not fit the data within the noise level. The choice of the threshold should therefore be carefully made based on the noise level, and its sensitivity should be assessed.

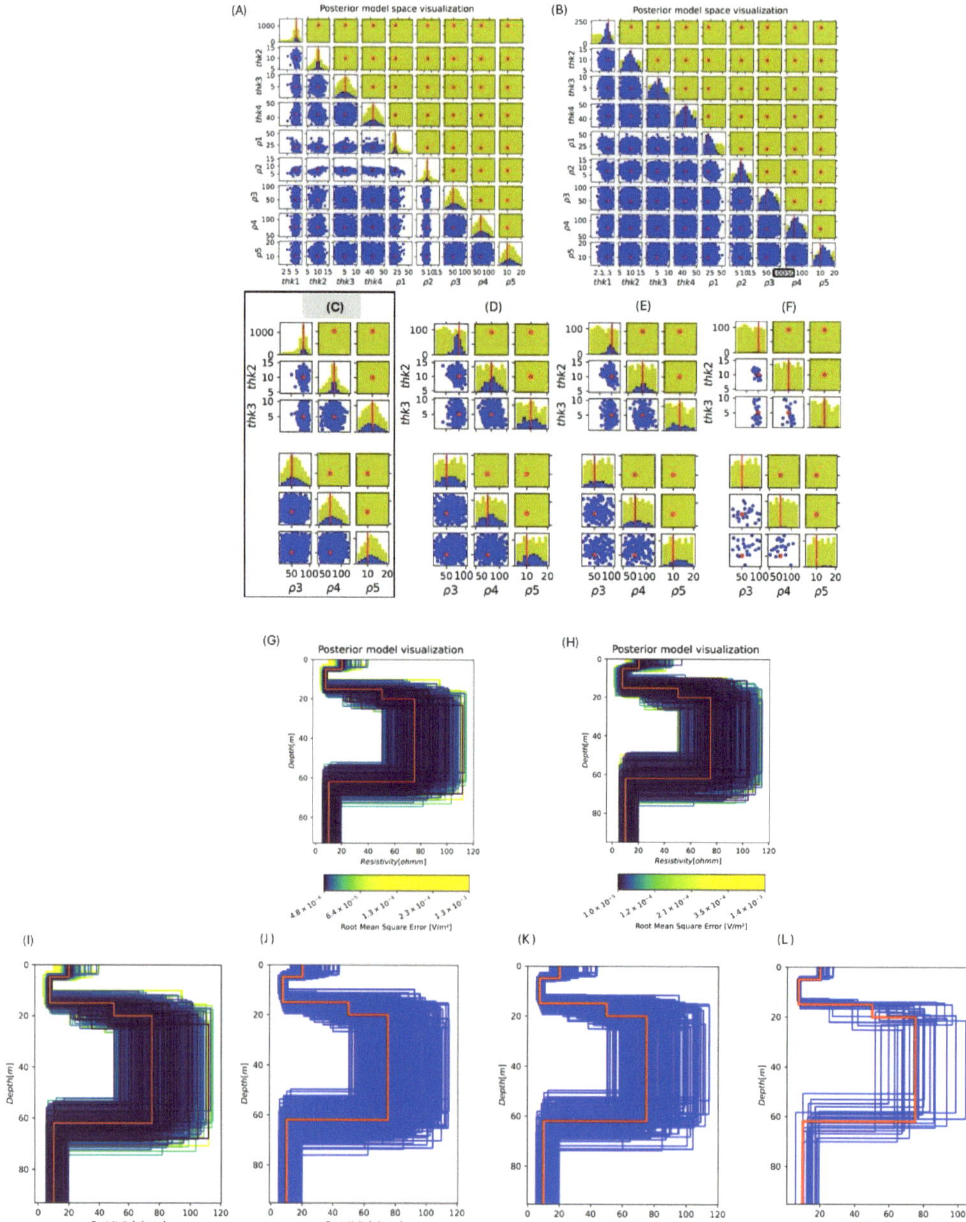

Figure 7. Posterior-model space visualization: yellow dots represent the prior distribution, blue dots show the posterior distribution, and the red line corresponds to the benchmark model. The panels represent the following: (**A**) the posterior-model space distribution after four iterations without a threshold (BEL1D-IPR); (**B**) the posterior-model space distribution after one iteration without threshold. Comparison between BEL1D-IPR (**C**) and three threshold values for BEL1D-T (0.18, 0.135 and 0.05 (**D–F**)). The x and y axes are equivalent to resistivity (ohm.m) and depth (m). Posterior-model distribution for BEL1D-IPR (**G**,**I**) and after one iteration without a threshold (**H**) (the color scale is based on the value of the RMSE) and BEL1D-T with threshold values (0.18, 0.135 and 0.05, **J–L**).

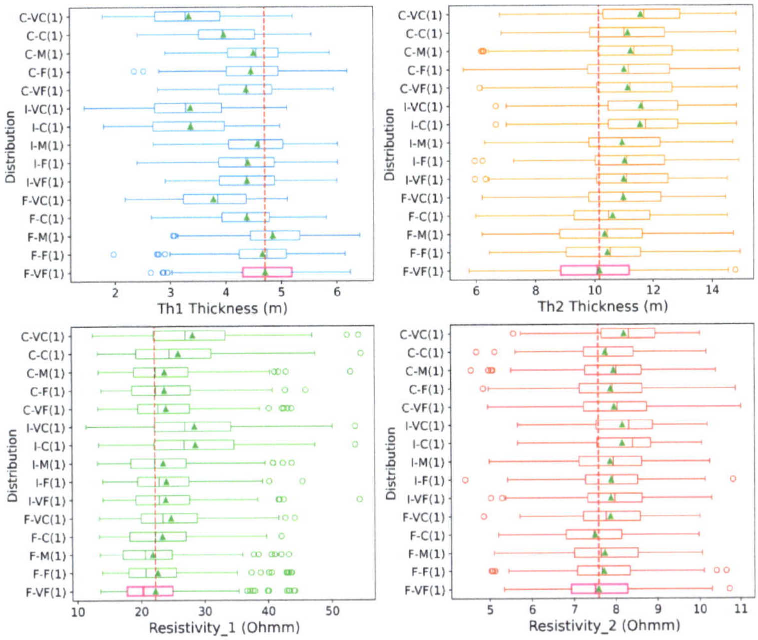

Figure 8. Box plot of thickness and resistivity for the two first layers after one iteration. Red line shows the benchmark F-VF (1). (1) represents the first iteration.

Since the choice of the threshold impacts the rejection rate, the number of samples to generate cannot be estimated a priori. An initial estimate can, however be derived from a limited set of posterior samples. For the selected threshold value of 0.135, only 166 models are retained after filtering, corresponding to a rejection rate of 83.4%. If more models are required in the posterior, it is necessary to generate new models, which is not computationally expensive in BEL1D. The only additional effort is to compute the resulting rRMSE. The total computational effort is therefore proportional to the efficiency of the forward solver (Table 4). For instance, generating 500 models in the posterior would require generating 3000 samples based on the same rejection rate, and therefore would take three- times longer. BEL1D-T is therefore equivalent to a smart sampler that quickly generates models only in the vicinity of the posterior distribution and can contribute to a first fast assessment of the posterior. If the generation of many models is required, we rather recommend using BEL1D-IPR.

In this case, the threshold value of 0.135 seems acceptable and close to the BEL1D-IPR posterior distribution after four iterations. A higher threshold seems to retain too many samples, resulting in an overestimation of the posterior. The threshold value of 0.05 corresponds to a very large rejection rate and would require generating more models to assess the posterior properly. In the remaining part of the paper, the threshold 0.135 is used. The visualization of model space encompassing all combinations of temporal and spatial discretization for the first two layers' thicknesses is illustrated in Figure S1 of the Supplementary Materials. Correspondingly, the depth-resistivity models are depicted in Figure S2 for the combinations of F-F, C-F, F-M, C-M, F-VC, and C-VC.

Figure 8 shows the boxplot results for BEL1D-T with the threshold 0.135 for various combinations of the discretization, and can be compared to the corresponding solution with BEL1D-IPR (Figure 6). Differences are less pronounced than with BEL1D-IPR. The F-VF and F-F and F-M discretization have similar posterior distributions as the reference for the thickness of the first two layers, while the uncertainty range for the resistivity is slightly underestimated. Figure 6 shows that the F-VF and F-F and F-M discretizations lead to results without bias for any parameters.

As with BEL1D-IPR, the very-coarse and coarse discretization are systematically biased. Most other combinations show a slight bias for the thickness of layer two, and—to a certain extent—also for layer one. Nonetheless, the difference with the reference for many combinations is less pronounced than for BEL1D-IPR. For example, the I-M and C-M combinations give relatively good approximations of the posterior. As in BEL1D-IPR the prior distribution is complemented with models sampled at the first iteration, without relying on their RMSE, an initial bias resulting from an error in the forward solver might be amplified in later iterations, leading to larger discrepancy between the response of the final model and the data. With BEL1D-T, the application of the threshold after iteration one prevents the solution deviating too much from the truth.

4.3. Impact of the Prior

In this section, we present some results obtained from the application of BEL1D to the TEM-fast dataset collected at sounding 2611, near project 22 (Figure 4). The measured signal can be seen in Figure 9, together with the standard deviation of the stacking error. A deterministic inversion of the data was carried out with SimPEG to have a first estimate of the electrical resistivity distribution (Figure 9). It shows a conductive zone at a shallow depth, likely corresponding to the saline part of the unconsolidated aquifer, while more resistive ground is found below 15 m, likely corresponding to the transition to the resistive bedrock. Below, a gradual decrease in resistivity can be observed.

In field cases, defining the prior distribution can be complicated, as the resistivity is not known in advance. We compare three possible prior combinations (obviously inconsistent prior range—case A, slightly inconsistent prior range—case B, acceptable prior range—case C) to better understand the impact of the choice of the prior. We apply BEL1D-T to bypass the additional computational time required in BEL1D-IPR, and use the F-F discretization.

The prior model consists of layers: the first five layers are characterized by their thickness and electrical resistivity, while the last layer has an infinite thickness. The prior distributions are shown in Figure 9 and Table 5. In case A, the prior is narrow, and was chosen to represent the main trend observed in the deterministic inversion. However, the first layers (upper 10 m) have a small resistivity range not in accordance with the deterministic inversion (red line in Figure 9). Similarly, the fourth layer underestimates the range of resistivity values expected from the deterministic inversion (60–70 Ohm.m). The prior for case B displays larger uncertainty: the first layer is forced to have larger resistivity values, and a strong transition is forced for the half-space. Finally, the last prior case C is very wide and allows a large overlap between successive layers, as well as a very large range of resistivity values.

Table 5. Prior distributions for the different cases: (**a**) obviously inconsistent prior range, (**b**) slightly inconsistent prior range and (**c**) acceptable prior range.

	Case A		Case B		Case C	
	Thickness (m)	Resistivity (ohmm)	Thickness (m)	Resistivity (ohmm)	Thickness (m)	Resistivity (ohmm)
Layer 1	0–10	2–5	0–10	10–25	0–10	10–55
Layer 2	5.0–10	0.5–6	5–10	0.5–5	5.0–10	0.5–15
Layer 3	0.5–10	20–100	0.5–10	20–50	0.5–10	20–100
Layer 4	35–50	60–70	35–50	50–100	35–50	50–600
Layer 5	45–60	5–10	45–60	0.2–0.5	45–60	0.2–10
layer 6	0–0	10–15	0–0	10–40	0–0	5–100

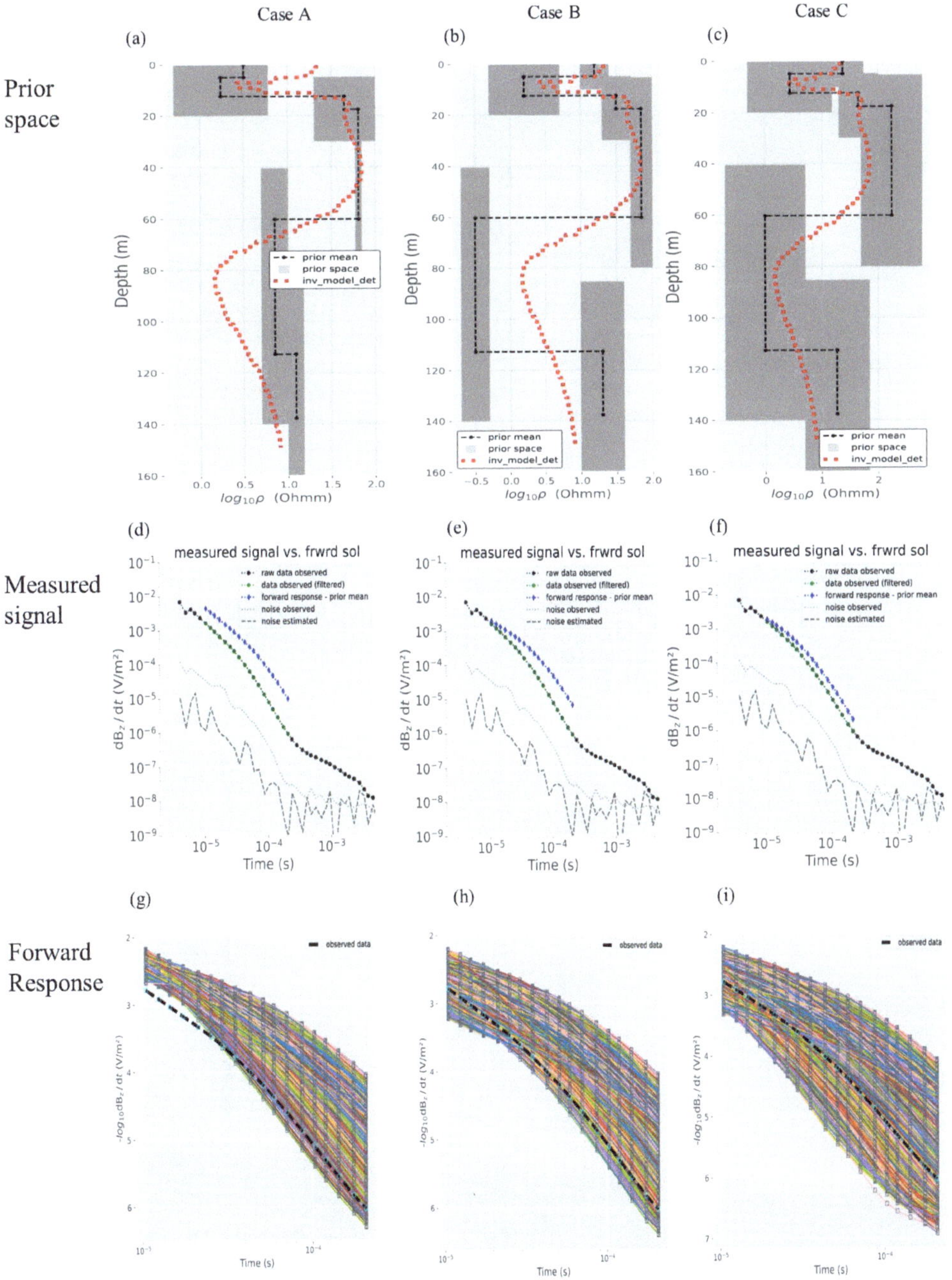

Figure 9. Prior distributions for the three cases of sounding 2611. Case A: obvious inconsistent prior range; case B: slightly inconsistent prior range; case C: acceptable prior range. (**a–c**) Prior range with deterministic inversion (red); (**d–f**) measured signal, noise and forward solution for the prior mean; (**g–i**) forward response of each prior model.

The forward responses of the mean prior model of each three cases are displayed in Figure 9d–f. We can see that the response of the prior is the following: (1) it largely deviates from the measured signal for case A, (2) it deviates at later times for case B, and (3) it has the lowest deviation in case C. We also display the range of the forward response for 4000 prior models (Figure 9g–i). Due to the poor selection of the prior, a large difference between the measured data and the prior data space can be seen for case A (Figure 9g). The prior is clearly not consistent with the data, as the latter lies outside of the prior range in the data space in the early time steps. On the other hand, for case B (Figure 9h), the prior data range now encompasses the observed data, although it is rather at the edge of the prior distribution. For case C (Figure 9i), the prior range in the data space encompasses the measured data which lie close to the response of the prior mean model (Figure 9).

However, visual inspection is not sufficient to verify the consistency of the prior. Indeed, it is necessary to ensure that specific behaviours of the measured data can be reproduced by the prior model. This can be carried out more efficiently in the reduced PCA and CCA space [72]. Indeed, as BEL1D relies on learning, it cannot be used for extrapolation, and should not be used if the data fall outside the range of the prior. To further support the argument, the PCA and (part of) the CCA spaces are shown in Figures 1 and 10, respectively. In Figure 10, the red crosses show the projection of the field data on every individual PCA dimension. It confirms that the prior for case A is inconsistent, with dimensions 2 and 3 lying outside, whereas the first PCA score lies at the edge of the prior distribution. For higher dimensions, the observed data lie within the range of the prior data space, but those dimensions represent only a limited part of the total variance. This is an indication that the prior is not able to reproduce the data and is therefore inconsistent. For cases B and C, no inconsistency is detected in the PCA space.

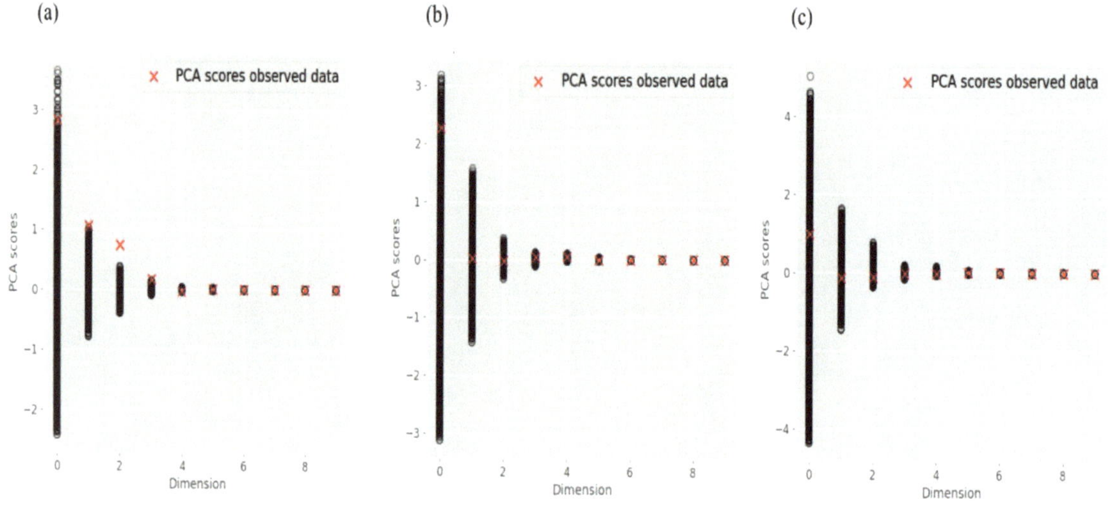

Figure 10. PCA space, (**a**) obvious inconsistent prior, (**b**) slightly inconsistent prior and (**c**) acceptable prior. The black dot represents the prior models and the red cross represents the observed data.

A similar exercise is then performed in the CCA space where the projection of the field data is marked by a red line. In Figure 11a the observed data (red line) are lying outside the zone covered by the sampled prior models for most dimensions (grey zone). In such a case, BEL1D returns an error message and does not provide any estimation of the posterior. For the sake of illustration, we deactivated this preventive action and, nevertheless, performed the inversion. The posterior models in Figure 12 (case A) show low uncertainty for layers 1, 2, 4, 5, and 6, because of the limited range provided in the prior. The posterior data space shows that the posterior models do not fit the data, as a result of the inability of BEL to extrapolate in this case. Note that the threshold was not applied in this case, as it would have left no sample in the posterior, since none of them fit the observed data.

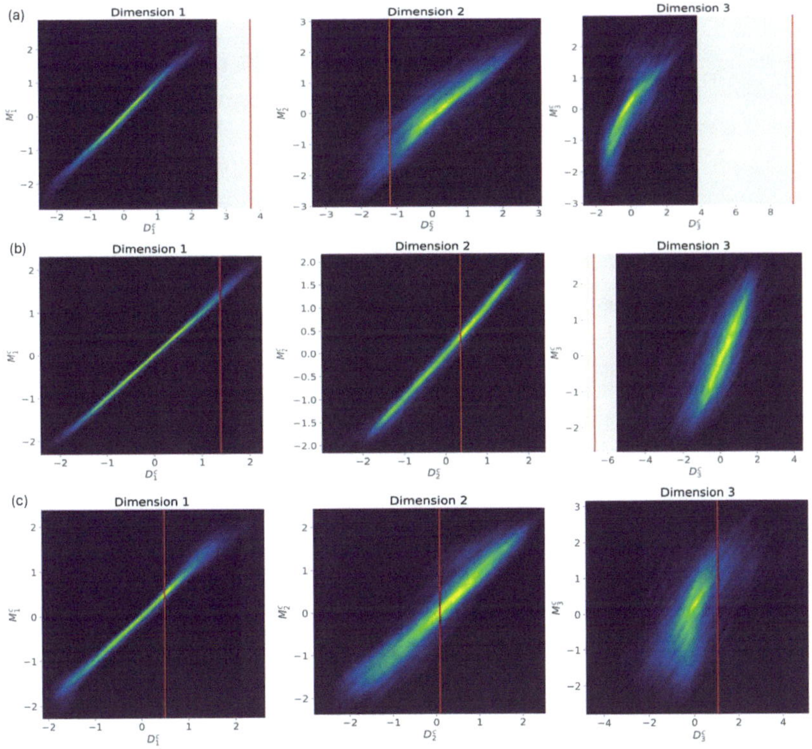

Figure 11. CCA space for the three first dimensions, (**a**) obviously inconsistent prior, (**b**) slightly inconsistent prior and (**c**) acceptable prior. The red line represents the observed data. The y-axis corresponds to the reduced models and the x-axis corresponds to the reduced data.

For case B, although it is apparently consistent in the PCA space, a similar occurrence of inconsistency appears in the CCA space (Figure 11b) for dimension three and some higher dimensions. Although apparently consistent with each individual dimension, the observed data do not correspond to combinations of dimensions contained in the prior, in which case they constitute an outlier for the proposed prior identified in the CCA space. However, in this case, the posterior models that are generated fit the data and have a relatively low RMSE (Figure 12c,d). The posterior-model visualization shows a limited uncertainty reduction for layers one to three and almost no uncertainty reduction for layers four, five and six (Figure 12c,d), likely pointing to a lack of sensitivity of the survey to these deeper layers. This indicates that BEL1D-T can overcome some inconsistency between the prior definition and the observed data, likely because the affected dimensions are only responsible for a small part of the total variance, to a level relatively similar to the noise level.

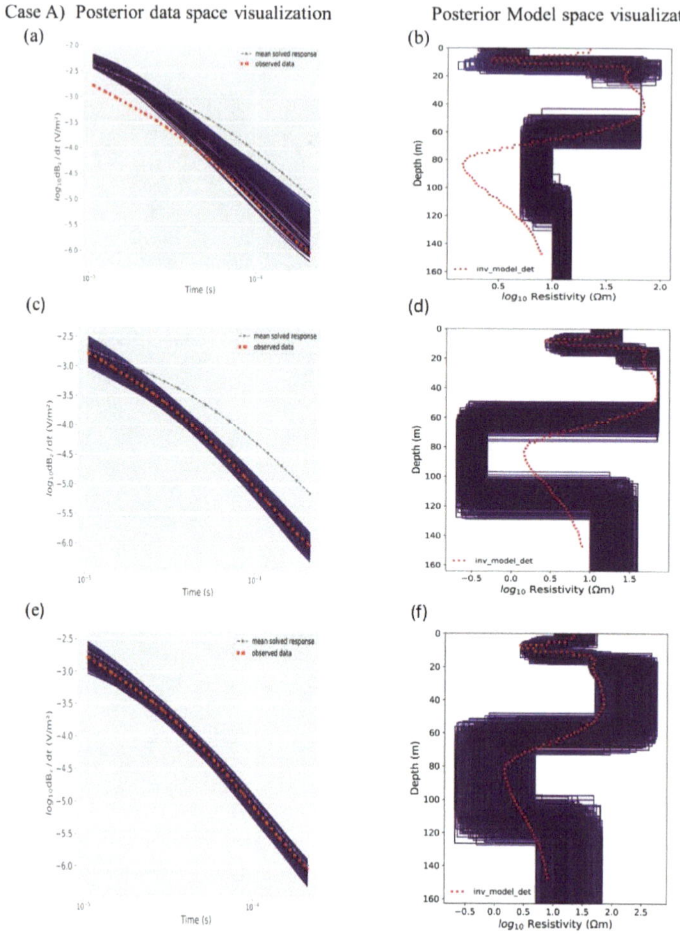

Figure 12. Response of both posterior data and model space for the three prior selections: (**a**,**b**) obviously inconsistent prior range (without application of the threshold, (**c**,**d**) slightly inconsistent prior range, (**e**,**f**) acceptable prior range.

In case C, no inconsistency is detected in the prior data space, the PCA and CCA space (Figures 10c and 11c). The posterior models do fit the data within the expected noise level and the deterministic inversion lies within the posterior (Figure 12e,f). The posterior uncertainty is large, especially for deeper layers (four, five and six). Therefore, in this case, BEL1D-T seems to correctly identify the posterior distribution of the model parameters. As the late times were filtered out, the data set is more sensitive to the shallow layers, and insensitive to the deeper layers. Increasing the prior range for those layers would also induce an increase in uncertainty in the posterior model.

4.4. Field Soundings

We selected two TDEM soundings, which are co-located with ERT profiles (red and yellow dots on Figure 4). The comparison with independent data can be used to evaluate the posterior solution from BEL1D-T. For the TDEM soundings (see Figure 13c,d), we compare the deterministic inversion, the BEL1D-T posterior distribution and a conductivity profile extracted from the ERT profile at the location of the sounding.

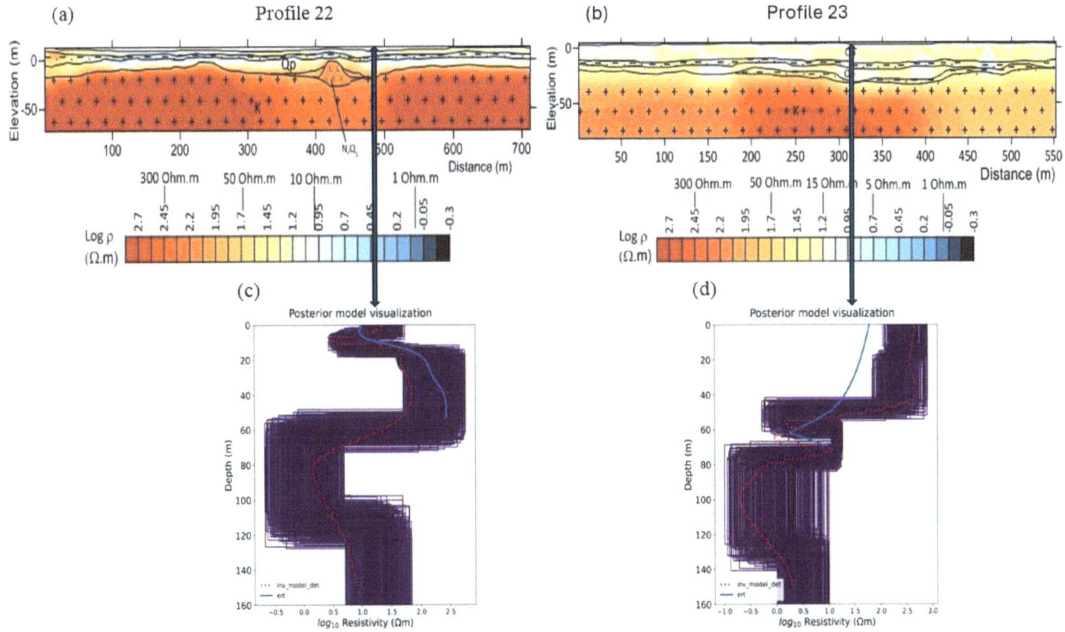

Figure 13. (**a**) ERT profile 22 near to the Luy River; (**b**) ERT profile 23 near the dunes. Posterior model visualization for TDEM soundings on profile 22 (**c**) and 23 (**d**). ERT inversion in blue and deterministic inversion of TDEM data in red.

The resistivity image and TDEM results of profile 22 show the same trend (Figure 13a,c). At a shallow depth between 5 and 15 m, less-resistive layers are observed, which indicates the presence of saltwater in the unconsolidated sediment (20 to 25 m thickness). At a greater depth, we have an increase in resistivity corresponding to the transition to the bedrock. The deterministic solution tends to show a decrease in resistivity at greater depths, which may be an artifact due to the loss of resolution. BEL1D-T is successful in providing a realistic uncertainty quantification, not resolved with the deterministic inversion. It can be observed that, except for the shallow layer, the reduction in uncertainty compared to the prior is relatively limited and concerns mostly the thickness and not the resistivity, illustrating the insensitivity of the survey set-up for depths below 60 m, where the solution is mostly driven by the definition of the prior distribution. The selection of an rRMSE threshold, however, ensures that all those models are consistent with the recorded data.

The results for profile 23 are different (Figure 13b,d). This site is at the foot of sand dunes, close to the sea, with an elevation level between 11 and 50 m. The shallow layer is relatively resistive, but the two methods do not agree on the value of the resistivity, with the TDEM resulting in higher values. BEL1D-T tends to predict a larger uncertainty towards low values of the resistivity for the shallow layers compared to the deterministic inversion. Below 50 m, the resistivity drops to 1–10 Ohm.m for both methods, which seems to show the presence of saltwater in the bedrock. The uncertainty range estimated by BEL1D-T seems to invalidate the presence of rapidly varying resistivity between 50 and 75 m, predicted by the deterministic TDEM inversion, which is quite coherent with the lack of sensitivity at this depth.

4.5. Summary and Discussion

Deterministic inversions are affected by the non-uniqueness of the solution preventing the quantification of uncertainty. Our approach using BEL1D-T allows us to retrieve not only the changes in resistivity with depth, but also to quantify the reliability of the model. We summarize the main outcomes of the sections above as the following:

(1) When using a numerical forward model, the temporal and spatial discretization have a significant effect on the retrieved posterior distribution. A semi-analytical approach is recommended when possible. Otherwise, a sufficiently fine temporal and spatial discretization must be retained and BEL1D-T constitutes an efficient and fast alternative for computing the posterior distribution.
(2) BEL1D-T is an efficient and accurate approach for predicting uncertainty with limited computational effort. It was shown to be equivalent to BEL1D-IPR but requires fewer forward models to be computed.
(3) As with any Bayesian approach, BEL1D-related methods are sensitive to the choice of the prior model. The consistency between the prior and the observed data is integrated, and the threshold approach allows for quickly identifying the inconsistent posterior model. We recommend running a deterministic inversion to define the prior model while keeping a wide range for each parameter, allowing for sufficient variability. Our findings illuminate the substantial uncertainty enveloping the deterministic inversion, highlighting the risk of disregarding such uncertainty, particularly in zones of low sensitivity at greater depths. We implement a threshold criterion to ensure all the models within the posterior distribution fit the observed data within a realistic error. Nonetheless, there exists a risk of underestimating uncertainty when the prior distribution is overly restrictive, as detailed in our prior analysis. Relying too much on the deterministic inversion is therefore dangerous, as it might not recover some variations occurring in the field because of the chosen inversion approach. To accommodate a broader prior, it may be imperative to resort to BEL1D-IPR or to increase the sample size significantly, ensuring a comprehensive exploration of the model space and a more accurate reflection of the inherent uncertainties.
(4) For the field case, the results are consistent with ERT and deterministic inversion. Our analysis reveals that the uncertainty reduction at depths greater than 60 m is almost non-existent. It is recommended to avoid interpreting the model parameters at that depth, as the solution is likely highly dependent on the prior.

5. Conclusions

In this paper, we introduce a new approach combining BEL1D with a threshold after the first iteration (BEL1D-T) as a fast and efficient stochastic inversion method for TDEM data. Although BEL1D-T only requires a limited number of forward runs, the computational time remains relatively important as we used the numerical solver of SimPEG to calculate the forward response. The proper selection of time-steps and space discretization is essential to limit the computation cost while keeping an accurate posterior distribution. Our numerical studies reveal that there is a compromise between the spatial and temporal discretization in the forward solver that minimizes the ricks of numerical errors in the posteriors generated, yet also reducing the computational cost. A fine temporal discretization seems to be important, as described in Table 1, yet a very fine spatial discretization does not seem mandatory. As this analysis is likely specific to every acquisition set-up and prior distribution, we suggest carefully evaluating the modelling error introduced by the forward model before starting the BEL1D-T inversion. The use of faster semi-analytical forward models is recommended when available. However, 2D and 3D effects when the 1D forward solver are used are expected to have a similar impact on the forward-model error. as observed in our work.

The application of a threshold on the rRMSE after one iteration is an efficient approach to limiting the computational costs. We select the threshold based on the estimated relative error in the data set, translated into an absolute value of the rRMSE calculated on a log scale. Selecting a too-selective threshold can result in overfitting and thus an underestimation of the uncertainty. We showed that selecting a threshold based on the expected noise level leads to a solution similar to the one obtained with the reference BEL1D-IPR. The proposed approach allows for partly mitigating the adverse effects of an inaccurate forward model, and therefore can be used to obtain a first fast assessment of the posterior distribution.

Moreover, it should be noted that, as with any stochastic methods, BEL1D is sensitive to the definition of the prior. We have experienced some prior distributions that might appear visually consistent in the data space resulting in inconsistencies in the low-dimensional spaces. It is thus crucial to verify the consistency of the prior also in the lower dimensional space. This feature is included by default in the pyBEL1D code [58], but it might be interesting to deactivate this feature in order to investigate the reasons and their impacts on the posterior. Beside the definition of the prior itself, the inconsistency can be attributed to the noisy nature of the field data [19].

In the case of large uncertainty, an iterative prior resampling approach is advised, as proposed by [19], but it comes at a larger computational cost. Therefore, we propose reducing the prior uncertainty by using the deterministic inversion as a guide, and limiting ourselves to the first iteration, while filtering the models based on their RMSE. However, care should be taken to avoid restricting the prior too much, as this might yield an underestimation of the uncertainty. In such cases, BEL1D-T acts more as a stochastic optimization algorithm, providing only a fast approximation of the posterior distribution, but still allowing for the rough estimation of the uncertainty of the solution, without requiring heavy computational power such as HPC facilities.

We validated the approach using TDEM soundings acquired in a saltwater intrusion context in Vietnam. The posterior distribution was consistent with both the deterministic inversion and the ERT profiles. The range of uncertainty was larger where TDEM and ERT deterministic inversions do not agree, which illustrates the intrinsic uncertainty of these type of data and the need for uncertainty quantification.

Supplementary Materials: The following supporting information can be downloaded at: https://www.mdpi.com/article/10.3390/w16071056/s1, Figure S1. (a) Posterior-model space visualization with one iteration and threshold (0.135), (b) posterior-model space visualization with four iterations; the above row is with fine-time discretization, whereas the other rows are with intermediate- and coarse-time discretization. From left to right, with spatial discretization (VF, F, M, C, and VC). Figure S2. Posterior-model visualization with respect to depth (m) vs resistivity (ohm.m); colour bar represents the RMSE values (a) with four iterations without a threshold and (b) with one iteration and a 0.135 threshold value.

Author Contributions: The conceptualization of this project was led by T.H., with the contribution of A.A., D.D. and A.F.O. The methods were developed by H.M. (implementation of BEL1D and BEL1D-IPR), A.A. (BEL1D-T link SimPEG forward solver), L.A. and W.D. (linking the SimPEG forward solver with BEL1D-IPR). A.A. ran all simulations (discretization, threshold, prior selection, and field data). The initial draft was written by A.A. with significant input from T.H. All authors edited and reviewed the draft. T.H. and D.D. provided supervision for A.A. throughout the whole research project. Project administration: T.H., and funding acquisition: A.A. and T.H. All authors have read and agreed to the published version of the manuscript.

Funding: AA: was funded by the Higher Education Commission (HEC) of Pakistan, with the reference number 50028442/PM/UESTPs/HRD/HEC/2022/December 5, 2022, and from the Bijzonder Onderzoeksfonds (BOF) of Ghent University, with grant number BOF.CDV.2023.0058.01. WD was funded by the Fund for Scientific Research (FWO) in Flanders, with grants 1113020N and 1113022N and the KU Leuven Postdoctoral Mandate PDMt223065. HM was funded by the F.R.S,-FNRS, with grants 32905391 and 40000777.

Data Availability Statement: The field data can be made available upon request.

Acknowledgments: We thank Robin Thibaut for helping in the analysis and plotting of the PCA space. We thank Robin Thibaut, Linh Pham Dieu, Diep Cong-Thi and all the team from the VIGMR for field data collection.

Conflicts of Interest: The authors declare no conflicts of interest.

References

1. Kemna, A.; Nguyen, F.; Gossen, S. On Linear Model Uncertainty Computation in Electrical Imaging. In Proceedings of the SIAM Conference on Mathematical and Computational Issues in the Geosciences, Santa Fe, NM, USA, 19–22 March 2007.
2. Hermans, T.; Vandenbohede, A.; Lebbe, L.; Nguyen, F. A Shallow Geothermal Experiment in a Sandy Aquifer Monitored Using Electric Resistivity Tomography. *Geophysics* **2012**, *77*, B11–B21. [CrossRef]
3. Parsekian, A.D.; Grombacher, D. Uncertainty Estimates for Surface Nuclear Magnetic Resonance Water Content and Relaxation Time Profiles from Bootstrap Statistics. *J. Appl. Geophys.* **2015**, *119*, 61–70. [CrossRef]
4. Aster, R.; Borchers, B.; Thurber, C. *Parameter Estimation and Inverse Problems*; Academic Press: Cambridge, MA, USA, 2013.
5. Sambridge, M. Monte Carlo Methods in Geophysical Inverse Problems. *Rev. Geophys.* **2002**, *40*, 3-1–3-29. [CrossRef]
6. Yeh, W.W.-G. Review of Parameter Identification Procedures in Groundwater Hydrology: The Inverse Problem. *Water Resour. Res.* **1986**, *22*, 95–108. [CrossRef]
7. McLaughlin, D.; Townley, L.R. A Reassessment of the Groundwater Inverse Problem. *Water Resour. Res.* **1996**, *32*, 1131–1161. [CrossRef]
8. Hidalgo, J.; Slooten, L.J.; Carrera, J.; Alcolea, A.; Medina, A. Inverse Problem in Hydrogeology. *Hydrogeol. J.* **2005**, *13*, 206–222. [CrossRef]
9. Zhou, H.; Gómez-Hernández, J.J.; Li, L. Inverse Methods in Hydrogeology: Evolution and Recent Trends. *Adv. Water Resour.* **2014**, *63*, 22–37. [CrossRef]
10. Sen, M.K.; Stoffa, P.L. Bayesian Inference, Gibbs' Sampler, and Uncertainty Estimation in Geophysical Inversion. *Geophys. Prospect.* **1996**, *44*, 313–350. [CrossRef]
11. Tarantola, A. *Inverse Problem Theory and Methods for Model Parameter Estimation*; Society for Industrial and Applied Mathematics: Philadelphia, PA, USA, 2005.
12. Irving, J.; Singha, K. Stochastic Inversion of Tracer Test and Electrical Geophysical Data to Estimate Hydraulic Conductivities. *Water Resour. Res.* **2010**, *46*, W11514. [CrossRef]
13. Trainor-Guitton, W.; Hoversten, G.M. Stochastic Inversion for Electromagnetic Geophysics: Practical Challenges and Improving Convergence Efficiency. *Geophysics* **2011**, *76*, F373–F386. [CrossRef]
14. De Pasquale, G.; Linde, N. On Structure-Based Priors in Bayesian Geophysical Inversion. *Geophys. J. Int.* **2017**, *208*, 1342–1358. [CrossRef]
15. Ball, L.B.; Davis, T.A.; Minsley, B.J.; Gillespie, J.M.; Landon, M.K. Probabilistic Categorical Groundwater Salinity Mapping from Airborne Electromagnetic Data Adjacent to California's Lost Hills and Belridge Oil Fields. *Water Resour. Res.* **2020**, *56*, e2019WR026273. [CrossRef]
16. Bobe, C.; Van De Vijver, E.; Keller, J.; Hanssens, D.; Van Meirvenne, M.; De Smedt, P. Probabilistic 1-D Inversion of Frequency-Domain Electromagnetic Data Using a Kalman Ensemble Generator. *IEEE Trans. Geosci. Remote Sens.* **2020**, *58*, 3287–3297. [CrossRef]
17. Tso, C.-H.M.; Iglesias, M.; Wilkinson, P.; Kuras, O.; Chambers, J.; Binley, A. Efficient Multiscale Imaging of Subsurface Resistivity with Uncertainty Quantification Using Ensemble Kalman Inversion. *Geophys. J. Int.* **2021**, *225*, 887–905. [CrossRef]
18. Michel, H.; Hermans, T.; Kremer, T.; Elen, A.; Nguyen, F. 1D Geological Imaging of the Subsurface from Geophysical Data with Bayesian Evidential Learning. Part 2: Applications and Software. *Comput. Geosci.* **2020**, *138*, 104456. [CrossRef]
19. Michel, H.; Hermans, T.; Nguyen, F. Iterative Prior Resampling and Rejection Sampling to Improve 1D Geophysical Imaging Based on Bayesian Evidential Learning (BEL1D). *Geophys. J. Int.* **2023**, *232*, 958–974. [CrossRef]
20. Scheidt, C.; Renard, P.; Caers, J. Prediction-Focused Subsurface Modeling: Investigating the Need for Accuracy in Flow-Based Inverse Modeling. *Math. Geosci.* **2015**, *47*, 173–191. [CrossRef]
21. Satija, A.; Scheidt, C.; Li, L.; Caers, J. Direct Forecasting of Reservoir Performance Using Production Data without History Matching. *Comput. Geosci.* **2017**, *21*, 315–333. [CrossRef]
22. Hermans, T.; Nguyen, F.; Klepikova, M.; Dassargues, A.; Caers, J. Uncertainty Quantification of Medium-Term Heat Storage from Short-Term Geophysical Experiments Using Bayesian Evidential Learning. *Water Resour. Res.* **2018**, *54*, 2931–2948. [CrossRef]
23. Athens, N.D.; Caers, J.K. A Monte Carlo-Based Framework for Assessing the Value of Information and Development Risk in Geothermal Exploration. *Appl. Energy* **2019**, *256*, 113932. [CrossRef]
24. Thibaut, R.; Compaire, N.; Lesparre, N.; Ramgraber, M.; Laloy, E.; Hermans, T. Comparing Well and Geophysical Data for Temperature Monitoring within a Bayesian Experimental Design Framework. *Water Resour. Res.* **2022**, *58*, e2022WR033045. [CrossRef]
25. Park, J.; Caers, J. Direct Forecasting of Global and Spatial Model Parameters from Dynamic Data. *Comput. Geosci.* **2020**, *143*, 104567. [CrossRef]
26. Yin, Z.; Strebelle, S.; Caers, J. Automated Monte Carlo-Based Quantification and Updating of Geological Uncertainty with Borehole Data (AutoBEL v1.0). *Geosci. Model Dev.* **2020**, *13*, 651–672. [CrossRef]
27. Tadjer, A.; Bratvold, R.B. Managing Uncertainty in Geological CO_2 Storage Using Bayesian Evidential Learning. *Energies* **2021**, *14*, 1557. [CrossRef]
28. Fang, J.; Gong, B.; Caers, J. Data-Driven Model Falsification and Uncertainty Quantification for Fractured Reservoirs. *Engineering* **2022**, *18*, 116–128. [CrossRef]

29. Thibaut, R.; Laloy, E.; Hermans, T. A New Framework for Experimental Design Using Bayesian Evidential Learning: The Case of Wellhead Protection Area. *J. Hydrol.* **2021**, *603*, 126903. [CrossRef]
30. Yang, H.-Q.; Chu, J.; Qi, X.; Wu, S.; Chiam, K. Bayesian Evidential Learning of Soil-Rock Interface Identification Using Boreholes. *Comput. Geotech.* **2023**, *162*, 105638. [CrossRef]
31. Hermans, T.; Oware, E.; Caers, J. Direct Prediction of Spatially and Temporally Varying Physical Properties from Time-Lapse Electrical Resistance Data. *Water Resour. Res.* **2016**, *52*, 7262–7283. [CrossRef]
32. Pradhan, A.; Mukerji, T. Seismic Bayesian Evidential Learning: Estimation and Uncertainty Quantification of Sub-Resolution Reservoir Properties. *Comput. Geosci.* **2020**, *24*, 1121–1140. [CrossRef]
33. Heagy, L.J.; Cockett, R.; Kang, S.; Rosenkjaer, G.K.; Oldenburg, D.W. A Framework for Simulation and Inversion in Electromagnetics. *Comput. Geosci.* **2017**, *82*, WB9–WB19. [CrossRef]
34. Cockett, R.; Kang, S.; Heagy, L.J.; Pidlisecky, A.; Oldenburg, D.W. SimPEG: An Open Source Framework for Simulation and Gradient Based Parameter Estimation in Geophysical Applications. *Comput. Geosci.* **2015**, *85*, 142–154. [CrossRef]
35. Chongo, M.; Christiansen, A.V.; Tembo, A.; Banda, K.E.; Nyambe, I.A.; Larsen, F.; Bauer-Gottwein, P. Airborne and Ground-Based Transient Electromagnetic Mapping of Groundwater Salinity in the Machile–Zambezi Basin, Southwestern Zambia. *Near Surf. Geophys.* **2015**, *13*, 383–396. [CrossRef]
36. Mohamaden, M.; Araffa, S.A.; Taha, A.; AbdelRahman, M.A.; El-Sayed, H.M.; Sharkawy, M.S. Geophysical Techniques and Geomatics-Based Mapping for Groundwater Exploration and Sustainable Development at Sidi Barrani Area, Egypt. *Egypt. J. Aquat. Res.* **2023**, *in press*. [CrossRef]
37. Rey, J.; Martínez, J.M.; Mendoza, R.; Sandoval, S.; Tarasov, V.; Kaminsky, A.; Morales, K. Geophysical Characterization of Aquifers in Southeast Spain Using ERT, TDEM, and Vertical Seismic Reflection. *Appl. Sci.* **2020**, *10*, 7365. [CrossRef]
38. Siemon, B.; Christiansen, A.V.; Auken, E. A Review of Helicopter-Borne Electromagnetic Methods for Groundwater Exploration. *Near Surf. Geophys.* **2009**, *7*, 629–646. [CrossRef]
39. Vilhelmsen, T.N.; Marker, P.A.; Foged, N.; Wernberg, T.; Auken, E.; Christiansen, A.V.; Bauer-Gottwein, P.; Christensen, S.; Høyer, A.-S. A Regional Scale Hydrostratigraphy Generated from Geophysical Data of Varying Age, Type, and Quality. *Water Resour. Manag.* **2018**, *33*, 539–553. [CrossRef]
40. Goebel, M.; Knight, R.; Halkjær, M. Mapping Saltwater Intrusion with an Airborne Electromagnetic Method in the Offshore Coastal Environment, Monterey Bay, California. *J. Hydrol. Reg. Stud.* **2019**, *23*, 100602. [CrossRef]
41. Auken, E.; Foged, N.; Larsen, J.J.; Lassen, K.V.T.; Maurya, P.K.; Dath, S.M.; Eiskjær, T.T. tTEM—A Towed Transient Electromagnetic System for Detailed 3D Imaging of the Top 70 m of the Subsurface. *Geophysics* **2019**, *84*, E13–E22. [CrossRef]
42. Lane, J.W., Jr.; Briggs, M.A.; Maurya, P.K.; White, E.A.; Pedersen, J.B.; Auken, E.; Terry, N.; Minsley, B.K.; Kress, W.; LeBlanc, D.R.; et al. Characterizing the Diverse Hydrogeology Underlying Rivers and Estuaries Using New Floating Transient Electromagnetic Methodology. *Sci. Total Environ.* **2020**, *740*, 140074. [CrossRef] [PubMed]
43. Aigner, L.; Högenauer, P.; Bücker, M.; Flores Orozco, A. A Flexible Single Loop Setup for Water-Borne Transient Electromagnetic Sounding Applications. *Sensors* **2021**, *21*, 6624. [CrossRef] [PubMed]
44. Viezzoli, A.; Christiansen, A.V.; Auken, E.; Sørensen, K. Quasi-3D Modeling of Airborne TEM Data by Spatially Constrained Inversion. *Geophysics* **2008**, *73*, F105–F113. [CrossRef]
45. Linde, N.; Ginsbourger, D.; Irving, J.; Nobile, F.; Doucet, A. On Uncertainty Quantification in Hydrogeology and Hydrogeophysics. *Adv. Water Resour.* **2017**, *110*, 166–181. [CrossRef]
46. Josset, L.; Ginsbourger, D.; Lunati, I. Functional Error Modeling for Uncertainty Quantification in Hydrogeology. *Water Resour. Res.* **2015**, *51*, 1050–1068. [CrossRef]
47. Scholer, M.; Irving, J.; Looms, M.C.; Nielsen, L.; Holliger, K. Bayesian Markov-Chain-Monte-Carlo Inversion of Time-Lapse Crosshole GPR Data to Characterize the Vadose Zone at the Arrenaes Site, Denmark. *Vadose Zone J.* **2012**, *11*, vzj2011.0153. [CrossRef]
48. Calvetti, D.; Ernst, O.; Somersalo, E. Dynamic Updating of Numerical Model Discrepancy Using Sequential Sampling. *Inverse Probl.* **2014**, *30*, 114019. [CrossRef]
49. Köpke, C.; Irving, J.; Elsheikh, A.H. Accounting for Model Error in Bayesian Solutions to Hydrogeophysical Inverse Problems Using a Local Basis Approach. *Adv. Water Resour.* **2018**, *116*, 195–207. [CrossRef]
50. Deleersnyder, W.; Dudal, D.; Hermans, T. Multidimensional Surrogate Modelling for Airborne TDEM Data Accepted Pending Revision. *Comput. Geosci.* **2024**. [CrossRef]
51. Khu, S.-T.; Werner, M.G.F. Reduction of Monte-Carlo Simulation Runs for Uncertainty Estimation in Hydrological Modelling. *Hydrol. Earth Syst. Sci.* **2003**, *7*, 680–692. [CrossRef]
52. Brynjarsdóttir, J.; O'Hagan, A. Learning About Physical Parameters: The Importance of Model Discrepancy. *Inverse Probl.* **2014**, *30*, 114007. [CrossRef]
53. Hansen, T.M.; Cordua, K.S.; Jacobsen, B.H.; Mosegaard, K. Accounting for Imperfect Forward Modeling in Geophysical Inverse Problems—Exemplified for Crosshole Tomography. *Geophysics* **2014**, *79*, H1–H21. [CrossRef]
54. Deleersnyder, W.; Maveau, B.; Hermans, T.; Dudal, D. Inversion of Electromagnetic Induction Data Using a Novel Wavelet-Based and Scale-Dependent Regularization Term. *Geophys. J. Int.* **2021**, *226*, 1715–1729. [CrossRef]
55. Deleersnyder, W.; Dudal, D.; Hermans, T. Novel Airborne EM Image Appraisal Tool for Imperfect Forward Modelling. *Remote Sens.* **2022**, *14*, 5757. [CrossRef]

56. Bai, P.; Vignoli, G.; Hansen, T.M. 1D Stochastic Inversion of Airborne Time-Domain Electromagnetic Data with Realistic Prior and Accounting for the Forward Modeling Error. *Remote Sens.* **2021**, *13*, 3881. [CrossRef]
57. Minsley, B.J.; Foks, N.L.; Bedrosian, P.A. Quantifying Model Structural Uncertainty Using Airborne Electromagnetic Data. *Geophys. J. Int.* **2021**, *224*, 590–607. [CrossRef]
58. Michel, H. *pyBEL1D: Latest Version of pyBEL1D*, version 1.1.0; Zenodo: Genève, Switzerland, 14 July 2022. [CrossRef]
59. Krzanowski, W.J. *Principles of Multivariate Analysis: A User's Perspective*; Oxford University Press: New York, NY, USA, 2000.
60. Thibaut, R. Machine Learning for Bayesian Experimental Design in the Subsurface. Ph.D. Dissertation, Ghent University, Faculty of Sciences, Ghent, Belgium, 2023. [CrossRef]
61. Wand, M.P.; Jones, M.C. Comparison of Smoothing Parameterizations in Bivariate Kernel Density Estimation. *J. Am. Stat. Assoc.* **1993**, *88*, 520. [CrossRef]
62. Devroye, L. An Automatic Method for Generating Random Variates with a Given Characteristic Function. *SIAM J. Appl. Math.* **1986**, *46*, 698–719. [CrossRef]
63. Werthmüller, D. An Open-Source Full 3D Electromagnetic Modeler for 1D VTI Media in Python: Empymod. *Geophysics* **2017**, *82*, WB9–WB19. [CrossRef]
64. Yee, K. Numerical Solution of Initial Boundary Value Problems Involving Maxwell's Equations in Isotropic Media. *IEEE Trans. Antennas Propag.* **1966**, *14*, 302–307. [CrossRef]
65. Haber, E. *Computational Methods in Geophysical Electromagnetics*; Society for Industrial and Applied Mathematics: Philadelphia, PA, USA, 2014. [CrossRef]
66. Hyman, J.; Morel, J.; Shashkov, M.; Steinberg, S. Mimetic Finite Difference Methods for Diffusion Equations. *Comput. Geosci.* **2002**, *6*, 333–352. [CrossRef]
67. Hyman, J.M.; Shashkov, M. Mimetic Discretizations for Maxwell's Equations. *J. Comput. Phys.* **1999**, *151*, 881–909. [CrossRef]
68. Diep, C.-T.; Linh, P.D.; Thibaut, R.; Paepen, M.; Ho, H.H.; Nguyen, F.; Hermans, T. Imaging the Structure and the Saltwater Intrusion Extent of the Luy River Coastal Aquifer (Binh Thuan, Vietnam) Using Electrical Resistivity Tomography. *Water* **2021**, *13*, 1743. [CrossRef]
69. Dieu, L.P.; Cong-Thi, D.; Segers, T.; Ho, H.H.; Nguyen, F.; Hermans, T. Groundwater Salinization and Freshening Processes in the Luy River Coastal Aquifer, Vietnam. *Water* **2022**, *14*, 2358. [CrossRef]
70. Cong-Thi, D.; Dieu, L.P.; Caterina, D.; De Pauw, X.; Thi, H.D.; Ho, H.H.; Nguyen, F.; Hermans, T. Quantifying Salinity in Heterogeneous Coastal Aquifers through ERT and IP: Insights from Laboratory and Field Investigations. *J. Contam. Hydrol.* **2024**, *262*, 104322. [CrossRef]
71. Aigner, L.; Werthmüller, D.; Flores Orozco, A. Sensitivity Analysis of Inverted Model Parameters from Transient Electromagnetic Measurements Affected by Induced Polarization Effects. *J. Appl. Geophys.* **2024**, *223*, 105334. [CrossRef]
72. Hermans, T.; Lesparre, N.; De Schepper, G.; Robert, T. Bayesian Evidential Learning: A Field Validation Using Push-Pull Tests. *Hydrogeol. J.* **2019**, *27*, 1661–1672. [CrossRef]

water

MDPI

Article

Coastal Groundwater Bodies Modelling Using Geophysical Surveys: The Reconstruction of the Geometry of Alluvial Plains in the North-Eastern Sicily (Italy)

Patrizia Capizzi [1,*], Raffaele Martorana [1], Alessandro Canzoneri [1], Alessandro Bonfardeci [1] and Rocco Favara [2]

[1] Department of Earth and Marine Sciences (DISTEM), Università degli Studi di Palermo, 90123 Palermo, Italy; raffaele.martorana@unipa.it (R.M.); alessandro.canzoneri@unipa.it (A.C.); alessandro.bonfardeci@unipa.it (A.B.)

[2] Istituto Nazionale di Geofisica e Vulcanologia, Sezione di Palermo, 90146 Palermo, Italy; rocco.favara@ingv.it

* Correspondence: patrizia.capizzi@unipa.it

Abstract: The integration of various geophysical methodologies is considered a fundamental tool for accurately reconstructing the extent and shape of a groundwater body and for estimating the physical parameters that characterize it. This is often essential for the management of water resources in areas affected by geological and environmental hazards. This work aims to reconstruct the pattern and extent of two groundwater bodies, located in the coastal sectors of the North-Eastern Sicily, through the integrated analysis and interpretation of several geoelectrical, seismic and geological data. These are the Sant'Agata-Capo D'Orlando (SCGWB) and the Barcelona-Milazzo (BMGWB) Groundwater Bodies, located at the two ends of the northern sector of the Peloritani geological complex. These two studied coastal plains represent densely populated and industrialized areas, in which the quantity and quality of the groundwater bodies are under constant threat. At first, the resistivity models of the two groundwater bodies were realized through the inversion of a dataset of Vertical Electrical Soundings (VES), constrained by stratigraphic well logs data and other geophysical data. The 3D resistivity models obtained by spatially interpolating 1D inverse VES models have allowed for an initial recognition of the distribution of groundwater, as well as a rough geological framework of the subsoil. Subsequently, these models were implemented by integrating results from active and passive seismic data to determine the seismic P and S wave velocities of the main lithotypes. Simultaneous acquisition and interpretation of seismic and electrical tomographies along identical profiles allowed to determine the specific values of seismic velocity, electrical resistivity and chargeability of the alluvial sediments, and to use these values to constrain the HVSR inversion. All this allowed us to recognize the areal extension and thickness of the various lithotypes in the two investigated areas and, finally, to define the depth and the morphology of the base of the groundwater bodies and the thickness of the filling deposits.

Keywords: Vertical Electrical Soundings; active and passive seismics; well log; groundwater body; coastal alluvial plain; 3D resistivity model

1. Introduction

The integration of geological and geophysical data for the reconstruction of subsurface models is a widespread practice nowadays. In numerous research fields, direct investigations, such as trenches and boreholes, are associated with data obtained by geophysical acquisitions to expand the available dataset and improve the subsurface characterization. Furthermore, the net of the actual cost of geophysical instrumentation, the relationship between the amount of data acquired and the economic evaluation of a geophysical survey make geophysical methodologies quite economical when compared to direct investigations.

This approach had been often used in many studies regarding stratigraphic [1–3] and structural geology [4–7], the analysis of depositional sequences [8–10], archaeological [11–13] and civil engineering research [14–16]. In these studies, acquired geophysical data constrained by the information derived from direct investigation provide a fundamental tool for the integration and definition of subsurface models. Even hydrogeology research exploits this workflow for numerous purposes: the description of groundwater depth and thickness [17–19], definition of the physical–chemical characteristics of an aquifer [20–22], assessment of marine ingression in coastal areas [23–25] and evaluation of water quality and degree of pollution [26–28].

In recent years, several studies based on integrated approaches of geological and geophysical datasets have focused on coastal areas that host floodplains [29–31]. These flat areas are often critically important for different reasons. They are usually densely populated and, considering their proximity to the sea, host strategic harbors; seaside tourism causes a clear increase in the population during the warm months. Moreover, the gentle morphology favors the development of numerous road and railway communication routes and, consequently, of industrialized areas adjacent to the main towns.

Unfortunately, coastal alluvial plains are often subjected to various risks, both natural, such as floods and paroxysmal events linked to wave motion, and those linked to the development of anthropic activities. Furthermore, the human impact on these areas causes excessive exploitation and contamination of aquifers. All this can lead to serious damage to infrastructure, negatively affecting industrial, tourism, fishing, and agricultural activities. Knowledge of the subsoil is, therefore, fundamental to effectively addressing all these problems.

This work presents the 2D modelling of two coastal alluvial plains, present in North-Eastern Sicily, hosting groundwater bodies: the Barcellona–Milazzo (BMGWB) and the S. Agata–Capo D'Orlando plain (SCGWB). The characterization of the alluvial geometries present in these two sectors have been defined by combining the boreholes data and several data collected by geophysical acquisition campaigns carried out at different times. Firstly, Vertical Electrical Soundings (VES) and Seismic Refraction (SR) profiles were realized in the 1970s by a project sponsored by a public entity named "Cassa per il Mezzogiorno" (CASMEZ), created by the Italian government to finance industrial initiatives aimed at the economic development of Southern Italy [32]. Subsequently, in 2019, these data were partly reinterpreted and integrated by other both active and passive seismic methods as part of a project conducted by the INGV (National Institute of Geophysics and Volcanology—Palermo Section) and DAR (Regional Water and Waste Department) [33]. Through the integration of all the data, the thickness of the alluvial deposit and the depth of the bottom were defined.

The subsoil models produced are a basic tool for evaluating the water resources of the area and the possibilities of exploiting them. Moreover, these study results should be a useful tool to define the geological hazards of the two sectors analyzed and to identify any areas that may be subject to environmental pollution. In conclusion, for all these reasons, the models created can also be considered fundamental to follow for basic studies in defining the city's master plans.

2. Geology and Hydrogeology

2.1. Geomorphological and Geological Setting

The Sant'Agata–Capo D'Orlando and Barcellona–Milazzo plains are in the Tyrrhenian coastal sector of the Peloritani Mountains (North-Eastern Sicily). In particular, the two areas studied are spread over an area of about 40.70 km^2 for the Sant'Agata–Capo D'Orlando plain and of about 1319 km^2 for the Barcellona–Milazzo one.

Several towns and villages lie in these two coastal plains, among which Sant'Agata di Militello, Capo D'Orlando, Barcellona Pozzo di Gotto and Milazzo are the main municipalities. Strategical infrastructures like harbors, highways, and railway lines, as well as industrial plants and several farms, are present in these sectors. Noteworthy is the site of national interest (SIN) represented by the petroleum refinery and chemical industries located east of Milazzo, at the mouth of the Corriolo Stream.

From a geo-structural point of view, the North-Eastern Sicily lies in a crucial sector of the central Mediterranean region in correspondence of the African and European plates boundary, characterizing the Sicilian foreland–foredeep–chain system (Figure 1c) [34–36]. The Peloritani Mountains' geological complex (hereafter, Peloritani Complex; PC) constitutes the innermost and structurally highest sector of the Sicilian Fold and Thrust Belt (SFTB) [36–38], (Figure 1b). In particular, the PC represents the southern termination of the Calabrian–Peloritan Arc (CPA), as well as the linking element between the Southern Appennines and the Sicilian Maghrebian Chain [39]. The PC is formed by different south-verging tectonic units (nappes), mainly composed by high-to-low-grade metamorphites, deriving from the deformation of an ancient Hercynian basement (Kabilo-Calabride Units) and related carbonate–clastic Meso-Cenozoic sedimentary covers [35,40–44]. The piling up of the tectonic units belonging to the PC and its incorporation into the SFTB began in the upper Oligocene and continued up to the Upper Miocene [45]. During the Plio-Pleistocene, extensional and transcurrent tectonic regimes affected the PC, producing an articulate dip-slip and strike-slip faults pattern, with NW-SE/W-E and NE-SW/N-S orientation [35,46–50]. Lastly, one of the most important tectonic structures of North-Eastern Sicily is represented by the fault zone known as "Taormina line", constituting the outer front of the PC [35,51]. Furthermore, through this latter tectonic structure, the PC overlaps on the Apennine–Maghrebidi Units (Figure 1c); more precisely, on those deriving from the deformation of the Sicilide complex, widely outcropping in the Nebrodi Mountains [40–44].

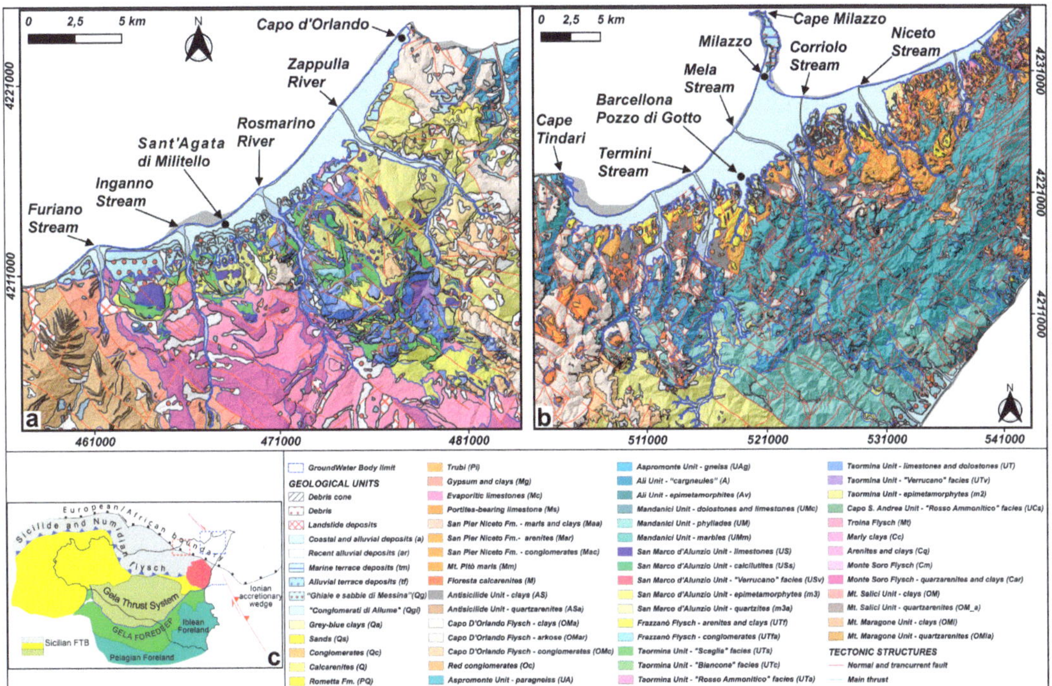

Figure 1. Geological sketch maps of the two investigated areas: the SCGWB on the left (**a**) and the BMGWB on the right (**b**). In the bottom left, (**c**) the structural map of Sicily, illustrating the main

elements of the foreland–foredeep–chain system, is shown (modified from [52]). The location of the study areas of SCGWB (red dashed line) and BMGWB (blue dashed line) is highlighted in this latter map. The geological sketch maps (**a**,**b**) were drawn with the QuantumGIS 3.16 software, provided by Open Source Geospatial Foundation (OSGeo), adopting a DEM image as a topographic base. Polygons and lines present in these two maps are referred to the outcropping geological units and tectonic structures, based on geological chart [53] and explanatory notes [40]. The description of the geological units and tectonic structures of the two geological sketch maps is shown in the bottom-right panel.

The tectono-stratigraphic sequence outcropping in the investigated sectors is composed, from the bottom to the top, by the Numidian Flysch, Sicilide and Kabilo–Calabride Units, unconformably covered by syn- and post-orogenic sedimentary sequences [40,44].

In particular, the Numidian Flysch and Sicilide (Figure 1a) complexes are composed of the following:

- the Upper Oligocene–Lower Miocene clays and quartzarenites of the Mt. Maragone and Mt. Salici Units;
- the clays and quartzarenites of the Monte Soro and the marly clays unit, Lower Cretaceous in age.

Tectonically superimposed to the Numidian Flysch and Sicilide Units, the Kabilo–Calabride Units and related Meso-Cenozoic sedimentary covers (Figure 1b) are present [40,44]. This latter sequence is composed of the following, from the bottom to the top:

- Capo Sant'Andrea, Longi-Taormina and San Marco d'Alunzio Units, comprising epimetamorphites deriving from an ancient Hercynian basement and Meso-Cenozoic carbonate covers;
- low- to high-grade metamorphic rocks belonging to the Mandanici and Aspromonte Units, derived from the deformation of the innermost sectors of the Kabilo–Calabride domain (Paleozoic);
- the Upper Oligocene–Lower Miocene syntectonic terrigenous deposit of the Capo d'Orlando Flysch;
- the Upper Cretaceous clays and quartzarenites of the Antisicilide Unit, overthrusting the Capo d'Orlando Flysch;
- the Lower-Middle Miocene Floresta calcarenites and the Mt. Pitò marls, unconformably lying on the deposits of the Antisicilide Unit;
- the Middle-Upper Miocene post-orogenic deposits belonging to the San Pier Niceto Fm. and to the evaporitic series upward.

The tectono-stratigraphic units described are unconformably covered by the Plio-Quaternary marine, fluvio-deltaic and alluvial sequences. These marine deposits, mainly outcropping toward the Tyrrhenian and Ionian coastal sectors, are composed of the Trubi unit (Lower Pliocene), the Upper Pliocene marls and sands, and the Rometta Fm. deposits (Upper Pliocene–Lower Pleistocene), passing laterally and upward to the conglomerates, sands and marly clays of Lower–Middle Pleistocene age. This sedimentary sequence ends with late Quaternary fluvio-deltaic ("Conglomerati di Allume" and "Ghiaie e sabbie di Messina"), marine and alluvial terrace, and alluvial and coastal plain deposits.

The Sant'Agata-Capo d'Orlando and Barcellona-Milazzo plains arise in two wide coastal sectors with peculiar NE-SW orientation. Several rivers and streams, arising further south in the central mountain areas of the PC, are characterized by articulated hydrographic patterns with predominant NW-SE and N-S direction of the river courses [41–43]. Furthermore, the orientation and development of the drainage direction of these streams appears orthogonal to the strike of the belts from which they arise [54]. Their northward flowing into the Sant'Agata-Capo d'Orlando and Barcellona-Milazzo coastal plains guarantees a considerable contribution in terms of freshwater recharge and sediments supply to these sectors.

Therefore, both the rivers and stream drainage patterns and the geometry and orientation of the two coastal plains appear directly controlled by the tectonic setting of these sectors of the PC. Normal and transtensional faults with predominant NW-SE and NE-SW trends characterize the geometry of the Sant'Agata-Capo d'Orlando and Barcellona-Milazzo coastal plains and the thickness of the sedimentary infill. The coastal plains infill is composed by the Middle-Late Pleistocene to Holocene fluvio-deltaic, marine and alluvial terrace deposits, recent alluvial and coastal deposits, reaching the thickness of even more than 100 m, as in the case of the central portion of the Barcellona-Milazzo plain [41,55–57].

2.2. Hydrogeological Setting

The lithology and stratigraphic setting of the sedimentary infill directly influence the groundwater circulation and its storage in the SCGWB and BMGWB. The Potential Infiltration Coefficient (CIP [19,58,59]) can be considered as the parameter through which the permeability of rocks and sedimentary deposits is quantified. As regards the Groundwater Bodies characterising the Sant'Agata-Capo d'Orlando and Barcellona-Milazzo coastal plains and related rivers and all tributaries, the CIP reaches the highest values (0.8–0.9%) estimated in the PC [33]. In these coastal sectors the groundwater bodies are composed by different Middle-Late Quaternary geological units mainly consisting of alternations of gravels, sands and clayey-silts, which are characterized by high permeability for primary porosity [60,61]. Therefore, due to the high porosity and thickness, generally increasing in correspondence of the river and stream mouths, and to the shoreline, SCGWB and BMGWB host significant unconfined aquifers [55,62,63]. The freshwater input on the SCGWB and BMGWB occurs with direct charging or, locally, through exchanges with the adjacent groundwater bodies and aquifers.

In particular, the SCGWB consists of the "Ghiaie e sabbie di Messina", marine and alluvial terrace deposits, alluvial and coastal plain [40,42,43], characterized by high-to-very high permeability for porosity. The SCGWB lower boundary is characterized by the presence of the flyschoid deposits of the Numidian Flysch and Monte Soro Flysch Units in the westernmost portion, the epimetamorphites and limestones of the San Marco d'Alunzio and Longi-Taormina Units in the central sector, and the Aspromonte Unit and Capo d'Orlando Flysch deposits in the north-easternmost portion. All these lithologies are characterized by medium-low-to-low fracture permeability [61].

The BMGWB is formed by the "Ghiaie e sabbie di Messina", marine and alluvial terrace deposits, alluvial and coastal plain [40,41,56,57], characterized by high-to-very high permeability for porosity. The BMGWB is superimposed on the Upper Pliocene–Pleistocene marly sandy sequences, the post- and syn-orogenic deposits in the western and central portion and the metamorphites of the Aspromonte Unit in the north-easternmost part of it. All the described units are characterized by medium-low-to-low permeability for porosity and fracturing [61].

3. Materials and Methods

3.1. Previous Geognostic and Geophysical Investigation

Litho-stratigraphic logs of several boreholes, lying in the two study areas, have been collected and analyzed to obtain information about the main sedimentological and stratigraphic features of the deposits. In total, 151 boreholes, lying in the two investigated sectors, were collected. In particular, as shown in Figure 2, 11 of these boreholes were located in the area of the SCGWB and 140 in the BMGWB one.

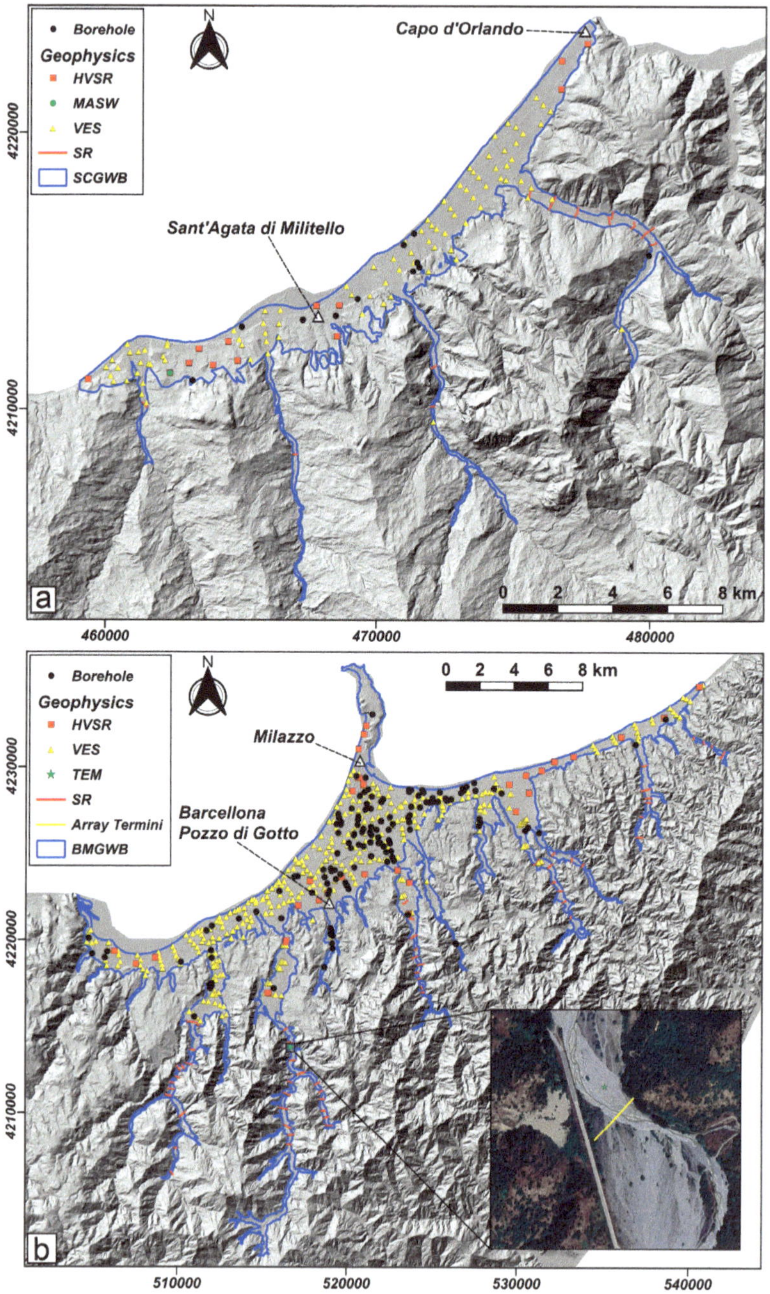

Figure 2. Charts illustrating the location of the boreholes and geophysical surveys analyzed for the study areas. DEM images were used as a topographic base. On the map at the top (**a**), the data available for the SCGWB are shown, while on the map at the bottom (**b**), the dataset for the BMGWB is described. As regards the (**b**) map, in the bottom-right satellite image, the location of the array (yellow line) carried out in the Termini Stream is shown. This array is referred to as the ERT and the SRT realized in this portion of the stream valley.

Most of the analyzed boreholes were drilled using the non-coring drilling method, mainly for groundwater research, while the remaining were drilled using the continuous coring method. The data obtained through the analysis of the boreholes have been reinterpreted and validated, homogenizing the lithologies described with those related to the litho-stratigraphic units present in the geological cartography [53] and the related explanatory notes [40] used for this work. The litho-stratigraphic information provided by these surveys has been used to constrain the inversion of the available geophysical data.

Both the previous SR and VES surveys, as shown in Figure 2, were carried out in the seventies in the two coastal plains for water research activities foreseen by a CASMEZ project [32]. In total, 90 VES and 12 SR profiles were selected within SCGWB, while BMGWB hosted 376 VES and 69 SR. The latter were carried out perpendicular to the axes of rivers present within the two alluvial plains.

3.2. New Processing and Interpretation of Vertical Electrical Soundings

Vertical Electrical Soundings (VES) are a geophysical methodology that usually provides a one-dimensional electrical resistivity model [64]. VES have often been used with good results for studying the physical properties of aquifers and their geometric characterization [65,66]. Despite all this, these surveys can also be useful to describe three-dimensional resistivity models. Indeed, within a studied area, several 1D resistivity data close to each other can be used to constrain the values of each VES. This approach, suitable when the studied area is characterized by a not-too-high resistivity gradient, is useful when large sectors need to be investigated, considering that electrical resistivity tomography appears impractical and expensive.

All vertical electrical soundings considered were performed using the Schlumberger array, with AB current dipole lengths increasing exponentially, so as to obtain roughly equally spaced values in a graph with logarithmic axes. Generally, 10 measurements were performed per logarithmic decade, with maximum AB/2 lengths ranging from 300 m for the shortest soundings up to 500 m for the longest ones, allowing maximum depths of investigation of about 250 m. Among all the VES carried out in the two groundwater bodies, we chose only those whose apparent resistivity vs. AB/2 curves presented a trend compatible with a one-dimensional layered modeling so as to be able to obtain from them 1D layered inverse models with a misfit not exceeding 5%.

For each VES, starting from analog data, the plot of the measured apparent resistivity ρ_a as a function of the half-distance AB/2 between the current electrodes has been digitalized and saved in ASCII files also containing topographic information and the array type.

The digitalized VES were inverted by ZondIP1D (v. 5.2), provided by Zond Software Ltd., Republic of Cyprus (EU) using the least squares method to constrain the inverse models with boreholes data where available. Considering that the studied areas present a fairly regular stratigraphy, a 1.5D inversion algorithm was used, according to which the deepest layer of the resistivity section is considered almost horizontal, while the more superficial layers may be affected by lateral variations in resistivity, albeit slight [67]. Therefore, given these assumptions, inverse models from contiguous VES are mutually constrained (Figure 3a). Consequently, the data were inverted by considering some VES alignments within the study areas, based on the mutual distance between VES points and the alleged geometric features of the geological bodies present in the substrate. The mutually constrained inverse models were laterally interpolated in order to construct two-dimensional electrical resistivity sections along the chosen alignments (Figure 3b). These vertical sections proved useful for providing an initial geological characterization of the two water bodies and as starting data for the subsequent creation of 3D electrical resistivity models.

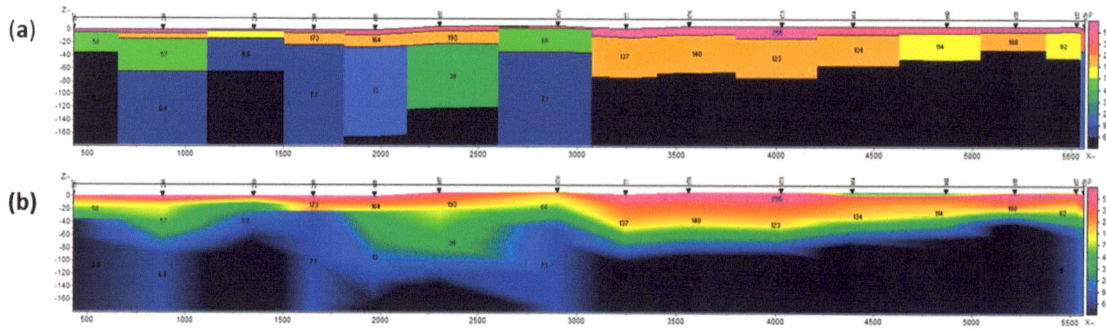

Figure 3. (**a**) Example of aligned 1D inverse electrical resistivity models obtained from the VES soundings; (**b**) corresponding 2D resistivity section by lateral interpolation.

3.3. New Geophysical Surveys

A new geophysical acquisition campaign was realized during the DAR project to better describe the geometrical features of SCGWB and BMGWB. For these reasons, both passive and active seismic surveys were performed with the aim of describing the volumes of deposits that could host groundwater.

In detail, 13 recordings of microtremor were acquired for the SCGWB and 50 for the BMGWB, as shown in Figure 2. The environmental noise records were acquired by a 3D velocimeter along the two plains and analyzed according to the HVSR (Horizontal-to-Vertical Spectral Ratio) methodology [68,69].

The inversion of an HVSR curve is subject to significant equivalence limitations and, thus, must be constrained by other geophysical data, mainly for shallower layers, to be considered reliable [70,71]. To this end, it was beneficial to consider the obtained 3D electrical resistivity models and the results of refraction seismic profiles. However, for a more detailed characterization of the vp and vs. values of lithotypes and the correlation with other observed geophysical parameters, a joint survey Seismic Refraction Tomography (SRT), Electrical Resistivity Tomography (ERT), and Induced Polarization Tomography (IPT) were carried out in coincident position [13]. Furthermore, a Transient Electromagnetic (TEM) survey was performed nearby. Finally, a Multichannel Analysis of Surveys Waves (MASW) survey [72] was performed in SCGWB, close to the Acquedolci Town. These surveys were situated in areas where geological knowledge allowed for their joint and detailed interpretation. The geophysical models derived from such surveys, along with other available geophysical models, were used to better constrain HVSR curve inversions, thereby obtaining layered models of seismic velocities. The latter, together with electrical resistivity models and seismic refraction sections, were crucial for determining the thickness of the affected alluvial sediments and the aquifer base.

The ERT, IPT and SRT surveys were carried out in coincident position within the Termini Stream (see Array Termini in Figure 2b). The SRT technique uses a series of aligned and equidistant geophones and a high number of shot records in order to obtain detailed sections of the trend of P-wave velocity in the subsurface [73]. Analogously, ERT and IPT methodologies allow us to obtain detailed 2D sections of electrical resistivity and chargeability using several aligned and equally spaced electrodes, and are often utilized in hydrogeological applications [74–76]. The aim of performing seismic and electrical tomographies along the same alignment in a noticeably area is to characterize the main lithologies in detail from a geophysical point of view, through the joint interpretation of the tomographic sections.

TEM is based on the study of a transient electromagnetic field artificially induced underground in order to reconstruct the trend of electrical conductivity in the investigated volumes. In addition to the mapping of aquifers, this methodology had been used for several applications and studies, such as the characterization of the thickness of volcanic covers [77], the identification of hydrocarbon reservoirs [78] and the identification of geothermal areas [79].

The MASW survey was performed close to the Acquedolci Town. A multichannel seismograph was used to acquire the data, using 48 vertical geophones (central frequency of 16 Hz) for the SRT and 24 vertical geophones (4.5 Hz frequency) for MASW.

4. Results and Discussion

4.1. Litho-Stratigraphic Interpretation of the Boreholes

The analysis of all the boreholes data for the SCGWB has allowed us to distinguish different kinds of geological substrates, characterized by different lithofacies and units, covered by the recent alluvial deposits. In the western part of the SCGWB, the alluvial deposits cover predominantly clayey–marly layers belonging to the Monte Maragone (OMi) and Monte Soro Flysch (Car) or the limestones and marly limestones of the San Marco (US) and Longi-Taormina Units (UTs). In the eastern part of the SCGWB, however, the geological substrate consists of the metamorphic lithofacies of the San Marco (m^3), Longi-Taormina (m^2) and Aspromonte Units (UA). Finally, in the North-Eastern sector of the SCGWB, under the coastal plain and alluvial deposits, the arkoses (OMar) and clays (OMa) of the Capo D'Orlando Flysch unit are present.

As regards the boreholes falling in the BMGWB, below the coastal plain and alluvial deposits, gray-blue clays (Qa), the calcarenites of the Rometta Fm. (PQ) and the clays (Maa), arenites (Mar) and conglomerates (Mac) belonging to the San Pier Niceto Fm. are present.

4.2. Results from SR Profiles and New Geophysical Surveys

As regards the SR profiles carried out during the CASMEZ project [32], although the original dataset is no longer available, several interpretative models have been evaluated. According to these models, the P-wave velocities in the SCGWB show values of about 2200–3000 m/s for the substrate; beneath the alluvial materials and altered portions of the substrate, the velocities are characterized by values of about 500 and 1300 m/s. The seismic velocity model concerning BMGWB describes a lower substrate and alluvial covers above characterized by 2200–3000 and 1000–1300 m/s, respectively.

The results of the joint-acquired SRT, ERT and IPT (Array Termini) carried out within the BMGWB are shown in Figures 4 and 5. SRT (Figure 4) shows a constant increase in velocities from the top to the bottom of the tomography. Values that characterize the upper 15–18 m vary from 400 to 1000 m/s. Velocities increasing from 1000 m/s to 3000 m/s are described from 20 to 40 m of depth. The deepest part of the section shows the highest P-wave velocity values by a maximum of 4500 m/s found at 60 m depth. Considering the data from the geologic map and from the boreholes located near the SRT, the lower velocities were attributed to recent alluvial sediments; the increase in the velocities of the intermediate layers could be related to the saturated part of these deposits. The highest values have been attributed to the metamorphic rocks of the Aspromonte Unit (UA) that constitute the substrate of this portion of the stream course, also outcropping at the banks.

Results of ERT and IPT are shown in Figure 5. The highest values of resistivity and chargeability have been identified mainly in the lower sections of the profile and are associated with the metamorphic basement. Low values in chargeability are, instead, present in the subsurface. These values, associated with medium-low resistivity values, should indicate the presence of alluvial deposits. These superficial deposits present lower values of resistivity where they are affected by the fluid presence.

The 1D resistivity model obtained by inverting the TEM survey provided a vertical model of resistivity values comparable with those derived from ERT near-realized. The TEM model shows resistivity values ranging from about 100 Ωm to 500 Ωm, according to the layers described in ERT as water-saturated.

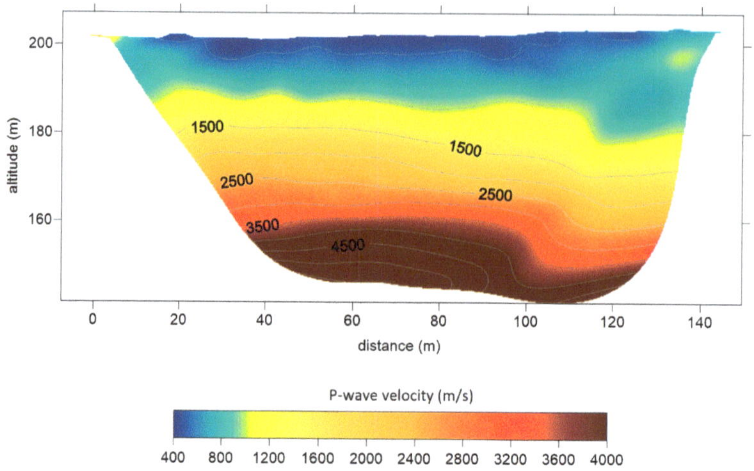

Figure 4. Results of seismic refraction tomography carried out inside the Barcelona–Milazzo Plain, within Termini Stream.

The simultaneous acquisition and interpretation of SRT, ERT and IPT allowed for characterizing the alluvial deposits for their typical values of P-wave velocity, electrical resistivity and chargeability. These values were used for lithotype recognition and the identification of their thickness in both 3D resistivity models and HVSR inverse models.

The MASW realized close to Acquedolci Town, within the SCGWB, provides a seismic–stratigraphic model (Figure 6) composed of a 250 m/s upper superficial cover followed by a 6 m thick layer characterized by a shear waves velocity of 480 m/s. Based on the outcropping rock and information taken from lithologic logs, these values are correlated to the alluvial deposits overlying the "Ghiaie e sabbie di Messina" (Qg) unit.

All the HVSR curves have been inverted, providing inverse shear wave velocity models constrained by all the geophysical and geological data discussed above [80]. Some example results are shown in Figures 7 and 8.

The HVSR inverse models regarding the SCGWB (Figure 7) are generally subdivided into layers varying from two to four. The shallower layers, characterized by an average value of 350 m/s and thickness varying from 5 to 30 m, should represent the alluvial and coastal deposits that predominantly outcrop along the plain. Several previous studies [19,81,82] attributed these velocities to sediments from coarse- to fine-size that usually fill the alluvial plains. The layers, with an average value of 600 m/s, correspond to the intermediate portions of the Vs-profiles. These show a maximum thickness of about 60 m and are associated with "Ghiaie e sabbie di Messina" (Qg). The deeper sections of the models represent high velocities, ranging from 800 to 1100 m/s. These values should describe the presence of different units that do not outcrop in the plain. In the west area of the SCGWB, the flyschoid deposits of Mt. Salici Unit (OM) and the limestones of San Marco d'Aluzio Unit (US) should cause these velocities. High values are, instead, attributed to the Frazzanò Flysch unit (UTf) and to the epimetamorphic rocks of the Longi-Taormina Unit (m2) for the easter sector. Finally, in the subsoil present between the Inganno and Rosmarino Streams, the overlapping of layers of the Marly clays (Cc) and the Flsych of Monte Soro Unit (Cm and Car) could be responsible for these elevated velocities computed.

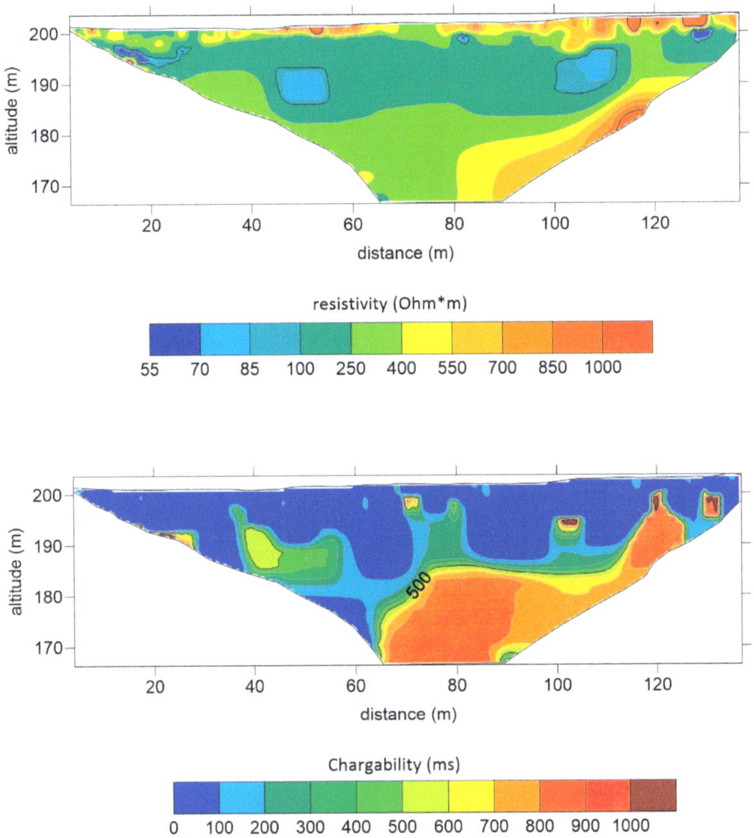

Figure 5. 2D models of resistivity (**above**) and chargeability (**below**) derived from the ERT and IPT carried out in the Plain of Barcelona–Milazzo, within Termini Stream (Array Termini).

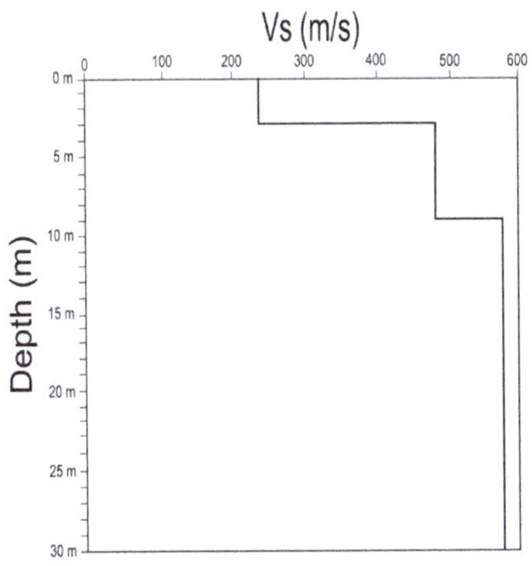

Figure 6. Inverse model resulting from the MASW survey performed in the S. Agata–Capo D'Orlando plain.

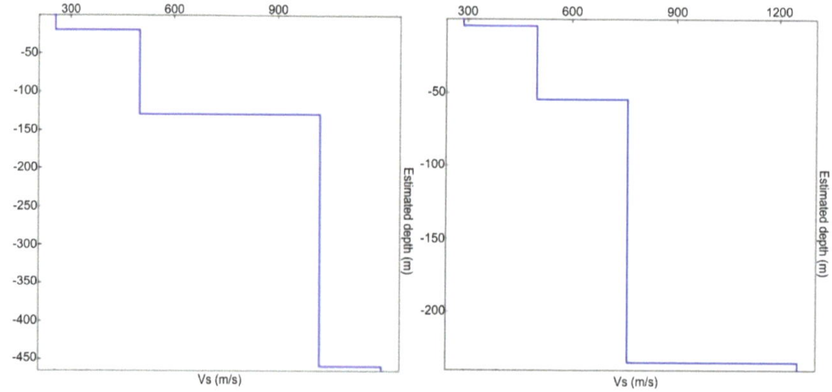

Figure 7. Examples of inverse models related to the HVSR carried out within the Sant'Agata–Capo D'Orlando Plain.

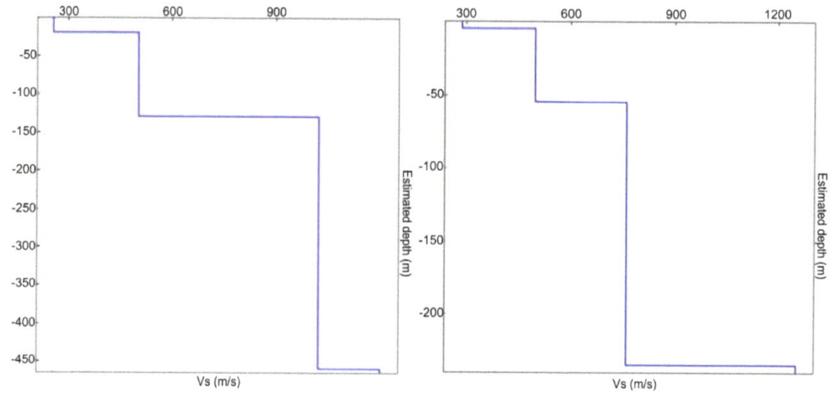

Figure 8. Examples of inverse models related to the HVSR carried out in the BMGWB.

The HVSR surveys realized in the BMGWB, shown in Figure 8, provide similar results about the near surface layers; inside also this waterbody, the superficial alluvial cover shows average Vs values of 300 m/s. Analyzing several single models, the increase of Vs within these first layers is noted, reaching values of 450 m/s, where alluvial deposits present the maximum thickness. This occurs, for example, south of the Milazzo Cape where 80 m of the alluvial deposits are described. Going deeper in the models, a Vs value from 500 to 700 m/s is often present; these values could both be referred to the following formation: "Ghiaie e sabbie di Messina" (Qg), Grey-blue clays (Qa), Rometta (PQ), Trubi (Pi), San Pier Niceto (Maa; Mar; Mac) and Capo d'Orlando Flysch (OMar; OMc). The highest values are found, also for BMGWB, at the base of the HVSR models. The Antisicilide Unit (AS) clays and the metamorphic rocks of the Peloritani Mountains, such as those from the Aspromonte units (UA), have been linked to velocities that exceed 800 m/s.

4.3. Tridimensional Models of the Electrical Resistivity

Three-dimensional models related to the variation of electrical resistivity were realized for SCGWB and BMGWB. These models were produced through the Voxler software v.4, provided by Golden Software (Golden Software, LLC, Golden, CO, USA), LLC, that allowed to laterally interpolate the inverse models of electrical resistivity, obtained from the new reanalyzes of the VES. The lower boundary of the models is given by the depth of investigation of VES while the lateral confinement is provided by the perimeters of the two groundwater bodies examined. The logarithm of electrical resistivity was chosen to be represented due to the high parameter contrasts found in the two areas.

The model realized for SCGWB, represented in Figure 9, shows areas characterized by different resistivity patterns. In the eastern side of the plain, higher values (200–500 Ω m) have been observed. Within the Furiano water stream, these values should be attributed to the presence of Monte Soro Flysch (Car) and Mt. Maragone Unit (OMia) at shallow depth, below the deposits of SCGWB. The alluvial deposits are, indeed, thin in this area and cover the aforementioned units, mainly composed of the alternations of quartzarenites and clays. Eastwards, in the mouth sector of the Rosmarino River, the resistivity values are high, probably due to the amount of fresh water flowing within alluvial deposits. Furthermore, in this sector of the Rosmarino River, these deposits reach the highest thickness. Lastly, around the Zappula River, the high values of resistivity can be linked to the presence of phyllades and metamorphosed arenites, rich in quartz. These deposits belong to the Longi-Taormina Unit (m^2) and are covered by thin alluvial bodies.

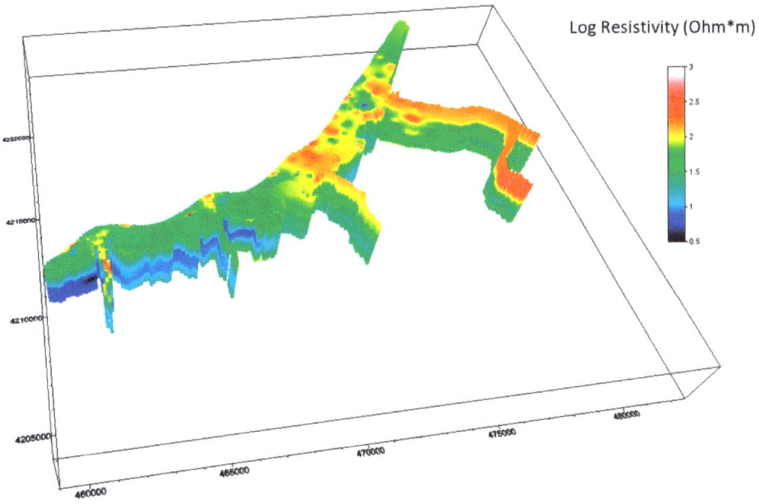

Figure 9. 3D model of the distribution of the logarithm of the electrical resistivity of the water body of the Sant'Agata–Capo D'Orlando Plain (SCGWB).

Medium resistivity values, comprising between 35 and 70 Ω m, are present in several areas of the SCGWB floodplain. In the area between the Furiano and Inganno streams, the abovementioned resistivity values are related to sands, silt and sandy matrix conglomerates, largely outcropping in this sector and also found in the borelogs. These latter sediments are ascribed to the "Ghiaie e sabbie di Messina" (Qg) unit and to the marine terrace deposits (tm). Moreover, the presence of gravels, pebbles and sands in a clayey matrix, identifying recent alluvial (ar), coastal, and stream deposits (a), may be the cause of the not elevated resistivity recognized along the coastline.

Low resistivity values (5–20 Ω m) are observable in the eastern sector of the SCGWB 3D model. The sea water intrusion could be considered one of the causes that produce this decrease in resistivity. Alluvial (ar) and coastal sediments (a) are partly composed of coarser deposits such as polygenic blocks, gravel, and coarse sands. An increase in grain size is often present in the lower portions of these deposits. Due to the sedimentological features, these deposits can be affected by strong sea water intrusion and the formation of a salt wedge below the plain. Moreover, not outcropping conductive rocks, such as those of the Marly clays (Cc) unit and the fractured limestones of the San Marco d'Alunzio Unit

(US), should be responsible for the decrease in the resistivity values, as observed in the deeper part of the model.

The 3D resistivity model of the BMGWB, shown in Figure 10, describes different resistivity trends. Resistivity values from 80 Ω m to 1000 Ω m are represented in different sectors of the BMGWB; most of the river arms, the central sector of the plain and the area of the Cape Milazzo promontory are characterized by these high values. The highest values, represented in white in Figure 10, are generally located in some sectors of the river arms in which the presence of fresh water causes an increase in resistivity, such as along the Termini and Niceto Streams. Resistivity values of about 80 Ω m are widely present in all river courses, both along the plain and in the areas located to the south, closer to the reliefs. In several areas around the mouths of the watercourses, these values are probably linked to the presence of thick alluvial deposits, such as in the coastal sectors of the Termini, Mela, Corriolo and Niceto Streams. Lastly, in the Cape Milazzo promontory, the outcropping metamorphic rocks of the Aspromonte Unit (UA; UAg) induce these medium-high values of resistivity.

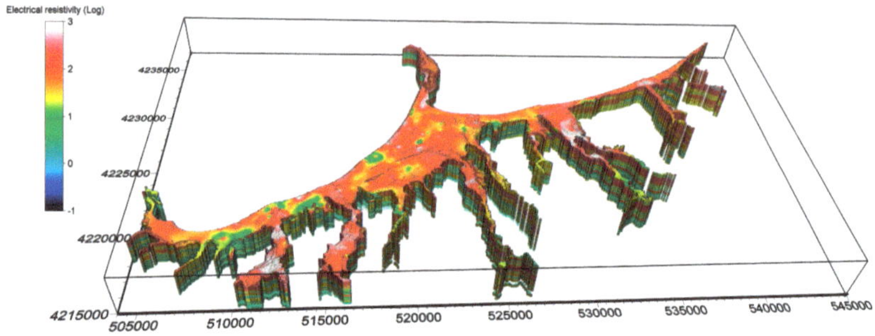

Figure 10. 3D model of the distribution of the logarithm of the electrical resistivity of the BMGWB.

Other areas of the BMGWB are affected by mean resistivity values of about 10–30 Ωm; these resistivities characterize the westernmost sector of the coastal plain, along some stream courses and in some areas between them. These values could be linked to portions of the BMGWB hosting alluvial sediments with less thickness. The lowest values described by the model are visible in the deepest parts of the plain, at depths between 40 and 120 m. As an example, the area between the towns of Barcelona P.G. and Milazzo, where the plain presents its greatest extension in N-S direction, is characterized by values less than 10 Ω m. This feature should be attributed to the presence of clays in the subsurface, as described by other studies performed in this area [57]. Moreover, for the lower values that appear quite shallow and close to the coast, marine ingression phenomena should not be excluded.

By analyzing the two resistivity models, we can notice the different contributions made by the floods and deposits that make up the water body. In the coastal plain areas, the resistivity model is, in fact, more influenced by the deposits of the SCGWB and BMGWB. In these areas, these deposits reach higher volumes. Within the mountain sector, represented in the models by the several arms of the rivers, the variation of resistivity is more linked to the characteristics of the rocks of the substrate, considering the thin thickness of the alluvial cover.

4.4. Estimate of the Bottom of the SCGWB and BMGWB

The thickness of the deposits that constituted SCGWB and BMGWB has been computed by a two-step workflow. Firstly, the deposits' thickness and depth were estimated at several points of the groundwater bodies, considering the results of HVSR inverse models. These were compared with 3D electrical resistivity models and other available geophysical data; additionally, this information was integrated, compiled and interpreted using data from several available boreholes in order to identify the main lithotypes and their thicknesses, as appropriate. Subsequently, all punctual thickness and depth data were spatially interpolated using a kriging algorithm available in the Surfer software v.16, provided by Golden Software, LCC (Golden Software, LLC, Golden, CO, USA). In this way, maps of both the bottom of the two water bodies and the thicknesses of the deposits were created (Figures 11 and 12).

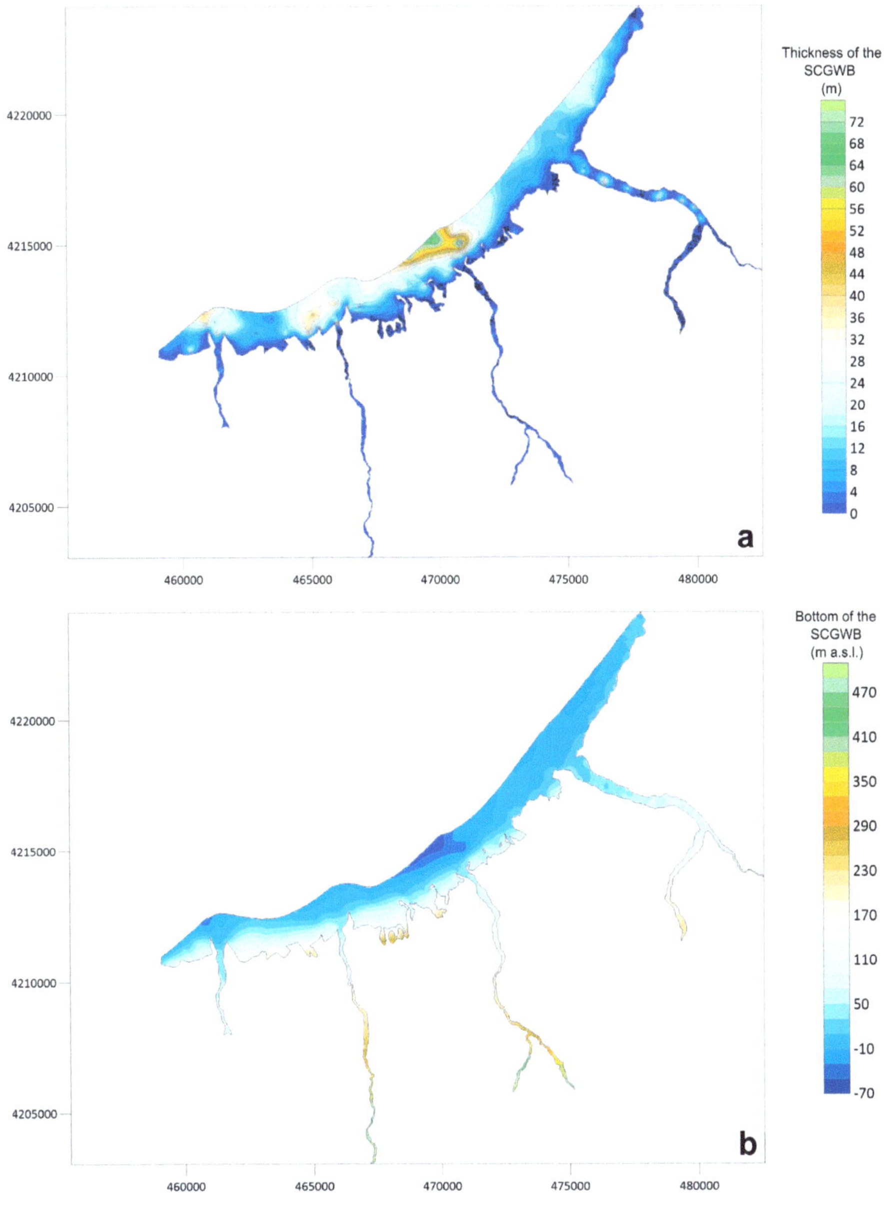

Figure 11. Maps of the thickness of the floods (**a**) and the depth of the flood bed (**b**) reconstructed through the integration of geophysical and geological data available within the water body of Sant'Agata–Capo D'Orlando (SCGWB).

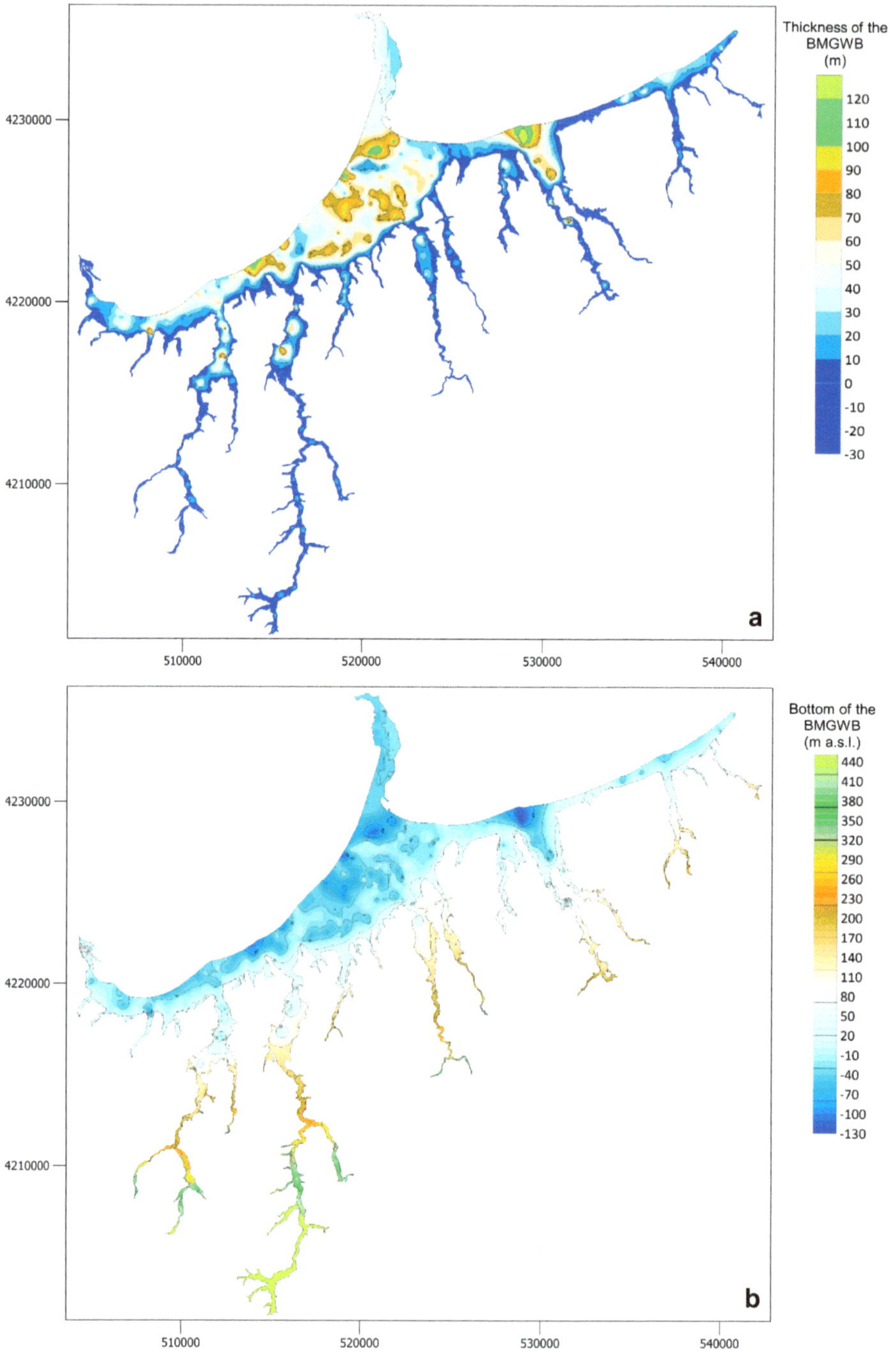

Figure 12. Maps of the thickness of the floods (**a**) and the depth of the flood bed (**b**) reconstructed through the integration of geophysical and geological data available within the water body of Barcellona P.G.–Milazzo (BMGWB).

Significant deposit volume variation in the different sectors of SCGWB and BMGWB can be noticed by analyzing the two thickness maps.

With regard to SCGWB (Figure 11a), the largest thicknesses are found in the coastal sector. A maximum depth of about 70 m b.s.l. is noted at the mouth of the Rosmarino river. Other large thicknesses are defined at the mouth of the Furiano Stream. About 30–40 m of floods are, instead, described on the left of the Inganno Stream, in a position slightly decentralized from the current estuary. In the eastern part of the SCGWB, the thicknesses are not large, as seen on the sides of the Zappula Stream. Along the mountain river paths of the SCGWB streams, the floods have been described with thicknesses that generally do not exceed 4–8 m. The same values are present in the plain areas close to the hilly and mountainous areas.

The area described within the BMGWB, shown in Figure 12a, shows the highest flood thicknesses in the northern parts of the plain. In particular, proceeding from West to East, a thickness of the surface coulters of more than 100 m and maximum values of about 120–130 m were obtained for these areas: near the mouth of the Termini Stream, in a coastal area close to the Mela Stream, inside the sector situated in the south of Milazzo Cape, in the area of the mouth of the Niceto stream. Moreover, 70–80 m of deposits were found within the central section of the plain, between Barcelona P.G. and Milazzo towns and in some small areas along the mountain river rods. This occurs, for example, along the Termini stream and along other river courses situated in the east sector of the coastal plain. Thicknesses of 30–50 m are widely present in the coastal areas of the plain, as well as in some small parts of the mountain arms of the rivers. These higher sectors of the BMGWB are mainly characterized by minimum thicknesses, which do not exceed 10 m.

5. Conclusions

The integrated analysis of geological and geophysical data allowed us to reconstruct the thickness of the alluvial deposits present in the two water bodies of the North-Eastern sector of Sicily, exploiting all the stratigraphic and geophysical data available for the two coastal floodplains. In particular, VES data were re-interpreted using a laterally constrained joint inversion algorithm and a 3D interpolation of the inverse models to provide 3D models of the electrical resistivity for the studied areas. Further active and passive seismic surveys, distributed in the study areas, allowed us to characterize the water bodies also with regard to the seismic P- and S-wave velocity of the different lithotypes and alluvial sediments.

The joint performing and interpretation of seismic and electrical tomographies along the same alignment helped to identify, in detail, the seismic velocity, electrical resistivity and chargeability values for the most common lithotypes.

Using all stratigraphic data collected and the geophysical models achieved, maps of the bottom and of the thickness of the deposits filling SCGWB and BMGWB were reconstructed.

These maps are tools that can be used to quantify the water flow rates possibly present in the subsoil. With an appropriate monitoring network, the hydrogeological flow model for these coastal plains could be indeed defined. The latter is a useful elaboration to indicate the opportunity for aquifer exploitation and to individuate the possible causes that could deplete the wealth of the aquifer by forced emulation or pollution.

Furthermore, the geophysical models obtained can be used for stratigraphic studies since the geometric characteristics of deposits can be better defined. Moreover, the variation of the thickness of the alluvial sediments can be correlated with the recent tectonic activity and the rates of uplift experienced by the northern coast of Sicily.

Finally, using such detailed subsurface models is of great interest for studies addressing the characterization of the geological hazards of the two areas that were often affected, in the past, by floods and earthquakes. Furthermore, the models and maps obtained can be a valid aid for identifying and analyzing areas subject to flooding or liquefaction phenomena, and can constitute the starting models for local seismic response studies.

The illustrated research approach, conducted on the North-Eastern coast of Sicily, can be used in similar hydrogeological contexts with a shallow groundwater body, mainly constituted by alluvial deposits. The study shows that the integration of the results of multiple geophysical surveys, supported by the constraint of boreholes, is a useful instrument to define more detailed and accurate subsoil models. Therefore, the different geophysical techniques adopted allow us to reduce the uncertainties related to the results and models derived from the use of geophysical individual surveys.

Author Contributions: Conceptualization, P.C., R.M. and R.F.; methodology, P.C. and R.M.; software, P.C., R.M., A.C. and A.B.; validation, P.C. and R.M.; formal analysis, P.C., R.M. and R.F.; investigation, P.C., A.C. and A.B.; resources, P.C., R.M. and R.F.; data curation, P.C. and R.M.; writing—original draft preparation, A.C. and A.B.; writing—review and editing, R.M.; visualization, P.C. and R.M.; supervision, P.C.; project administration, R.F.; funding acquisition, R.F. All authors have read and agreed to the published version of the manuscript.

Funding: This research was funded by Regione Siciliana, Commissario Delegato per l'Emergenza Bonifiche e la tutela delle Acque in Sicilia.

Data Availability Statement: The data presented in this study are available on request from the corresponding author. The data are not publicly available due to funder restrictions.

Conflicts of Interest: The authors declare no conflicts of interest.

References

1. Cacas, M.C.; Laigle, J.M.; Albouy, E.; Sulzer, C. Integrating stratigraphic modelling and seismic interpretation: Two practical examples. *Pet. Geosci.* **2008**, *14*, 253–261. [CrossRef]
2. Villa, V.; Virmoux, C.; Chaussé, C.; Degeai, J.P.; Robert, V.; Kuzucuoglu, C.; Boschian, G.; Agostini, S.; Aureli, D.; Pagli, M.; et al. The long sedimentary succession of the Valle Giumentina basin (Abruzzo, Central Italy): New evidence from stratigraphic studies and Electrical Resistivity Tomography (ERT). *Alp. Mediterr. Quat.* **2016**, *29*, 193–199.
3. Martorana, R.; Agate, M.; Capizzi, P.; Cavera, F.; D'Alessandro, A. Seismo-stratigraphic model of "La Bandita" area in the Palermo Plain (Sicily, Italy) through HVSR inversion constrained by stratigraphic data. *Ital. J. Geosci.* **2018**, *137*, 73–86. [CrossRef]
4. Caputo, R.; Salviulo, L.; Piscitelli, S.; Loperte, A. Late Quaternary activity along the Scorciabuoi Fault (Southern Italy) as inferred from electrical resistivity tomographies. *Ann. Geophys.* **2007**, *50*, 213–224. [CrossRef]
5. Imposa, S.; Mele, G.; Corrao, M.; Coco, G. Borehole seismic surveys for the mechanical characterization of a calcarenite cliff in the area of Ispica (southern Sicily). *Bull. Eng. Geol. Environ.* **2015**, *74*, 971–980. [CrossRef]
6. Drahor, M.G.; Berge, M.A. Integrated geophysical investigations in a fault zone located on southwestern part of İzmir city, Western Anatolia, Turkey. *J. Appl. Geophys.* **2017**, *136*, 114–133. [CrossRef]
7. Wang, Z.; Cai, X.; Yan, J.; Wang, J.; Liu, Y.; Zhang, L. Using the integrated geophysical methods detecting active faults: A case study in Beijing, China. *J. Appl. Geophys.* **2018**, *156*, 82–91. [CrossRef]
8. Abbas, A.; Zhu, H.; Zeng, Z.; Zhou, X. Sedimentary facies analysis using sequence stratigraphy and seismic sedimentology in the paleogene Pinghu Formation, Xihu depression, east China sea shelf basin. *Mar. Pet. Geol.* **2018**, *93*, 287–297. [CrossRef]
9. Ni, J.; Zhao, D.; Liao, X.; Li, X.; Fu, L.; Chen, R.; Xia, Z.; Liu, Y. Sedimentary Architecture Analysis of Deltaic Sand Bodies Using Sequence Stratigraphy and Seismic Sedimentology: A Case Study of Jurassic Deposits in Zhetybay Oilfield, Mangeshrak Basin, Kazakhstan. *Energies* **2022**, *15*, 5306. [CrossRef]
10. Todaro, S.; Sulli, A.; Spatola, D.; Basilone, G.; Aronica, S. Seismic stratigraphy of the north-westernmost area of the Malta Plateau (Sicily Channel): The Middle Pleistocene-Holocene sedimentation in a tidally influenced shelf. *Mar. Geol.* **2022**, *445*, 106740. [CrossRef]
11. Abu Zeid, N.; Corradini, E.; Bignardi, S.; Nizzo, V.; Santarato, G. The Passive Seismic Technique 'HVSR' as a Reconnaissance Tool for Mapping Paleo-soils: The Case of the Pilastri Archaeological Site, Northern Italy. *Archaeol. Prospect.* **2017**, *24*, 245–258. [CrossRef]
12. Bottari, C.; Martorana, R.; Scudero, S.; Capizzi, P.; Cavallaro, D.; Pisciotta, A.; D'Alessandro, A.; Coltelli, M.; Lodato, L. Coseismic damage at an archaeological site in sicily, italy: Evidence of roman age earthquake surface faulting. *Surv.Geophys.* **2018**, *39*, 1263–1284. [CrossRef]
13. Cafiso, F.; Canzoneri, A.; Capizzi, P.; Carollo, A.; Martorana, R.; Romano, F. Joint interpretation of electrical and seismic data aimed at modelling the foundation soils of the Maredolce monumental complex in Palermo (Italy). *Archaeol. Prospect.* **2023**, *30*, 69–85. [CrossRef]
14. Butchibabu, B.; Khan, P.K.; Jha, P.C. Foundation evaluation of underground metro rail station using geophysical and geotechnical investigations. *Eng. Geol.* **2019**, *248*, 140–154. [CrossRef]
15. Liu, B.; Wang, C.; Liu, Z.; Xu, Z.; Nie, L.; Pang, Y.; Wang, N.; Feng, S. Cascade surface and borehole geophysical investigation for water leakage: A case study of the Dehou reservoir, China. *Eng. Geol.* **2021**, *294*, 106364. [CrossRef]

16. Polonia, A.; Albertazzi, S.; Bellucci, L.G.; Bonetti, C.; Bonetti, J.; Giorgetti, G.; Giuliani, S.; López Correa, M.; Mayr, C.; Peruzza, L.; et al. Decoding a complex record of anthropogenic and natural impacts in the Lake of Cavazzo sediments, NE Italy. *Sci. Total Environ.* **2021**, *787*, 147659. [CrossRef]
17. Chaudhuri, A.; Sekhar, M.; Descloitres, M.; Godderis, Y.; Ruiz, L.; Braun, J.J. Constraining complex aquifer geometry with geophysics (2-D ERT and MRS measurements) for stochastic modelling of groundwater flow. *J. Appl. Geophys.* **2013**, *98*, 288–297. [CrossRef]
18. Kumar, D.; Rao, V.A.; Sarma, V.S. Hydrogeological and geophysical study for deeper groundwater resource in quartzitic hard rock ridge region from 2D resistivity data. *J. Earth Syst. Sci.* **2014**, *123*, 531–543. [CrossRef]
19. Canzoneri, A.; Capizzi, P.; Martorana, R.; Albano, L.; Bonfardeci, A.; Costa, N.; Favara, R. Geophysical Constraints to Reconstructing the Geometry of a Shallow Groundwater Body in Caronia (Sicily). *Water* **2023**, *15*, 3206. [CrossRef]
20. Gondwe, B.R.; Lerer, S.; Stisen, S.; Marín, L.; Rebolledo-Vieyra, M.; Merediz-Alonso, G.; Bauer-Gottwein, P. Hydrogeology of the south-eastern Yucatan Peninsula: New insights from water level measurements, geochemistry, geophysics and remote sensing. *J. Hydrol.* **2010**, *389*, 1–17. [CrossRef]
21. Zouhri, L.; Lutz, P. Hydrogeophysical characterization of the porous and fractured media (chalk aquifer in the Beauvais, France). *Environ. Earth. Sci.* **2016**, *75*, 343. [CrossRef]
22. Muhammad, S.; Ehsan, M.I.; Khalid, P.; Sheikh, A. Hydrogeophysical modeling and physio-chemical analysis of quaternary aquifer in central part of Bari Doab, Punjab, Pakistan. *Model. Earth Syst. Environ.* **2023**, *9*, 1427–1443. [CrossRef]
23. Hasan, M.; Shang, Y.; Jin, W.; Shao, P.; Yi, X.; Akhter, G. Geophysical Assessment of Seawater Intrusion into Coastal Aquifers of Bela Plain, Pakistan. *Water* **2020**, *12*, 3408. [CrossRef]
24. Sendrós, A.; Urruela, A.; Himi, M.; Alonso, C.; Lovera, R.; Tapias, J.C.; Rivero, L.; Garcia-Artigas, R.; Casas, A. Characterization of a shallow coastal aquifer in the framework of a subsurface storage and soil aquifer treatment project using electrical resistivity tomography (Port de la Selva, Spain). *Appl. Sci.* **2021**, *11*, 2448. [CrossRef]
25. Tarallo, D.; Alberico, I.; Cavuoto, G.; Pelosi, N.; Punzo, M.; Di Fiore, V. Geophysical assessment of seawater intrusion: The Volturno Coastal Plain case study. *Appl. Water. Sci.* **2023**, *13*, 234. [CrossRef]
26. Matias, M.S.; Da Silva, M.M.; Ferreira, P.; Ramalho, E. A geophysical and hydrogeological study of aquifers contamination by a landfill. *J. Appl. Geophys.* **1994**, *32*, 155–162. [CrossRef]
27. Ciampi, P.; Esposito, C.; Cassiani, G.; Deidda, G.P.; Flores-Orozco, A.; Rizzetto, P.; Chiappa, A.; Bernaberi, M.; Gardon, A.; Papini, M.P. Contamination presence and dynamics at a polluted site: Spatial analysis of integrated data and joint conceptual modeling approach. *J. Contam. Hydrol.* **2022**, *248*, 104026. [CrossRef] [PubMed]
28. Mohammed, M.A.; Szabó, N.P.; Szűcs, P. Exploring hydrogeological parameters by integration of geophysical and hydrogeological methods in northern Khartoum state, Sudan. *Ground. Sustain. Dev.* **2023**, *20*, 100891. [CrossRef]
29. Dochartaigh, B.E.O.; Archer, N.A.L.; Peskett, L.; MacDonald, A.M.; Black, A.R.; Auton, C.A.; Merritt, J.E.; Gooddy, D.C.; Bonell, M. Geological structure as a control on floodplain groundwater dynamics. *Hydrogeol. J.* **2019**, *27*, 703–716. Available online: https://nora.nerc.ac.uk/id/eprint/521927 (accessed on 23 February 2024). [CrossRef]
30. Cianflone, G.; Vespasiano, G.; De Rosa, R.; Dominici, R.; Apollaro, C.; Vaselli, O.; Pizzino, L.; Tolomei, C.; Capecchiacci, F.; Polemio, M. Hydrostratigraphic framework and physicochemical status of groundwater in the Gioia Tauro Coastal Plain (Calabria—Southern Italy). *Water* **2021**, *13*, 3279. [CrossRef]
31. von Suchodoletz, H.; Pohle, M.; Khosravichenar, A.; Ulrich, M.; Hein, M.; Tinapp, C.; Shultz, J.; Ballasus, E.; Veit, U.; Ettel, P.; et al. The fluvial architecture of buried floodplain sediments of the Weiße Elster River (Germany) revealed by a novel method combination of drill cores with two-dimensional and spatially resolved geophysical measurements. *Earth Surf. Process. Landf.* **2022**, *47*, 955–976. [CrossRef]
32. CASMEZ (CASsa per il MEZzogiorno)—Direzione Generale Progetti Speciali. Progetto Speciale n. 30: Utilizzazione delle acque degli schemi idrici intersettoriali della Sicilia. In *Indagini Idrogeologiche e Geofisiche per il Reperimento di Acque Sotterranee per L'approvvigionamento Idrico del Sistema IV Zona Nord Orientale Della Sicilia (Messinese)*; CASMEZ: Palermo, Italy, 1978.
33. Granata, A.; Castrianni, G.; Pasotti, L.; Favara, R.; Gagliano Candela, E.; Scaletta, C.; Madonia, P.; Morici, S.; Bellomo, S.; La Pica, L.; et al. Studio per la definizione dei modelli concettuali dei corpi idrici sotterranei di Peloritani, Nebrodi e ragusano e indagini geofisiche correlate. In Proceedings of the 37 Convegno del Gruppo Nazionale di Geofisica della Terra Solida-GNGTS, Roma, Italy, 19–21 November 2018; Volume 37, pp. 104–108. Available online: https://hdl.handle.net/10447/316230 (accessed on 10 January 2024).
34. Catalano, R.; Di Stefano, P.; Sulli, A.; Vitale, F. Paleogeography and structure of the central Mediterranean: Sicily and its offshore area. *Tectonophysics* **1996**, *260*, 291–323. [CrossRef]
35. Giunta, G.; Luzio, D.; Agosta, F.; Calò, M.; Di Trapani, F.; Giorgianni, A.; Oliveri, E.; Orioli, S.; Perniciaro, M.; Vitale, M.; et al. An integrated approach to investigate the seismotectonics of northern Sicily and southern Tyrrhenian. *Tectonophysics* **2009**, *476*, 13–21. [CrossRef]
36. Catalano, R.; Valenti, V.; Albanese, C.; Accaino, F.; Sulli, A.; Tinivella, U.; Gasparo Morticelli, M.; Zanolla, C.; Giustiniani, M. Sicily's fold/thrust belt and slab rollback: The SI.RI.PRO. seismic crustal transect. *J. Geol. Soc.* **2013**, *170*, 451–464. [CrossRef]
37. Morticelli, M.G.; Valenti, V.; Catalano, R.; Sulli, A.; Agate, M.; Avellone, G.; Albanese, C.; Basilone, L.; Gugliotta, C. Deep controls on foreland basin system evolution along the Sicilian fold and thrust belt. *BSG-Earth Sci. Bull.* **2015**, *186*, 273–290. [CrossRef]

38. Basilone, L.; Bonfardeci, A.; Romano, P.; Sulli, A. Natural Laboratories for Field Observation About Genesis and Landscape Effects of Palaeo-Earthquakes: A Proposal for the Rocca Busambra and Monte Barracù Geosites (West Sicily). *Geoheritage* **2018**, *11*, 821–837. [CrossRef]
39. Amodio-Morelli, L.; Bonardi, G.; Colonna, V.; Dietrich, D.; Giunta, G.; Ippolito, F.; Liguori, V.; Lorenzoni, F.; Paglionico, A.; Perrone, V.; et al. L'Arco Calabro-Peloritano nell'orogene Appenninico-Maghrebide. *Mem. Soc. Geol. Ital.* **1976**, *17*, 1. Available online: https://arpi.unipi.it/handle/11568/1278 (accessed on 10 January 2024).
40. Lentini, F.; Catalano, S.; Carbone, S. Note Illustrative Della Carta Geologica Della Provincia di Messina, Scala 1:50.000: S.EL.CA, Firenze. 2000; pp. 1–70. Available online: https://www.isprambiente.gov.it/Media/carg/note_illustrative/601_Messina_Reggio_Calabria.pdf (accessed on 1 February 2024).
41. Carbone, S.; Messina, A.; Lentini, F. *Note Illustrative dei F. 587–600 "Milazzo–Barcellona Pozzo di Gotto" della Carta Geologica d'Italia Alla Scala 1:50.000*; S.EL.CA., 2011; 262, pp. 146 ff., 5 tavv., 4 tabb. ISPRA—Regione Siciliana—Università di Catania; Servizio Geologico d'Italia: Rome, Italy.
42. Servizio Geologico d'Italia. *Carta Geologica d'Italia Alla Scala 1:50.000, F. 599 Patti*; Servizio Geologico d'Italia: Rome, Italy, 2023. [CrossRef]
43. Giunta, G.; Giorgianni, A. *Note Illustrative del F.598 "S. Agata di Militello" Della Carta Geologica d'Italia Alla Scala 1:50.000*; SystemCart; ISPRA—Università degli Studi di Palermo: Rome, Italy, 2013. Available online: https://www.isprambiente.gov.it/Media/carg/note_illustrative/598_S_Agata_di_Militello.pdf (accessed on 1 February 2024).
44. Lentini, F.; Carbone, S. Geologia della Sicilia (Geology of Sicily). *Mem. Descr. Carta Geol. d'It.* **2014**, *XCV*, 7–414.
45. Giunta, G.; Nigro, F. Tectono-sedimentary constraints to the Oligocene-to-Miocene evolution of the Peloritani thrust belt (NE Sicily). *Tectonophysics* **1999**, *315*, 287–299. [CrossRef]
46. Nigro, F.; Sulli, A. Plio-Pleistocene extensional tectonics in the Western Peloritani area and its offshore (northeastern Sicily). *Tectonophysics* **1995**, *252*, 295–305. [CrossRef]
47. Giunta, G.; Nigro, F.; Renda, P. Extensional tectonics during Maghrebides chain building since Late Miocene: Examples from Northern Sicily. *Ann. Soc. Geol.* **2000**, *70*, 81–89.
48. Giunta, G.; Nigro, F.; Renda, P.; Giorgianni, A. The Sicilian–Maghrebides Tyrrhenian Margin: A neotectonic evolutionary model. *Mem. Soc. Geol.* **2000**, *119*, 553–565.
49. Renda, P.; Tavarnelli, E.; Tramutoli, M.; Gueguen, E. Neogene deformations of Northern Sicily, and their implications for the geodynamics of the Southern Tyrrhenian Sea margin. *Mem. Soc. Geol.* **2000**, *55*, 53–59.
50. Nigro, F.; Renda, P. Plio-Pleistocene stike-slip deformation in NE Sicily: The example of the area between Capo Calavà and Capo Tindari. *Boll. Soc. Geol.* **2005**, *124*, 377–394.
51. Giunta, G.; Bellomo, D.; Carnemolla, S.; Pisano, A.; Profeta, R.; Runfola, P. La "Linea di Taormina": Residuo epidermico di una paleostruttura crostale del fronte cinematico maghrebide? *Acts 8○ GNGTS Congr.* **1989**, *8*, 1197–2013.
52. Gasparo Morticelli, M.; Avellone, G.; Sulli, A.; Agate, M.; Basilone, L.; Catalano, R.; Pierini, S. Mountain building in NW Sicily from the superimposition of subsequent thrusting and folding events during Neogene: Structural setting and tectonic evolution of the Kumeta and Pizzuta ridges. *J. Maps* **2017**, *13*, 276–290. [CrossRef]
53. Lentini, F. Carta Geologica della Provincia di Messina, Scala 1:50.000: S.EL.CA, Firenze. 2000. Available online: https://www2.regione.sicilia.it/beniculturali/dirbenicult/bca/ptpr/documentazione%20tecnica%20messina/CARTOGRAFIA/ANALISI/03_Geologia.pdf (accessed on 1 February 2024).
54. Arisco, G.; Arnone, G.; Favara, R.; Nigro, F.; Perricone, M.; Renda, P.; Mondello, C. Integrated neotectonic and morphometric analysis of northern Sicily. *Boll. Soc. Geol.* **2006**, *125*, 221–244. Available online: http://hdl.handle.net/2122/2497 (accessed on 10 January 2024).
55. Ferrara, V. Vulnerabilità all'inquinamento degli Acquiferi dell'area Peloritana (Sicilia Nord-Orientale), Studi Sulla Vulnerabilità Degli Acquiferi 14. In *Quaderni di Tecniche di Protezione Ambientale: Edizione Pitagora*; Pitagora Editrice Srl.: Bologna, Italy, 1999.
56. Cangemi, M.; Madonia, P.; Albano, L.; Bonfardeci, A.; Di Figlia, M.G.; Di Martino, R.M.R.; Nicolosi, M.; Favara, R. Heavy Metal Concentrations in the Groundwater of the Barcellona-Milazzo Plain (Italy): Contributions from Geogenic and Anthropogenic Sources. *Int. J. Environ. Res. Public Health* **2019**, *16*, 285. [CrossRef] [PubMed]
57. Capizzi, P.; Martorana, R.; Favara, R.; Albano, L.; Bonfardeci, A.; Catania, M.; Costa, N.; Gagliano, A. Geophysical Contribution to the Reconstruction of the Hydrological Model of "Barcellona-Milazzo Plain" Groundwater Body, Northen Sicily. In Proceedings of the 25th European Meeting of Environmental and Engineering Geophysics; European Association of Geoscientists & Engineers, Moscow, Russia, 8–12 September 2019; pp. 1–5. [CrossRef]
58. Celico, P. *Prospezioni Idrogeologiche, Vol. I e Vol. II*; Liguori Editore Napoli: Naples, Italy, 1988.
59. Allocca, V.; De Vita, P.; Manna, F. La variabilità climatica in Italia meridionale e gli effetti sulla ricarica degli acquiferi carbonatici. *EHE Geol.* **2012**, *14*, 9–10. Available online: https://hdl.handle.net/11588/425079 (accessed on 23 February 2024).
60. Castany, G. Idrogeologia. In *Principi e Metodi: Edizione Flaccovio Dario, Palermo*; Pitagora Editrice Srl.: Bologna, Italy, 1985.
61. Civita, M. *Idrogeologia Applicata e Ambientale*; CEA Editore: Rozzano, Italy, 2005; p. 800. ISBN 8808087417.
62. Cimino, A.; Cosentino, C.; Oieni, A.; Tranchina, L. A geophysical and geochemical approach for seawater intrusion assessment in the Acquedolci coastal aquifer (Northern Sicily). *Environ. Geol.* **2007**, *55*, 1473–1482. [CrossRef]
63. Pantaleone, D.V.; Vincenzo, A.; Fulvio, C.; Silvia, F.; Cesaria, M.; Giuseppina, M.; Ilaria, M.; Vincenzo, P.; Rosa, S.A.; Gianpietro, S.; et al. Hydrogeology of continental southern Italy. *J. Maps* **2018**, *14*, 230–241. [CrossRef]

64. Kosinski, W.K.; Kelly, W.E. Geoelectric Soundings for Predicting Aquifer Properties. *Groundwater* **1981**, *19*, 163–171. [CrossRef]
65. Heigold, P.C.; Gilkeson, R.H.; Cartwright, K.; Reed, P.C. Aquifer Transmissivity from Surficial Electrical Methods. *Groundwater* **1979**, *17*, 338–345. [CrossRef]
66. Van Overmeeren, R.A. Aquifer boundaries explored by geoelectrical measurements in the coastal plain of Yemen: A case of equivalence. *Geophysics* **1989**, *54*, 38–48. [CrossRef]
67. Auken, E.; Christiansen, A.V. Layered and laterally constrained 2D inversion of resistivity data. *Geophysics* **2004**, *69*, 752–761. [CrossRef]
68. Nogoshi, M.; Iragashi, T. On the propagation characteristic of the microtremors. *J. Seismol. Soc. Jpn.* **1970**, *24*, 24–40.
69. Nakamura, Y. A method for dynamic characteristics estimation of subsurface using microtremor on the ground surface. *Quat. Rep. Railw.* **1989**, *30*, 25–33. Available online: https://worldcat.org/oclc/3127232 (accessed on 23 February 2024).
70. Castellaro, S.; Mulargia, F. VS30 estimates using constrained H/V measurements. *Bull. Seismol. Soc. Am.* **2009**, *99*, 761–773. [CrossRef]
71. Castellaro, S. The complementarity of H/V and dispersion curves. *Geophysics* **2016**, *81*, T323–T338. [CrossRef]
72. Park, C.B.; Miller, R.D.; Xia, J. Multichannel analysis of surface waves. *Geophysics* **1999**, *64*, 800–808. [CrossRef]
73. White, D.J. Two-Dimensional Seismic Refraction Tomography. *Geophys. J. Int.* **1989**, *97*, 223–245. [CrossRef]
74. Koch, K.; Wenninger, J.; Uhlenbrook, S.; Bonell, M. Joint interpretationof hydrological and geophysical data: Electrical resistivityn tomography results from a process hydrological research site in the Black Forest Mountains, Germany. *Hydrol. Process. Int. J.* **2009**, *23*, 1501–1513. [CrossRef]
75. Robert, T.; Dassargues, A.; Brouyère, S.; Kaufmann, O.; Hallet, V.; Nguyen, F. Assessing the contribution of electrical resistivity tomography (ERT) and self-potential (SP) methods for a water well drilling program in fractured/karstified limestones. *J. Appl. Geophys.* **2011**, *75*, 42–53. [CrossRef]
76. Martorana, R.; Lombardo, L.; Messina, N.; Luzio, D. Integrated geophysical survey for 3D modelling of a coastal aquifer polluted by seawater. *Near Surf. Geophys.* **2014**, *12*, 45–59. [CrossRef]
77. Frischknecht, F.C.; Raab, P.V. Time-domain electromagnetic soundings at the Nevada Test Site, Nevada. *Geophysics* **1984**, *49*, 981–992. [CrossRef]
78. Wightman, W.E.; Kaufman, A.A.; Hoekstra, P. Mapping gas-water contacts in shallow producing formations with transient EM. In SEG Technical Program Expanded Abstracts. *Soc. Explor. Geophys.* **1983**, *0059*, 59–60. [CrossRef]
79. Fitterman, D.V.; Stanley, W.D.; Bisdorf, R.J. Electrical structure of Newberry volcano, Oregon. *J. Geophys. Res. Solid Earth* **1988**, *93*, 10119–10134. [CrossRef]
80. Martorana, R.; Capizzi, P.; Avellone, G.; D'Alessandro, A.; Siragusa, R.; Luzio, D. Assessment of a geological model by surface wave analyses. *J. Geophys. Eng.* **2017**, *14*, 159–172. [CrossRef]
81. Di Capua, G.; Peppoloni, S.; Amanti, M.; Cipolloni, C.; Conte, G. Site classification map of Italy based on surface geology. *Geol. Soc. Lond. Eng. Geol. Spec. Publ.* **2016**, *27*, 147–158. [CrossRef]
82. Forte, G.; Chioccarelli, E.; De Falco, M.; Cito, P.; Santo, A.; Iervolino, I. Seismic soil classification of Italy based on surface geology and shear-wave velocity measurements. *Soil Dyn. Earthq. Eng.* **2019**, *122*, 79–93. [CrossRef]

water

MDPI

Article

Environmental Monitoring of Pig Slurry Ponds Using Geochemical and Geoelectrical Techniques

Ximena Capa-Camacho [1], Pedro Martínez-Pagán [2], José A. Acosta [1], Marcos A. Martínez-Segura [2,*], Marco Vásconez-Maza [3] and Ángel Faz [1]

[1] Sustainable Use, Management and Reclamation of Soil and Water Research Group, Universidad Politécnica de Cartagena, Paseo Alfonso XIII, 48, 30203 Cartagena, Murcia, Spain; ximena.capa@upct.es (X.C.-C.); ja.acosta@upct.es (J.A.A.); angel.fazcano@upct.es (Á.F.)

[2] Applied Near-Surface Geophysics Research Group, Departamento de Ingeniería Minera y Civil, Universidad Politécnica de Cartagena, Paseo Alfonso XIII, 52, 30203 Cartagena, Murcia, Spain; p.martinez@upct.es

[3] British Geological Survey, Keyworth, Nottinghamshire NG12 5GG, UK; mvm@bgs.ac.uk

* Correspondence: marcos.martinez@upct.es

Abstract: The efficient management of slurry, which is a by-product rich in nutrients derived from feces, urine, cleaning water, and animal waste that stands out for its high concentration of nutrients such as nitrogen, phosphorus, and potassium, is of vital importance, highlighting the importance of slurry management in storage ponds, which. The Murcia–Spain region has an important number of pig farms. Hence, infrastructures dedicated to managing by-products are necessary to prevent environmental pollution and eutrophication of groundwater. The aim of a recent study was to evaluate the relationship between electrical values and geochemical parameters of pig slurry stored in a pond using ERT and geochemical analysis. In addition, the study was designed to monitor the pond to determine the geochemical characteristics of the slurry and to assess the risk of lateral contamination. The study results indicate a noticeable decrease in electrical resistivity values at 0.4 and 1.6 m depth in surveys 1 and 2. The reduction ranges from 50 to 100 percent. This paper presents a new method for monitoring slurry ponds using electrical resistivity tomography. This non-invasive method provides detailed information on the distribution and characteristics of the fluids, as well as a clear picture of the electrical resistivity of the subsurface.

Keywords: electrical resistivity tomography (ERT); waste characterization; time-lapse; hydrogeophysical studies; groundwater

1. Introduction

Pig slurry is a nutrient-rich byproduct composed of pig feces, urine, cleaning water, feed remnants, and other animal waste [1]. This by-product material has crucial nutrients, including nitrogen, phosphorus, potassium, and various inorganic compounds (ammonia (NH_3), carbon dioxide (CO_2)) and organic compounds (methane (CH_4), oil acids, phenolic compounds, proteins and peptides, organic phosphorus, humic and fulvic acids, and volatile organic compounds (VOCs)) [2]. The elevated concentration of these nutrients (N, P, K) underscores the highest importance of effective management of pig slurry, serving as a critical measure to mitigate environmental pollution concerns, notably the risk of groundwater quality deterioration due to eutrophication [3]. In the 2022 Global Pig Production ranking, Spain ranked third place due to a remarkable number of 88,437 pig farms located in the country [4]. Notably, the region of Murcia emerged as a prominent contributor, housing a notable 7% of the total pig farms, with a comprehensive count of 1444 farms and an approximate production of 72,000 m^3 of slurry per year. The substantial presence of pig farms in this region underlines the essential need for dedicated infrastructure to efficiently manage and store the by-products generated by these porcine operations.

Presently, different legislation has been established to construct slurry storage ponds, such as the Water Resources (Control of Pollution. Silage, Slurry and Fuel Oil, England, Regulations 2010) [5], which aims to minimize water pollution risks. According to these regulations, slurry stores must be constructed at a minimum distance of 10 m from watercourses. They must also be adequately waterproofed and have sufficient capacity to accommodate four months of slurry production. An additional United States Department of Agriculture (USDA) regulation requires that the construction of slurry ponds be located in soils that meet permeability standards to prevent groundwater contamination, according to the Natural Resources Conservation Service—Conservation Practice Standard Code 313 (NRCS-CPS) [6]. To accomplish this, it is necessary to perform a thorough analysis of the pond sealing or lining through various geotechnical tests, including compacted soil treatment (Code 520), geomembrane or geosynthetic clay liner (Code 521), or concrete (Code 522). In Spain, RD 306/2020 [7] regulations require that storage facilities be closed and waterproofed by their natural structure or artificial conditioning. Similarly, the Council Directive of 12 December 1991 (91/676/EEC) [8] states that storage containers for pig slurry must have a capacity more significant than the amount of waste produced on the farm during its most extended period. The calculation of this capacity depends on the location of the farm and whether it is in an area vulnerable to nitrate pollution caused by nitrates used in agriculture. However, in the past, waste storage consisted of depositing by-products in open pits excavated in the ground, using in situ material as a sealing method. The environmental risks associated with possible slurry seepage from storage ponds are closely related to nitrate (NO_3^-) leaching, which can contaminate surface and groundwater in case of infiltration [9,10]. Non-destructive techniques are required to make measurements to ensure the stability and security of the slurry pond. These techniques should also provide a representation of the distribution of pig slurry in the storage pond and its infiltration into the surrounding soil, which can be obtained through the use of electrical resistivity tomography (ERT).

The electrical resistivity tomography (ERT) method is a geophysical technique with many applications, including marine [11], geotechnical, environmental [12], and composition of the soil, including building foundation prospecting, archaeological prospecting [13], landfill delineation [14], contamination [15], and chemical tracer studies [16,17]. This technique is based on measuring the electrical resistivity of soil [18], which is dependent on the geometry, pore size, and total porosity of the soil [19]. Therefore, ERT provides fast, low-cost, and accurate results in the subsurface [20]. Several studies have highlighted the effectiveness of ERT in identifying and determining the extent of leachate contamination in urban waste [21,22]. For instance, Akiang et al. [23] applied ERT to identify the infiltration zones of leachate and map out the surface areas contaminated by leachate in an urban waste site. Similarly, Morita et al. (2022) [24] demonstrated the successful use of ERT in identifying covered and uncovered zones and boundary zones within an exposed landfill. Moreover, Zaini et al. (2022) [25] established that ERT is one of the most effective methods for evaluating the existence and extent of leachate in urban landfills. Additionally, Udosen et al. (2022) [26] conducted geoelectrical modeling using ERT to determine the extent of leachate contamination at a landfill site in southeastern Nigeria. A few studies have explored the impact of slurry on soil and groundwater. Capa-Camacho et al. [27,28] used geophysical and geochemical techniques to identify areas affected by slurry and found that the most significant accumulation of contaminants occurs at two meters depth. Martínez-Pagán et al. [29,30] demonstrated the effectiveness of electrical resistivity tomography (ERT) in assessing the impact of slurry ponds on the subsoil. These studies show the potential of ERT to delimit the subsoil areas affected by slurry ponds and to characterize the physicochemical properties of these areas. Conversely, monitoring the development of the possible infiltration plume in slurry ponds has not been carried out.

With ERT, one of the ways to corroborate the results can be by geochemical techniques, which are crucial to studying the composition of fluids, soils, and rocks in various environmental contexts [31]. However, conventional geochemical methods to analyze the possibility of slurry seepage into a storage pond intended for containment require the collection of samples at specific locations in the pond. These methods, while valuable, do not provide a global understanding of the subsurface conditions throughout the pond. In contrast, tomographic techniques, such as the ERT method, offer a unique advantage by allowing continuous and noninvasive global subsurface monitoring of the subsoil. ERT generates a subsurface image consisting of spatially distributed apparent electrical resistivity values, facilitating a more holistic and dynamic assessment of the entire storage pond [32,33]. This approach is advantageous when investigating potential slurry infiltration, as the tomographic technique can effectively identify preferential pathways and areas vulnerable to slurry infiltration without the limitations of discrete sampling points [27,29,34].

Using the ERT method to monitor a slurry pond is a novel approach. This non-invasive technique provides detailed information on the distribution and characteristics of the fluids in the pond. It offers a clear image of the electrical resistivity of the subsurface materials, which helps to identify the composition and concentration of fluids present in the pond. This is crucial for understanding slurry dynamics and making informed decisions around waste management. Furthermore, ERT allows for the detection of changes in the electrical resistivity of the soil. This information can be used to identify the presence and movement of liquids, which is essential in predicting and preventing environmental issues related to slurry management. Implementing ERT is essential in promoting sustainability and minimizing negative impacts on the surrounding environment [35].

The main objective of this work was to evaluate the relationship between the electrical values and geochemical parameters of pig slurry stored in a pond using ERT and geochemical analysis. ERT was used as a proxy to understand the different conditions of the pig slurry inside the slurry pond. Additionally, the analysis aimed to monitor the pond to determine the geochemical characteristics of the pig slurry and assess the risk of contamination through lateral and vertical seepage.

2. Materials and Methods

2.1. Study Area

Fuente Álamo is a municipality (Region of Murcia, SE Spain) (Figure 1a) with 244 pig farms annually producing around 12,000 cubic meters of slurry. Moreover, Fuente Álamo is located near the Mar Menor coastal lagoon, where nitrates from agriculture and cattle raising can cause contamination [36,37]. This zone has an average annual temperature of 17.3 °C and receives a yearly rainfall of 321 mm [38]. The geological composition of the Fuente Álamo municipality can be found on sheet 955 of the National Geological Map (MAGNA) 1:50,000 [39]. The northern part is characterized by gravel, conglomerates, sands, and silts, while the southern region is composed of undifferentiated Quaternary. Regarding hydrogeological characteristics, Fuente Álamo is located in the "Campo de Cartagena" region, a complex and huge hydrogeological unit comprising several aquifers. These aquifers are characterized by Neogene Quaternary materials, primarily consisting of loamy soils with dendritic and calcareous intercalations from the Miocene to the Quaternary, resulting in various aquifer levels [40]. Remarkably, the "Quaternary" aquifer is prominent within the study area, composed of 20–150 m of gravels, sands, silts, clays, and caliches deposited on tertiary marls, which act as an impermeable base.

Consequently, a slurry pond in this municipality was chosen as a point to employ the ERT method to monitor potential slurry (Figure 1a). The year of construction of the slurry pond corresponds to the year 2001; the type of production system of the farm is intensive (fattening) with a maximum capacity of 4000 animals per year with an average weight of 120 kg and an approximate annual slurry production of 2830 m^3/year. The type of waterproofing used in the slurry pond construction is natural, i.e., soil from the area was used. The pig slurry pond has a storage capacity of 2867 m^3 with dimensions of 64 m in length, 28 m in width, and 1.6 m in depth (Figure 1a).

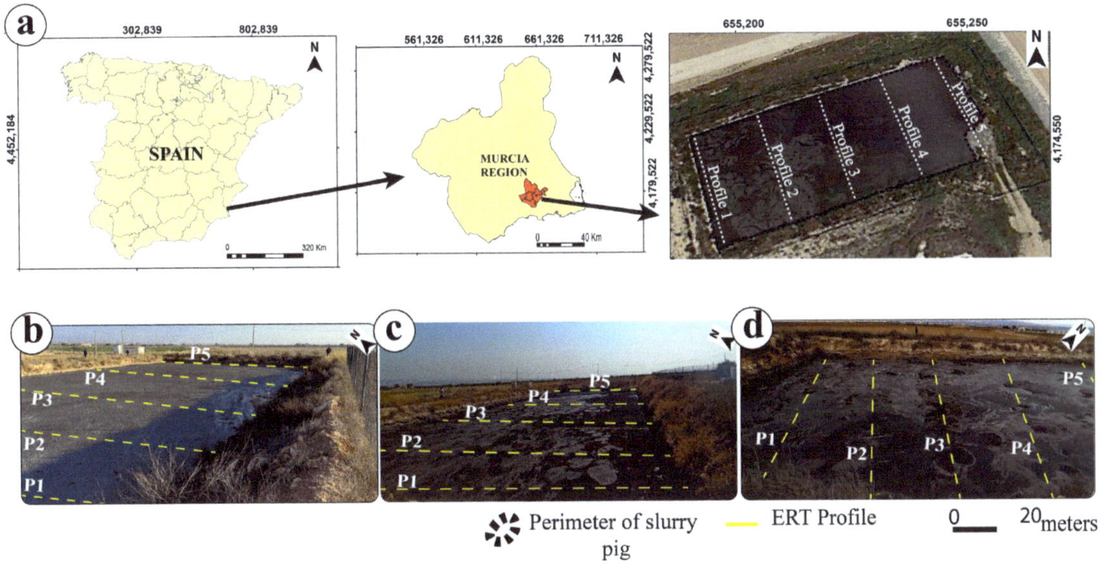

Figure 1. Location and schema of the selected slurry pond. (**a**) Geographical location of the studied pond in southeast Spain; layout of ERT profiles in the slurry pond (**b**) in survey 1, (**c**) in survey 2, (**d**) in survey 3.

2.2. Electrical Resistivity Tomography

During the ERT measurement process, an electrical current (I) is injected into the soil through a pair of electrodes, and the resulting difference of potential (V) is measured between another pair of potential electrodes. By comparing the voltage measured at the potential electrodes to the current input at the current electrodes, an apparent resistivity value (ρ_a) is obtained in Ohm·m. The resistivity mapping of the area of study is presented in 2D and 3D pseudo sections after an inversion process [16]. Different electrode arrangements can be used to determine the apparent resistivity at varying depths and lateral positions [41].

ERT was conducted within the slurry pond using a modified marine cable from Advanced Geosciences Inc (AGI). The modifications involved adding polyethylene floats and plastic clamps to the line, which allowed the graphite electrodes to float on the surface of the slurry pond. We employed a 28-electrode dipole–dipole array with one-meter spacing between electrodes for the measurements, providing acceptable penetration and horizontal resolution. This array also enabled 2D modeling by producing pseudo sections that yielded insightful information [18,42]. The AGI SuperSting R4 resistivity meter was used for field data collection. Five profiles were performed to monitor the slurry pond with a 15 m separation between each profile (Figure 1a). The pond was measured three times—the control measurement was conducted in December 2020 (Figure 1b), named survey 1, while the second and third field measurements were carried out in July 2021 (Figure 1c) and January 2022 (Figure 1d), named survey 2 and 3, respectively. EarthImager 2D v. 2.4.2 software was employed to process the raw electrical data. Firstly, a post-processing stage was undertaken to eliminate outliers and perform static correction to normalize the resistivity variations caused by the difference in electrode elevation. Then, smooth inversion was carried out on the apparent electrical resistivity values obtained from the control pond. Smooth model inversion, also known as inversion of Occam, is a mathematical technique determining the smoothest possible model that can fit the data while adhering to an a priori Chi-square statistic. This approach is predicated on assuming a Gaussian distribution of data errors. By minimizing the roughness of the model, smooth model inversion aims to generate a model that is most representative of the data without overfitting. The percentage value of the root mean square error (RMS), which indicates the mismatch between the field measurements and model data [43], was less than 10% for the modeling of each ERT profile section. The interpolation of 2D sections to 2D models at different depths was calculated using the commercial software Surfer v. 25.4.320 by Golden Software.

2.3. Pig Slurry Sampling

The selected slurry pond was examined three times, with three samples collected during each survey. The first survey was conducted in December 2020, with samples taken at different points along the pond. The second and third surveys were conducted in July 2021 and January 2022, respectively. As soon as the laboratory received the slurry samples, they were promptly stored at 4 °C to prevent any probable chemical or biological reactions.

The sample pH and electrical conductivity (EC) measurements were carried out using the HANNA Instrument portable equipment model HI 9025(Hanna Instruments S.L. Eibar, Spain). The measurement is taken directly from a homogenized sample. The pH and EC values are read once the reading stabilizes in the values parameter with a standard temperature of 25 °C [41]. To determine the total suspended solids (TSS), 1 mL of the homogenized sample was filtered through a glass fiber filter and dried at 60 °C for 24 h. The sample was filtered using a Watman filter with a 0.45 μm pore size and a Vacuum Brand vacuum pump (method 2440-D, APHA-AWWA-WEF, 2012) [44]. The TSS units of measurement are expressed in g L^{-1}. Total nitrogen (TN) is the collective amount of all forms of nitrogen in a sample. Furthermore, Kjeldahl nitrogen (NK) is a subset of TN encompassing organic and ammoniacal nitrogen and nitrates and nitrites. Measuring NK involves treating the sample with a combination of sulfuric acid and catalyst at 400 °C for 40 min. The distilled sample is then titrated using 0.1 N hydrochloric acid [45]. Ammoniacal nitrogen ($N\text{-}NH_4^+$) is measured without treatment, while organic nitrogen (NO) is calculated by subtracting $N\text{-}NH_4^+$ from NK. Lastly, nitrate (NO_3^-) and nitrite (NO_2) are measured separately using ion exchange chromatography.

2.4. Statistical Analyses

The IMB SPSS 23 program was used to analyze the data for descriptive statistics. To identify significant differences in the chemical compositions of the slurry during the three surveys, a one-way analysis of variance (ANOVA) was performed, followed by Tukey's post hoc test at $p < 0.05$. The different letters assigned by Tukey's post hoc test (a, b) indicate statistically significant differences among the means of each parameter. The same test was also used to determine statistical differences between the resistivity values obtained through ERT during each survey and at different depths.

3. Results and Discussion

3.1. Electrical Resistivity Tomography of Pig Slurry

The average resistivity of the slurry during the three surveys was 0.97 Ohm·m, 1.42 Ohm·m, and 3.18 Ohm·m for surveys 1 in 2020, 2 in 2021, and 3 in 2022, respectively. A comparison between the 2D ERT sections from the ERT in each of the periods in the selected slurry storage pond shows the changes in the resistivity of the slurry accumulated in the pond and the variations of these resistivities at depth (Figure 2). The depth of the slurry pond is 1.6 m, which was confirmed by using ERT during each survey. The 2D ERT sections showed variations in the resistivity values from survey to survey, evidencing the aging of the slurry reflected in the three layers that were distinguished in each survey. The formation of these three layers within the slurry pond is consistent with the pig slurry behavior, which, in natural decantation, separates 45–57% of the suspended particles with a diameter greater than 400 μm, forming three distinct layers: the crust zone, the most liquid part, and the sedimentation zone [46].

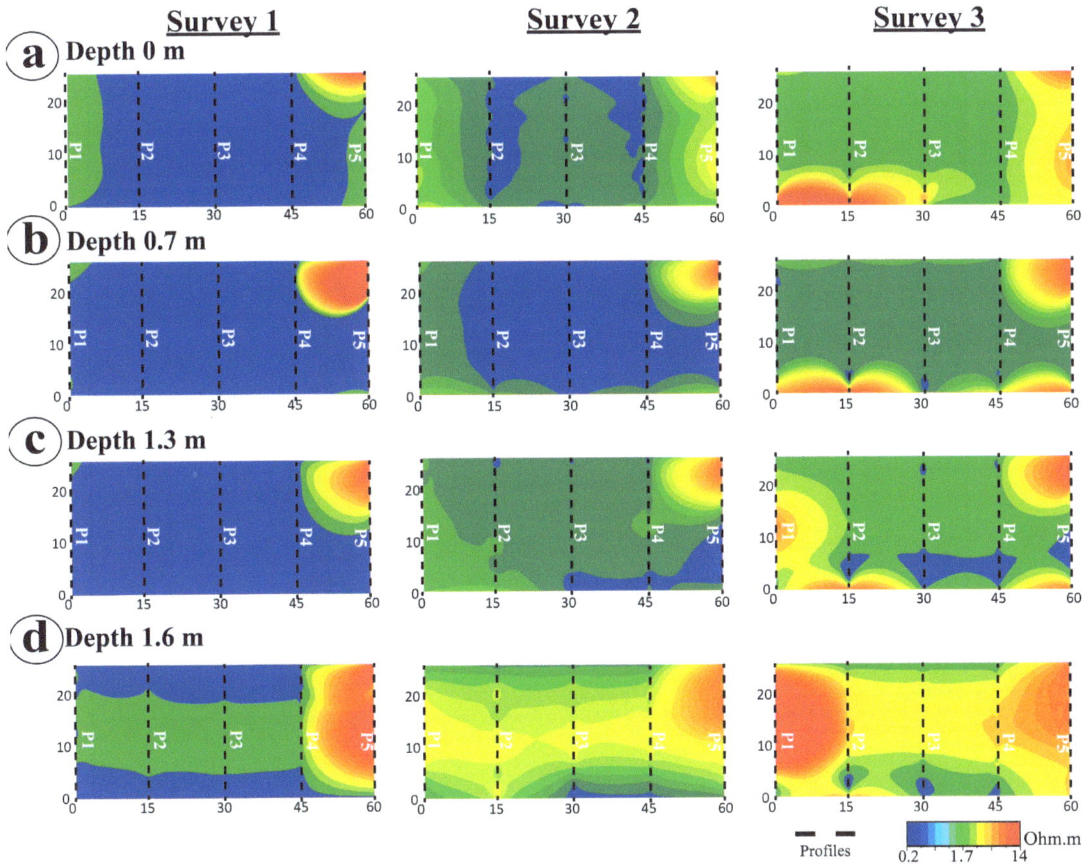

Figure 2. Horizontal sections of the resistivity model of the pig manure pond. The resistivity distribution is represented for the following depths: (**a**) surface (0 m), (**b**) 0.7 m, (**c**) 1.3 m, (**d**) 1.6 m. The depth of 1.6 m represents the bottom of the pond.

Figure 3 shows an increase in resistivity values from one survey to the next. Furthermore, the resistivity values also exhibit variability based on the depth within the slurry pond, which can be categorized into three distinct layers. In survey 1, the average resistivity value is 0.97 Ohm·m. However, at a depth of 0.7 m, the average resistivity value rises to 1.97 Ohm·m, and at the bottom of the pond, it further increases to 3.08 Ohm·m. For survey 2, the average surface value of the slurry increases to 1.42 Ohm·m; this is attributed to the formation of the crust. In the center of the pond, the average value of the slurry decreases to 0.81 Ohm·m, corresponding to the liquid part of the slurry, and at depth, this value increases to an average value of 3.02 Ohm·m. The same behavior is observed in survey 3, where the surface crust zone shows a value of 3.18 Ohm·m, the liquid slurry zone shows an average value of 1.65 Ohm·m, and the bottom of the pond shows a remarkable increase with an average value of 7.13 Ohm·m (Figure 2).

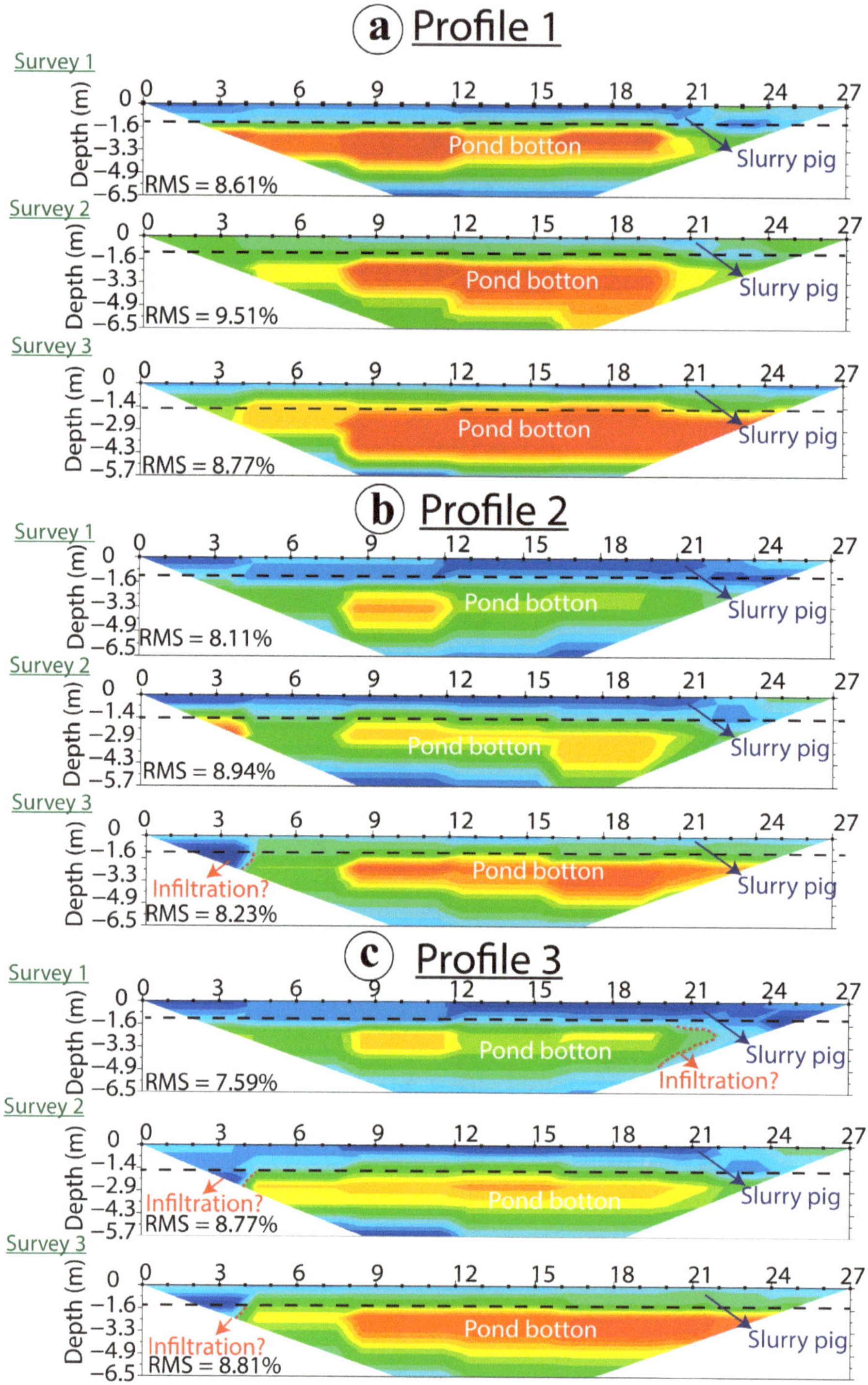

Figure 3. *Cont.*

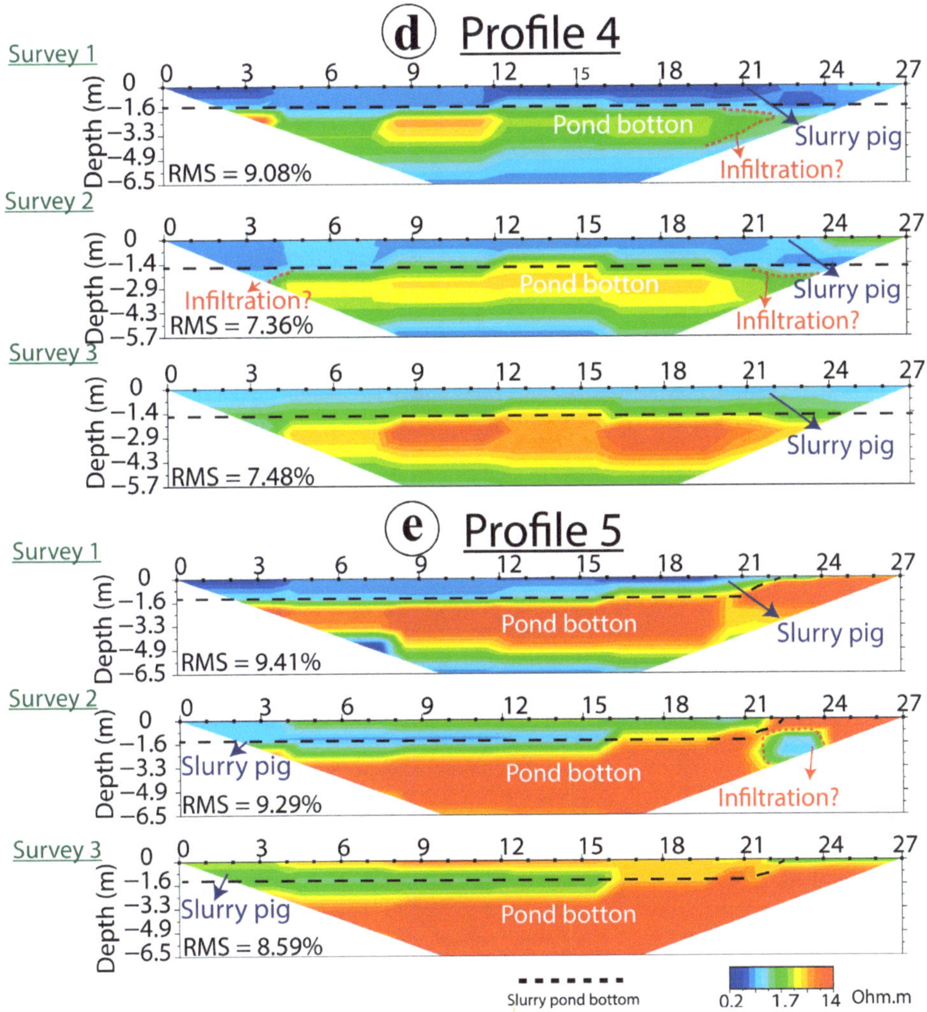

Figure 3. ERT 2D electrical section comparison of (**a**) profile 1, (**b**) profile 2, (**c**) profile 3, (**d**) profile 4, and (**e**) profile 5; retrieved from surveys 1, 2, and 3.

The resistivity values obtained in the three ERT surveys are consistent with similar values reported in other studies for pig slurry [27,29,34]. The superficial electrical resistivity values of each profile with respect to survey 1 (0.97 Ohm·m) indicated an increase of 146% (1.42 Ohm·m) for survey 2 and a further increase of 328% (3.18 Ohm·m) for survey 3 (Figure 2). The increase is attributed to the elevated crustal content at the surface. The crustal content corresponds to the solid surface layer that develops on top of pig slurry ponds, commonly referred to as the pig slurry crust. This crust results from several physical and chemical factors and environmental conditions, such as temperature. However, the increase in resistivity values was higher at the bottom of the pond. All profiles from the first survey showed a lower resistivity at depth (3.08 Ohm·m), indicating less sediment accumulation of the slurry at the time of measurement at the bottom of the pond. In contrast, the resistivity value in the pig slurry pond profiles from the second survey at the base was similar with a value of 3.02 Ohm·m, and the slurry in this survey had a higher crust concentration on the slurry surface (Figure 3). Likewise, survey 3 revealed an increase in resistivity values compared to previous surveys, indicating a higher accumulation of slurry sediment at the bottom of the pond than the first and second. Survey 3 was conducted six months after the second and eleven months after the first. The 2D ERT sections at the depth of the bottom showed a 236% (7.13 Ohm·m) increase in ground electrical resistivity compared to the second survey (3.02 Ohm·m) and an increase of 232% (7.13 Ohm. m) compared to the first survey (3.08 Ohm·m).

The results of the three surveys conducted on the slurry pond are displayed in Figure 3. The profiles show that the slurry had a more liquid consistency in the first survey, resulting in lower resistivity values. However, the resistivity values increased in subsequent surveys, leading to higher values. In fact, we also noticed more crust on the surface of the pig slurry than in previous surveys (Figure 1). This is attributed to the fact that during storage in the slurry pond, the particles settle by natural gravity, resulting in a solid–liquid separation process. Thus, this separation process causes an increase in the resistivity value at the bottom of the pond by sediment accumulation [47]. These waste materials, with higher resistivity values, may suggest lower water content [40], while on the surface, the slurry forms that hard crust covering the liquid part, which is observed in contrast obtained in the 2D ERT sections for the first, second, and third surveys.

Overall, the resistivity values tended to increase during the ERT analysis. However, interestingly, the ERT identified specific areas in the pond at a depth of 1.6 m where the resistivity values decreased. Figure 3 depicts this resistivity difference, which was primarily concentrated in profiles 2, 3, 4, and 5, specifically between electrodes 1 and 3, located at the edge of the pond.

The resistivity values of profile 1 increased from one survey to the next. However, the third survey showed a decrease in resistivity values to less than 1 Ohm·m for the first three electrodes at depth. This behavior is consistent across profiles 2, 3, and 4 and coincides with electrodes 1 and 3 (Figure 3). This decrease in the resistivity value could be attributed to the different degrees of saturation of the soil, probably due to the introduction of the slurry, which includes the infiltration of salts prevalent in slurry [48], allowing a reduction in the resistivity, making the soil more conductive [49,50]. In addition, the findings above indicate that the slurry may have spread laterally due to the soil composition in this region [51], because the decrease in electrical resistivity values was observed under the electrodes situated adjacent to the slurry pond side.

Also, analyzing surveys 1 and 2, it becomes apparent that the electrical resistivity values differ from those of survey 3. This disparity is especially noticeable in the electrodes where no decrease in resistivity value was detected during survey 3. To highlight, the resistivity values at the bottom for surveys 1 and 2 closely align with the 1 Ohm·m range commonly associated with pig slurry (Figure 4). This can be attributed to the composition of the slurry. The slurry in surveys 1 and 2 exhibited a more fluid consistency, which may have facilitated throwing into the soil. This finding suggests that the fluidity of the slurry plays a crucial role in the ability to penetrate the soil and also emphasizes the significance of monitoring the pond from the interior and the perimeter of the pond for potential external contamination [51]. In cases where low resistivity values are observed, it could be an indication of pig slurry seeping into the soil. This is because the presence of dissolved salts in the pig slurry initially causes an increase in electrical conductivity, which consequently lowers the resistivity of the soil [52]. However, as the soil dries up and the salts become more concentrated, the resistivity may increase due to a higher concentration of dissolved salts in the dry soil [53]. This is evidenced by the increase in the resistivity value from survey to survey at the bottom of the pond.

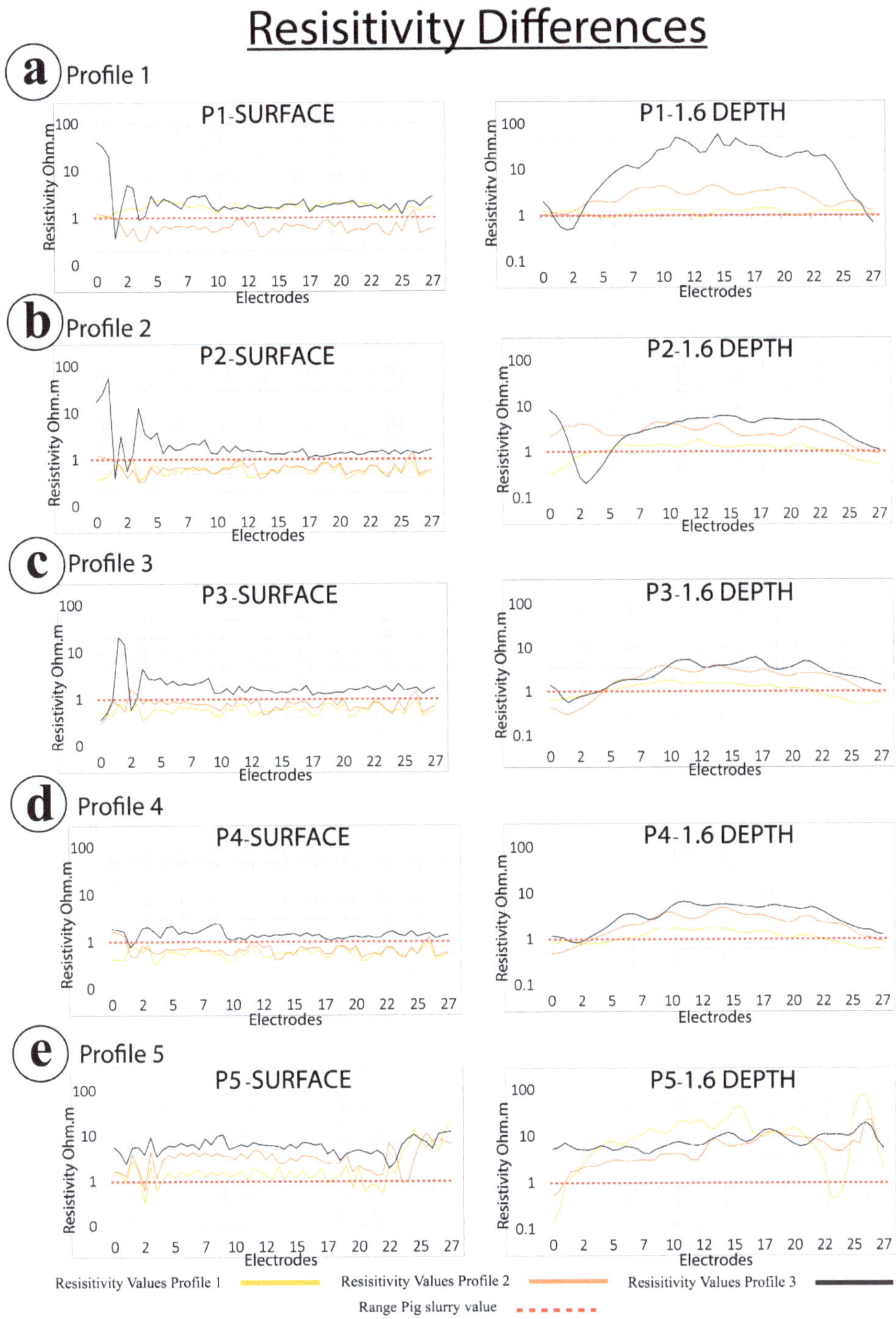

Figure 4. Differences in the electrical resistivity values on the surface and at the bottom of the slurry pond measured at each electrode between studies 1, 2, and 3 for (**a**) profile 1, (**b**) profile 2, (**c**) profile 3, (**d**) profile 4, and (**e**) profile 5.

In addition, a noticeable decrease in electrical resistivity values was observed under the electrodes located outside the pond, specifically in profile 5, from electrodes no. 22 to no. 27, at a depth of 0.4 m and 1.6 m in surveys 1 and 2 (Figure 3). This decrease in resistivity values was more pronounced in Figure 4, which showed values below 1 Ohm in survey 2. This reduction can be attributed to an external infiltration that may compromise the efficacy of the slurry containment system [54].

3.2. *Pig Slurry Characterization and Statistical Analysis*

As a result of the differences obtained by the ERT for each level within the slurry pond, the average resistivity values at three different depths were examined. The surface, middle, and depth resistivity values of the slurry showed no significant differences in profiles 1 to 5; therefore, the mean value of the pig slurry was analyzed for each survey. The analysis determined that surveys 1 and 2 differ statistically from survey 3 (Figure 4).

For the chemical characterization of the pig slurry, the initial survey yielded an average pH of 7.3, and subsequent surveys showed a marginal increase that did not reach statistical significance. The fluctuations in pH levels across the analyzed slurry samples can be attributed to variances in salt intake through the feed and changes in water volume added during cleaning procedures [1]. The EC values ranged from 37 dS/m to 17.9 dS/m; the results indicated significant differences in the values obtained from surveys 1 and 2 compared to survey 3, consistent with the resistivity values (Figure 4). This could be attributed to the slurry storage time, since an influential factor in the electrical conductivity value is the organic matter content [55], which can be reduced by the storage time of the pig slurry in the pond [56]. Over time, the TSS values in the surveys increased, with notable disparities between the first and second surveys compared to the third. The gradual rise in TSS levels suggested a progressive buildup of solid particles in the pond; this could be attributed to the fact that the TSS amount was considerably more significant for the third survey, which increased the slurry resistivity value, as suspended particles within the fluid act as insulators, increasing resistivity [57].

Additionally, significant differences were observed between the PO_4^{-3} values of surveys 1 and 2 compared to survey 3 (Figure 5). The variances in the PO_4^{-3} concentrations found in the slurry pond result from biogeochemical occurrences [58]. Inorganic phosphate is one of the different types of phosphorus present in pig slurry. This is because around 50–60% of the phosphorus in their food is excreted by pigs through their feed and urine, as their digestive system cannot fully absorb it [59]. As organic matter decomposes in the slurry, specific chemical reactions may cause phosphate to be released or retained on the surface. According to research by Masse et al. [60] and Christensen et al. [61], approximately 70% of the undissolved phosphorus in swine manure is bound to particles ranging from 0.45 µm to 10 µm or colloids; this is because these small particles contain a significant portion of the total phosphorus in swine manure, which could indicate a relationship between the decrease in PO_4^{-3} and the increase in the amount of solids in the slurry. This suggests that the solid particles in the slurry may contain a significant amount of PO_4^{-3}, which could lead to a lower PO_4^{-3} value during laboratory analysis in liquid slurry. Additionally, it implies that more PO_4^{-3} might be trapped in the sediment at the bottom of the pond, explaining why the PO_4^{-3} level decreases from one survey to another at the surface [62]. Organic nitrogen showed significant differences between survey 1, survey 2, and survey 3 (Figure 5). The NO concentrations were highest during survey 1 compared to surveys 2 and 3. This can be attributed to the fact that organic nitrogen in pig slurry is usually associated with organic matter [56]. Organic matter is the primary nitrogen source in the slurry that is broken down by microorganisms through biological decomposition processes. This results in the breakdown of nitrogen into different forms, meaning that some organic nitrogen can be converted into ammonia nitrogen through microbial activity [63]. Moreover, it is essential to note that the amount of time slurry is stored can impact the levels of organic nitrogen. When slurry is stored in an anaerobic environment, ammonia nitrogen levels increase while NO levels decrease. This can explain why longer storage times result in lower NO levels, as observed in all three surveys. Nonetheless, external environmental conditions, like temperature and precipitation, can also affect nutrient dynamics in the slurry pond. Temperature fluctuations can influence microbial activity rates, affecting nutrient transformation processes [64]. But also, the significant fluctuations observed in the TN and PO_4^{-3} values in the slurry storage pond highlight the complex interplay of biogeochemical cycles, microbial activity, and external environmental factors in shaping the dynamic nature of nutrient content within the system.

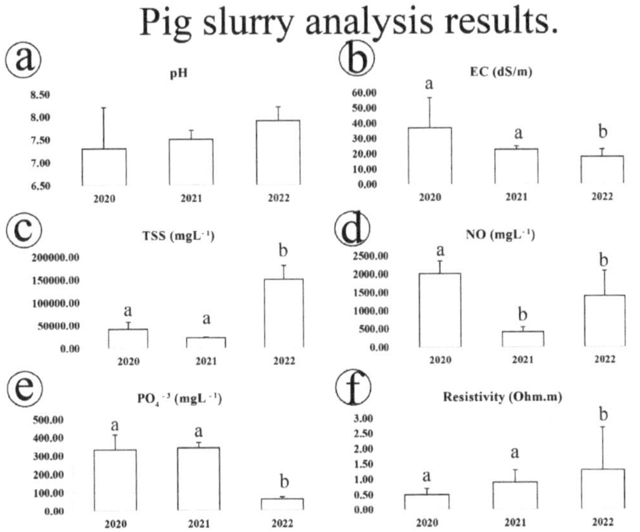

Figure 5. Bar graph of the slurry analysis results. (**a**) pH results, (**b**) EC (electrical conductivity) results, (**c**) TSS (total solid suspension) results, (**d**) NO (organic nitrogen) results, (**e**) PO_4^{-3} (Phosphate ion) results, (**f**) resistivity values. Different letters (a, b) indicate statistically significant differences ($p < 0.05$) among the means of each parameter.

4. Conclusions

This study revealed that slurry composition varied significantly over time. This variation can be attributed to several factors, including changes in diet, environmental conditions such as temperature and humidity, farm management practices, and the type and amount of cleaning water used for slurry handling and storage.

Regarding the pig slurry, the electrical resistivity values obtained from the ERT method showed significant differences from those values obtained for EC, TSS, PO_4^{-3}, and NO. This suggests that the use of ERT in pig slurry ponds can be used as a proxy for estimating these concentrations; however, a more exhaustive study should be carried out to confirm if there is a direct correlation between slurry composition and electrical resistivity values.

The variations in resistivity values observed during the time-lapse studies were found to be associated with the consistency of the slurry. Therefore, we can conclude that electrical resistivity tomography (ERT) can be used to accurately identify different layers of the slurry, such as the crust zone, more fluid part, and sedimentation zone. The research findings indicate that the utilization of electrical resistivity can be a valuable means of monitoring slurry storage systems in real time while also providing insight into the internal variability of the slurry.

The effectiveness of ERT In detecting and monitoring possible horizontal and vertical infiltrations in a slurry pond has been demonstrated. This highlights the ability of ERT to pinpoint areas that may be susceptible to the migration of slurry components into the surrounding environment. Consequently, ERT can be a valuable tool in mitigating environmental risks by aiding in informed decision-making regarding the location and design of storage ponds and implementing practices that reduce the likelihood of contamination, therefore helping to ensure more effective environmental management of pig slurry.

Author Contributions: Conceptualization, X.C.-C.; methodology, J.A.A.; software, M.A.M.-S.; validation, P.M.-P. and M.V.-M.; data curation, J.A.A.; writing—original draft preparation, X.C.-C.; writing—review and editing, J.A.A., P.M.-P. and M.V.-M.; visualization, M.A.M.-S. and Á.F. All authors have read and agreed to the published version of the manuscript.

Funding: This work was supported by the research grant 21583/FPI/21. Fundación Séneca. Región de Murcia (Spain), and the Sustainable Use, Management, and Reclamation of Soil and Water research group.

Data Availability Statement: The data that support the findings of this study are available upon request from the corresponding author M.A.M.-S.

Conflicts of Interest: The authors declare no conflicts of interest.

References

1. Suresh, A.; Choi, H.L.; Oh, D.I.; Moon, O.K. Prediction of the nutrients value and biochemical characteristics of swine slurry by measurement of EC–Electrical conductivity. *Bioresour. Technol.* **2009**, *100*, 4683–4689. [CrossRef] [PubMed]
2. Martinez, J.; Burton, C.H.; Sneath, R.W.; Farrent, J.W. A study of the potential contribution of sedimentation to aerobic treatment processes for pig slurry. *J. Agric. Eng. Res.* **1995**, *61*, 87–96. [CrossRef]
3. Venglovsky, J.; Sasakova, N.; Gregova, G.; Papajova, I.; Toth, F.; Szaboova, T. Devitalisation of pathogens in stored pig slurry and potential risk related to its application to agricultural soil. *Environ. Sci. Pollut. Res.* **2018**, *25*, 21412–21419. [CrossRef] [PubMed]
4. Ministerio de Agricultura Pesca y Alimentación. El Sector de la Carne de Cerdo en Cifras: Principales Indicadores Económicos. In *The Pig Sector in Numbers: Main Economic Indicators*; Subdirección General de Producciones Ganaderas y Cinegéticas, Dirección General de Producciones y Mercados Agrarios: Madrid, Spain, 2022.
5. Williams, J.; Jones, R. Work Package 1. The Contribution of Slurry Management Practices to Diffuse Pollution from Agriculture. Available online: https://assets.publishing.service.gov.uk/media/5a7ca2dc40f0b6629523ad27/pb14045-slurry-management-storage-report-appendix.pdf#page=4 (accessed on 26 October 2023).
6. Nrcs, U. Conservation Practice Standard Waste Storage Facility (Code 313). Available online: https://www.nrcs.usda.gov/ (accessed on 26 October 2023).
7. *Real Decreto 306/2020, de 11 de febrero*; Boletín Oficial del Estado: Madrid, Spain, 2019; Volume 2014, pp. 18987–19106.
8. Council Directive 91/676/EEC of 12 December 1991 Concerning the Protection of Waters against Pollution Caused by Nitrates from Agricultural Sources. Available online: https://eur-lex.europa.eu/legal-content/EN/ALL/?uri=celex:31991L0676 (accessed on 26 October 2023).
9. Díez López, J.A.; Hernáiz, P.J.; Muñoz, M.J.; Torre, A.; Vallejo, A. Impact of pig slurry on soil properties, water salinization, nitrate leaching and crop yield in a four-year experiment in Central Spain. *Soil Use Manag.* **2004**, *20*, 444–450.
10. Frey, L.; Tanunchai, B.; Glaser, B. Antibiotics residues in pig slurry and manure and its environmental contamination potential. A meta-analysis. *Agron. Sustain. Dev.* **2022**, *42*, 31. [CrossRef]
11. Alcalá, F.J.; Paz, M.C.; Martínez-Pagán, P.; Santos, F.M. Integrated Geophysical Methods for Shallow Aquifers Characterization and Modelling. *Appl. Sci.* **2022**, *12*, 2271. [CrossRef]
12. Martín-Crespo, T.; Gómez-Ortiz, D.; Martínez-Pagán, P.; De Ignacio-San José, C.; Martín-Velázquez, S.; Lillo, J.; Faz, A. Geoenvironmental characterization of riverbeds affected by mine tailings in the Mazarrón district (Spain). *J. Geochem. Explor.* **2012**, *119*, 6–16. [CrossRef]
13. Capozzoli, L.; Giampaolo, V.; De Martino, G.; Perciante, F.; Lapenna, V.; Rizzo, E. ERT and GPR prospecting applied to unsaturated and subwater analogue archaeological site in a full scale laboratory. *Appl. Sci.* **2022**, *12*, 1126. [CrossRef]
14. Babaiwa, D.A.; Ikponmwen, M.O. Application of 2D and 3D Electrical Resistivity Tomography (ERT) in Predicting Soil Erodibility in Oredide Village, Auchi in Etsako West LGA of Edo State, Southern Nigeria. *J. Appl. Sci. Environ. Manag.* **2021**, *25*, 1073–1079.
15. Yan, Z.; Song, X.; Wu, Y.; Gao, C.; Wang, Y.; Yang, Y. Fingerprinting Organochlorine Groundwater Plumes Based on Non-Invasive ERT Technology at a Chemical Plant. *Appl. Sci.* **2022**, *12*, 2816. [CrossRef]
16. Bondo Medhus, A.; Klynkby, L. *Engineering Geophysics*, 1st ed.; CRC Press: London, UK, 2022; p. 324.
17. Farzamian, M.; Santos, F.A.M.; Khalil, M.A. Application of EM38 and ERT methods in estimation of saturated hydraulic conductivity in unsaturated soil. *J. Appl. Geophys.* **2015**, *112*, 175–189. [CrossRef]
18. Butler, D.K. *Near-Surface Geophysics*; Society of Exploration Geophysicists: Tulsa, OK, USA, 2005; Volume 1.
19. Rosales, R.M.; Martínez-Pagán, P.; Faz, A.; Bech, J. Study of subsoil in former petrol stations in SE of Spain: Physicochemical characterization and hydrocarbon contamination assessment. *J. Geochem. Explor.* **2014**, *147*, 306–320. [CrossRef]
20. Song, W.; Wang, Y.; Shao, Z. Categorical modeling on electrical anomaly of room-and-pillar coal mine fires and application for field electrical resistivity tomography. *J. Appl. Geophys.* **2017**, *136*, 474–483. [CrossRef]

21. Ezeh, C.C.; Maike, S.M. Using vertical electrical sounding and 2D resistivity tomography to investigate Osogbo central waste dumpsite, Osun State, Nigeria. *Int. J. Phys. Sci.* **2023**, *18*, 25–37.
22. Arifin, M.H.; Kayode, J.S.; Ismail MK, I.; Abdullah, A.M.; Embrandiri, A.; Nazer NS, M.; Azmi, A. Environmental hazard assessment of industrial and municipal waste materials with the applications of RES2-D method and 3-D Oasis Montaj modeling: A case study at Kepong, Kuala Lumpur, Peninsula Malaysia. *J. Hazard. Mater.* **2021**, *406*, 124282. [CrossRef]
23. Akiang, F.B.; Emujakporue, G.O.; Nwosu, L.I. Leachate delineation and aquifer vulnerability assessment using geo-electric imaging in a major dumpsite around Calabar Flank, Southern Nigeria. *Environ. Monit. Assess.* **2023**, *195*, 123. [CrossRef] [PubMed]
24. Morita, A.K.; Pelinson, N.S.; Bastianon, D.; Saraiva, F.A.; Wendland, E. Using Electrical Resistivity Tomography (ERT) to assess the effectiveness of capping in old unlined landfills. *Pure Appl. Geophys.* **2023**, *180*, 3599–3606. [CrossRef]
25. Zaini MS, I.; Hasan, M.; Zolkepli, M.F. Urban landfills investigation for leachate assessment using electrical resistivity imaging in Johor, Malaysia. *Environ. Chall.* **2022**, *6*, 100415. [CrossRef]
26. Udosen, N.I. Geo-electrical modeling of leachate contamination at a major waste disposal site in south-eastern Nigeria. *Model. Earth Syst. Environ.* **2022**, *8*, 847–856. [CrossRef]
27. Capa-Camacho, X.; Martínez-Pagán, P.; Martínez-Segura, M.A.; Gabarrón, M.; Faz, Á. Delimiting pig slurry affected subsurface areas by combining geophysical and geochemical techniques. *Water* **2022**, *14*, 1872. [CrossRef]
28. Capa-Camacho, X.; Martínez-Pagán, P.; Martínez-Segura, M.; Gabarrón, M.; Faz, Á. Electrical resistivity tomography (ERT) and geochemical analysis dataset to delimit subsurface affected areas by livestock pig slurry ponds. *Data Brief* **2022**, *45*, 108684. [CrossRef] [PubMed]
29. Martínez-Pagín, P.; Faz, A.; Aracil, E. The use of 2D electrical tomography to assess pollution in slurry ponds of the Murcia region, SE Spain. *Near Surf. Geophys.* **2009**, *7*, 49–61. [CrossRef]
30. Martínez-Pagán, P.; Cano, Á.F.; Ramos da Silva, G.R.; Olivares, A.B. 2-D electrical resistivity imaging to assess slurry pond subsoil pollution in the southeastern region of Murcia, Spain. *J. Environ. Eng. Geophys.* **2010**, *15*, 29–47. [CrossRef]
31. Gabarrón, M.; Martínez-Pagán, P.; Martínez-Segura, M.A.; Bueso, M.C.; Martínez-Martínez, S.; Faz, Á.; Acosta, J.A. Electrical resistivity tomography as a support tool for physicochemical properties assessment of near-surface waste materials in a mining tailing pond (El Gorguel, SE Spain). *Minerals* **2020**, *10*, 559. [CrossRef]
32. Chen, W.; Nover, D.; Xia, Y.; Zhang, G.; Yen, H.; He, B. Assessment of extrinsic and intrinsic influences on water quality variation in subtropical agricultural multipond systems. *Environ. Pollut.* **2021**, *276*, 116689. [CrossRef] [PubMed]
33. Martinez-Pagan, P.; Gomez-Ortiz, D.; Martin-Crespo, T.; Martin-Velazquez, S.; Martinez-Segura, M. Electrical resistivity imaging applied to tailings ponds: An overview. *Mine Water Environ.* **2021**, *40*, 285–297. [CrossRef]
34. Martínez Pagán, P. "Aplicación de diferentes técnicas no destructivas de prospección geofísica a problemas relacionados con contaminación ambiental producidas por diferentes actividades antrópicas en la Región de Murcia." (2006).
35. Alshuwaikhat, H.M.; Abubakar, I. An integrated approach to achieving campus sustainability: Assessment of the current campus environmental management practices. *J. Clean. Prod.* **2008**, *16*, 1777–1785. [CrossRef]
36. BORM-s-2021-90328 Decreto-Ley 5/2021, de 27 de Agosto, de Modificación de la Ley 3/2020, de 27 de Julio, de Recuperación y Protección del Mar Menor. Available online: https://www.boe.es/buscar/doc.php?id=BORM-s-2021-90328 (accessed on 21 November 2023).
37. Boletín Oficial de la Región de Murcia. Publicación Número 8097 del BORM Número 298 de 27/12/2019. Available online: https://www.borm.es/services/anuncio/ano/2019/numero/8097/pdf?id=782214 (accessed on 21 November 2023).
38. Fuente Álamo de Murcia Climate: Temperature Fuente Álamo de Murcia & Weather by Month. Available online: https://en.climate-data.org/europe/spain/region-de-murcia/fuente-alamo-de-murcia-31093/ (accessed on 20 September 2023).
39. Marín Lechado, C.; Roldán García, F.J.; Pineda Velasco, A.; Martinez Zubieta, P.; Rodero Pérez, J.; Diaz Pinto, G. Mapa Geológico Continuo de España1:50 000 Zonas Internas de las Cordilleras Béticas (Zona, -2.2.1.0.). in, G.E.O.D.E. Mapa Geológico Digital Continuo de España. [En Linea]. 2008. Available online: https://info.igme.es/cartografiadigital/geologica/geodezona.aspx?Id=Z1800 (accessed on 23 November 2023).
40. Romero Díaz, A.; Belmonte Serrato, F.; Hernández Bastida, J.A. *El Campo de Cartagena una visión global. Recorridos por el Campo de Cartagena. Control de la degradación y uso sostenible del suelo*; Instituto Mediterráneo del Agua: Murcia, Spain, 2011; pp. 17–48.
41. Everett, M.E. *Near-Surface Applied Geophysics*; Cambridge University Press: Cambridge, UK, 2013.
42. de la Vega, M.; Osella, A.; Lascano, E. Joint inversion of Wenner and dipole–dipole data to study a gasoline-contaminated soil. *J. Appl. Geophys.* **2003**, *54*, 97–109. [CrossRef]
43. I. (AGI) Advanced Geosciences," *Instruction Manual for EarthImager 2D Version 2.4.0 Resistivity and IP Inversion Software"*, no. 512, p. 139, 2009. Available online: https://geophysicalequipmentrental.com/files/2020/01/EarthImager2DManual.pdf. (accessed on 6 December 2023).
44. APHA; AWWA; WEF. *Standard Methods for Examination of Water and Wastewater*, 22nd ed.; American Public Health Association: Washington, DC, USA, 2012; p. 1360.
45. Hoegger, R. *Training Papers Nitrogen Determination According to Kjeldahl*; Büchi Labortechnik AG: Flawil, Switzerland, 1998; Volume 1, p. 18.

46. González Martínez, A. Problemática Medioambiental de los Distintos Tipos de Explotaciones Ganaderas; 1995. Available online: https://helvia.uco.es/bitstream/handle/10396/3831/08-1995-04.pdf?sequence=1&isAllowed=y. (accessed on 14 December 2023).
47. Popovic, O.; Hjorth, M.; Stoumann Jensen, L. Phosphorus, copper, and zinc in solid and liquid fractions from full-scale and laboratory-separated pig slurry. *Environ. Technol.* **2012**, *33*, 2119–2131. [CrossRef]
48. Turki, N.; Elaoud, A.; Gabtni, H.; Trabelsi, I.; Khalfallah, K.K. Agricultural soil characterization using 2D electrical resistivity tomography (ERT) after direct and intermittent digestate application. *Arab. J. Geosci.* **2019**, *12*, 423. [CrossRef]
49. Dumont, G.; Pilawski, T.; Dzaomuho-Lenieregue, P.; Hiligsmann, S.; Delvigne, F.; Thonart, P.; Hermans, T. Gravimetric water distribution assessment from geoelectrical methods (ERT and EMI) in municipal solid waste landfill. *Waste Manag.* **2016**, *55*, 129–140. [CrossRef] [PubMed]
50. Bichet, V.; Grisey, E.; Aleya, L. Spatial characterization of leachate plume using electrical resistivity tomography in a landfill composed of old and new cells (Belfort, France). *Eng. Geol.* **2016**, *211*, 61–73. [CrossRef]
51. Rey, J.; Martínez, J.; Hidalgo, M.C.; Mendoza, R.; Sandoval, S. Assessment of tailings ponds by a combination of electrical (ERT and IP) and hydrochemical techniques (Linares, southern Spain). *Mine Water Environ.* **2021**, *40*, 298–307. [CrossRef]
52. Kusim, A.S.; Abdullah, N.E.; Hashim, H.; Kutty, S.B. Effects of salt content on measurement of soil resistivity. In Proceedings of the 2013 IEEE 7th International Power Engineering and Optimization Conference (PEOCO), Langkawi, Malaysia, 3–4 June 2013.
53. Zhang, D.; Liu, S.; Fan, L.; Deng, Y. Effect of salt concentrations on the electrical resistivity of cement-treated soils. In *GeoCongress 2012: State of the Art and Practice in Geotechnical Engineering*; ASCE: Reston, VA, USA, 2012; pp. 1016–1025.
54. Igbal, M.Z.; Krothe, N.C. Infiltration mechanisms related to agricultural waste transport through the soil mantle to karst aquifers of southern Indiana, USA. *J. Hydrol.* **1995**, *164*, 171–192. [CrossRef]
55. Lukas, V.; Neudert, L.; Novák, J.; Kren, J. Estimation of Soil Physico-chemical Properties by On-the-go Measurement of Soil Electrical Conductivity. *Agric. Conspec. Sci.* **2018**, *83*, 93–98.
56. Petersen, S.O.; Lind, A.M.; Sommer, S.G. Nitrogen, and organic matter losses during storage of cattle and pig manure. *J. Agric. Sci.* **1998**, *130*, 69–79. [CrossRef]
57. Helene LP, I.; Moreira, C.A.; Bovi, R.C. Identification of leachate infiltration and its flow pathway in landfill by means of electrical resistivity tomography (ERT). *Environ. Monit. Assess.* **2020**, *192*, 249. [CrossRef] [PubMed]
58. Hanief, A. Phosphorus Fractions in Biosolids, Biosolid-Amended Soils, Runoffs, and Its Impact on Primary Productivity in Aquatic Ecosystems. Master's Thesis, Ryerson University, Toronto, ON, Canada, 2011.
59. Samanta, P.; Schwark LV, U.S.; Horn, H.; Saravia, F. Nutrient recovery and ammonia-water production by MF-vacuum evaporation treatment of pig manure. *J. Environ. Chem. Eng.* **2022**, *10*, 106929. [CrossRef]
60. Masse, L.; Massé, D.I.; Beaudette, V.; Muir, M. Size distribution and composition of particles in raw and anaerobically digested swine manure. *Trans. ASAE* **2005**, *48*, 1943–1949. [CrossRef]
61. Christensen, M.L.; Hjorth, M.; Keiding, K. Characterization of pig slurry with reference to flocculation and separation. *Water Res.* **2009**, *43*, 773–783. [CrossRef] [PubMed]
62. Peters, K.; Hjorth, M.; Jensen, L.S.; Magid, J. Carbon, nitrogen, and phosphorus distribution in particle size–fractionated separated pig and cattle slurry. *J. Environ. Qual.* **2011**, *40*, 224–232. [CrossRef] [PubMed]
63. FOG, K. The effect of added nitrogen on the rate of decomposition of organic matter. *Biol. Rev.* **1988**, *63*, 433–462. [CrossRef]
64. Zhu, B.; Wang, T.; Dzakpasu, M.; Li, X. Nutrient dynamics and retention in a vegetated drainage ditch receiving nutrient-rich sewage at low temperatures. *Sci. Total Environ.* **2020**, *741*, 140268.

water

MDPI

Article

A Real-Time Prediction Approach to Deep Soil Moisture Combining GNSS-R Data and a Water Movement Model in Unsaturated Soil

Xiaotian Luo [1,2], Cong Yin [2,3], Yueqiang Sun [2,3], Weihua Bai [2,3], Wei Li [2,3] and Hongqing Song [1,2,*]

[1] School of Civil and Resource Engineering, University of Science and Technology Beijing, Beijing 100083, China; d202210002@xs.ustb.edu.cn

[2] Joint Lab for Application of GNSS Atmospheric Remote Sensing Data, Beijing 100083, China

[3] National Space Science Center, Chinese Academy of Sciences (NSSC/CAS), Beijing 100190, China

* Correspondence: songhongqing@ustb.edu.cn

Abstract: Deep soil moisture data have wide applications in fields such as engineering construction and agricultural production. Therefore, achieving the real-time monitoring of deep soil moisture is of significant importance. Current soil monitoring methods face challenges in conducting the large-scale, real-time monitoring of deep soil moisture. This paper innovatively proposes a real-time prediction approach to deep soil moisture combining GNSS-R data and a water movement model in unsaturated soil. This approach, built upon surface soil moisture data retrieved from GNSS-R signal inversion, integrates soil–water characteristics and soil moisture values at a depth of 1 m. By employing a deep soil moisture content prediction model, it provides predictions of soil moisture at depths from 0 to 1 m, thus realizing the large-scale, real-time dynamic monitoring of deep soil moisture. The proposed approach was validated in a study area in Goodwell, Texas County, Oklahoma, USA. Predicted values of soil moisture at a randomly selected location in the study area at depths of 0.1 m, 0.2 m, 0.5 m, and 1 m were compared with ground truth values for the period from 25 October to 19 November 2023. The results indicated that the relative error (δ) was controlled within the range of $\pm$14%. The mean square error (MSE) ranged from 2.90×10^{-5} to 1.88×10^{-4}, and the coefficient of determination (R^2) ranged from 82.45% to 89.88%, indicating an overall high level of fitting between the predicted values and ground truth data. This validates the feasibility of the proposed approach, which has the potential to play a crucial role in agricultural production, geological disaster management, engineering construction, and heritage site preservation.

Keywords: deep soil moisture; soil–water characteristics; mathematical model; prediction; GNSS-R

1. Introduction

Water in soil exists in various forms, including structural water, bound water, free water, solid-state water, and gaseous water. Investigating the quantity of water in soil is of paramount importance [1,2]. The amount of water in soil is typically expressed as moisture content, which fundamentally represents the ratio of water, excluding structural water, to the mass or volume of the solid or soil body. This can be expressed in two primary methods: mass moisture content and volumetric moisture content. Buckingham and Gardner conducted initial research on the amount of water with respect to the energy level with which water is held in the soil, and the relationship is known as the soil water characteristic curve (SWCC) [3,4]. SWCC is a relationship between the moisture content in the soil and soil suction (soil moisture potential), and it is unique for each soil type. SWCC is used to predict soil moisture storage and field capacity and to understand the drying and wetting characteristics of the soil and its pore structure [5,6]. The water movement equation in unsaturated soil is a mathematical model describing the process of water movement within soil, which holds significant importance in understanding soil moisture distribution, hydrological cycles, and agricultural irrigation practices. Water movement in soil is typically influenced by soil properties, initial moisture conditions, boundary conditions, and environmental factors. In the early 20th century, Richards proposed the renowned Richards equation, which has been widely employed to describe water movement in unsaturated soil. This equation, grounded on Darcy's law and the principle of mass conservation, incorporates factors such as soil moisture, soil water potential, and hydraulic conductivity, and it has been extensively utilized to investigate various scenarios of soil water movement [7]. As research progressed, scholars recognized limitations in the Richards equation when describing certain situations, such as nonlinearity and a lack of physical interpretability. Consequently, to address these issues, several improved models have been proposed, including the Brooks–Corey model, the Van Genuchten model, and others [8,9]. These models take into account factors such as soil pore structure and capillary pressure curves, providing a more accurate description of the soil moisture movement process. In determining soil engineering properties, controlling the quality of compacted soil construction, monitoring and forecasting geological hazards, managing agricultural production with precision, and preserving cultural relics, it is essential to conduct the testing and monitoring of soil moisture conditions [10,11]. Therefore, achieving the real-time monitoring of deep soil moisture holds significant importance.

Surface soil moisture can be retrieved through GNSS-R signals, meeting the demand for all-weather autonomous monitoring. GNSS, which stands for Global Navigation Satellite System, primarily includes the United States GPS system, China's BDS system, Russia's GLONASS system, and the European Union's GALILEO system [12]. These navigation satellites not only provide navigation positioning and timing information to users in real-time but also offer L-band microwave signals suitable for remote sensing detection characterized by global coverage, strong penetration, and a high temporal resolution [13]. Retrieving soil moisture data from GNSS-R signals involves capturing both direct and reflected satellite signals, analyzing the time delay or power changes of the surface-reflected signals, and deducing relevant parameters reflecting surface features based on the geometric relationships among GNSS satellites, ground receivers, reflection points, and the variations in reflection signal characteristics and surface soil properties. The technology for retrieving surface soil moisture data through GNSS-R signals has become relatively mature. In 2002, the University of Colorado and others, under the leadership of NASA, conducted a series of soil moisture retrieval experiments, and experimental data validated the accuracy of the monitoring results [14]. The Starlab Institute in Spain has designed a soil moisture detection device based on L-band GNSS signal observations. This device analyzes the GNSS signals after interference to obtain relevant information about soil moisture [15]. Utilizing the Advanced Integrated Equation Model (AIEM), a method for soil moisture monitoring was derived, and the accuracy of the GNSS-R soil moisture monitoring model was validated based on existing experimental data [16].

Current soil monitoring methods face challenges in achieving the large-scale, real-time monitoring of deep soil moisture. Various methods are available for soil moisture monitoring, including the drying–weighing method, Time Domain Reflectometry (TDR), Ground Penetrating Radar (GPR), the soil resistance method, and the capacitance method for single-point or small-scale soil moisture monitoring. Remote sensing technology is commonly used for spatial and temporal distribution and changes in soil moisture over large areas [17–41]. However, several issues persist in real-time soil moisture monitoring. For instance, the applicability and accuracy of testing methods are often influenced and constrained due to soil characteristics. Furthermore, instruments and sensors commonly suffer from issues such as a large size, high energy consumption, and a high cost, resulting in high real-time monitoring expenses [17–19]. The L-band signals carried by Global Navigation Satellite Systems (GNSSs) are highly sensitive to soil moisture, making them particularly suitable for monitoring soil moisture variations [42]. By utilizing signal power or delay as attributes and actual soil moisture values as labels, inversion models based on navigation signals can be established through the combination of empirical dielectric constant models, Support Vector Machines (SVMs), Random Forest algorithms, and neural network methods such as BP and Deep Belief Networks [43–45]. Therefore, the introduction of soil moisture data retrieved from GNSS-R signals largely overcomes the limitations of traditional soil moisture measurement methods in terms of small effective measurement area and a lack of representativeness, meeting the demand for all-weather autonomous monitoring. However, the penetration depth of GNSSs' L-band signals is limited to only 10 cm. As a result, soil moisture data within 10 cm depth can be retrieved through GNSS-R signal inversion, and soil moisture data within 5 cm depth can be more accurately inverted through GNSS-R signals. In summary, current soil monitoring methods face challenges in achieving the large-scale, real-time monitoring of deep soil moisture. However, deep soil moisture data are crucial in agriculture, geological hazard monitoring, engineering construction, and site preservation.

Therefore, by integrating soil–water characteristics and soil moisture at a depth of 1 m with surface soil moisture data retrieved from GNSS-R signal inversion, deep soil moisture prediction models can be developed. These models enable the prediction of soil moisture values in the 0–1 m depth range, facilitating the large-scale, real-time dynamic monitoring of deep soil moisture.

This paper addresses the current challenge of the inability of existing soil moisture monitoring methods to achieve the large-scale, real-time dynamic monitoring of deep soil moisture. It innovatively proposes a real-time prediction approach to deep soil moisture combining GNSS-R data and water movement model in unsaturated soil. This approach, relying on soil moisture data retrieved from GNSS-R signal inversion, largely overcomes the limitations of traditional soil moisture measurement methods, such as small effective measurement areas and a lack of representativeness, thereby meeting the demand for all-weather autonomous monitoring. By integrating soil–water characteristics and soil moisture values at a depth of 1 m with surface soil moisture data retrieved from GNSS-R signal inversion, the approach can predict soil moisture values in the 0–1 m depth range. This approach overcomes the limitation of GNSS-R signal inversion, which is restricted to the soil surface. The proposed method realizes the large-scale, real-time dynamic monitoring of deep soil moisture and is expected to play a crucial role in agriculture, geological hazard monitoring, engineering construction, and site preservation.

2. Methodology

2.1. Method Design

The core of this real-time prediction approach to deep soil moisture combining GNSS-R data and a water movement model in unsaturated soil lies in the prediction model for deep soil moisture based on GNSS-R detection data. The deep soil moisture prediction model combines the unsaturated soil infiltration model, Darcy's law, and a soil–water characteristic model. Soil moisture data obtained from humidity sensors in the study area serve as boundary conditions for fitting the soil–water characteristic model parameters. Surface soil moisture data retrieved from GNSS-R signal inversion serve as input variables for the deep soil moisture prediction model, thereby obtaining the monitored deep soil moisture. Section 2.2 describes the construction process of the deep soil moisture prediction model, Section 2.3 describes the data acquisition process, and Section 2.4 describes the validation process of the deep soil moisture prediction model. After validation, the deep soil moisture prediction model can be applied to obtain real-time deep soil moisture data through the real-time surface soil moisture data retrieved from GNSS-R signal inversion, thus facilitating its widespread application. The flowchart of the deep soil moisture prediction method is illustrated in Figure 1 shown below.

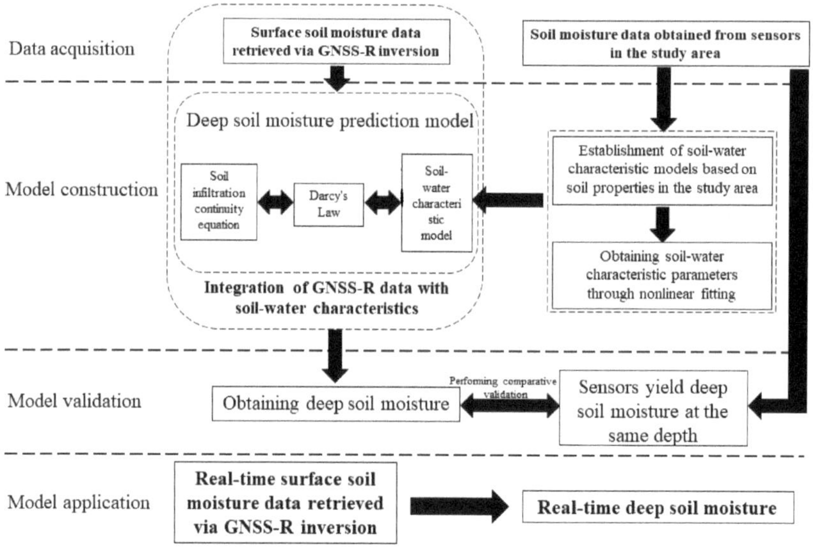

Figure 1. Flowchart of the deep soil moisture prediction method.

2.2. Mathematical Model for Predicting Deep Soil Moisture

2.2.1. Governing Equation of Water Transport in Unsaturated Soil

The Richards equation serves as the governing equation for water movement in unsaturated soil, and it is expressed as a nonlinear partial differential equation. By assuming soil porosity to be uniform and neglecting anisotropy, according to mass conservation, the one-dimensional vertical soil infiltration continuity equation is represented as Equation (1):

$$\frac{\partial \theta}{\partial t} = -\frac{\partial q}{\partial z} \tag{1}$$

where θ is soil moisture, t is the time, q is the vertical infiltration rate perpendicular to the soil surface, and z is the soil depth, with a positive direction oriented vertically downwards (with z = 0 at the soil surface).

Richards introduced Darcy's law into the equation governing unsaturated soil water movement, which can be represented as Equation (2):

$$q = -K\frac{\partial H}{\partial z} = -K\frac{\partial (h-z)}{\partial z} = -K\left(\frac{\partial h}{\partial z} - 1\right) \tag{2}$$

where K is the permeability coefficient, H is the vertical total water potential, and h is the matric suction head.

Furthermore, the expression of the infiltration control equation, known as the Richards equation, can be obtained as shown in Equation (3):

$$\frac{\partial \theta}{\partial t} = \frac{\partial}{\partial z}\left[K\left(\frac{\partial h}{\partial z} - 1\right)\right] \tag{3}$$

To unify the variables, two new variables, C and D, are introduced in Equation (4), representing soil moisture θ and the matric suction head h, respectively:

$$\begin{cases} C(\theta) = \frac{\partial \theta}{\partial h} \\ D(\theta) = \frac{K}{C(\theta)} = \frac{K}{\frac{\partial \theta}{\partial h}} = K\frac{\partial h}{\partial \theta} \end{cases} \tag{4}$$

where C represents the specific water capacity, reflecting the rate at which soil moisture changes with matric suction head and describing the quantitative indicator of the soil's water release capacity, D represents the soil moisture diffusion coefficient, which reflects the soil porosity, the pore size distribution, and its hydraulic conductivity, thereby influencing the soil moisture movement conditions.

By incorporating the specific water capacity, C, and soil moisture diffusion coefficient, D, into the infiltration control equation, as shown in Equation (3) for the Richards equation, we obtain a one-dimensional unsaturated soil water movement control equation with soil moisture θ as the single variable, as shown in Equation (5):

$$\frac{\partial \theta}{\partial t} = \frac{\partial}{\partial z}\left(D(\theta)\frac{\partial \theta}{\partial z}\right) - \frac{\partial K}{\partial z} \tag{5}$$

By defining $V = dK/d\theta = (K(\theta_s) - K(\theta_0))/(\theta_s - \theta_0)$, the one-dimensional unsaturated soil water movement control equation can be transformed into Equation (6):

$$\frac{\partial \theta}{\partial t} = D(\theta)\frac{\partial^2 \theta}{\partial z^2} - V\frac{\partial \theta}{\partial z} \tag{6}$$

2.2.2. Soil–Water Characteristic Model

The soil–water characteristic model is primarily used in soil science to reflect properties such as water retention, moisture movement, and changes in suction. It portrays the functional relationship between soil water energy and soil moisture. From the modern perspective of soil mechanics, the constitutive model of water retention characteristics is one of the constitutive models of unsaturated soil. Therefore, it serves as an important indicator for representing the basic hydraulic properties of soil, and it plays a crucial role in studying soil water retention and movement. The deep soil moisture prediction model proposed in this paper adopts the Van Genuchten model to characterize the soil–water characteristic model, which represents the relationship between soil moisture and suction. The Van Genuchten model is depicted as Equation (7):

$$\begin{cases} h = \frac{1}{\alpha}\left(S_e^{-\frac{1}{m}} - 1\right)^{\frac{1}{n}} \\ K = K_s \cdot K_r = K_s \cdot S_e^{0.5}\left[1 - \left(1 - S_e^{1/m}\right)^m\right]^2 \end{cases} \tag{7}$$

where the following applies: S_e is the soil–water saturation, $S_e = (\theta - \theta_r)/(\theta_s - \theta_r)$; θ is the soil moisture; θ_r is the soil residual moisture; θ_s is the soil saturated moisture; h is the soil suction head with the unit in meters (m); α is a fitting parameter related to the soil–air entry value, approximately equal to the reciprocal of the air entry value, with units in meters to the power of negative one (m^{-1}); n is a fitting parameter related to the soil pore size distribution, $n > 1$; m is a fitting parameter associated with the overall symmetry of the soil characteristic curve, $m = 1 - 1/n$ $(0 < m < 1)$; K represents the soil permeability coefficient with units in meters per second (m/s); K_s represents the saturated hydraulic conductivity of the soil with units in meters per second (m/s); and K_r represents the soil relative permeability coefficient, $K_r = K/K_s$.

The parameters α, n, and m in the Van Genuchten model are obtained through the nonlinear fitting of soil moisture at different depths, thereby establishing the soil–water characteristic model.

2.2.3. Model for Predicting Deep Soil Moisture

The present study integrates the equation governing water movement in unsaturated soil with the soil–water characteristic model and further establishes the deep soil moisture prediction model.

In unsaturated infiltration, the relationship between the permeability coefficient, the matric suction head, h, and the variable soil moisture, θ, is a significant factor influencing the nonlinearity of the Richards equation. Therefore, we integrate the soil–water characteristic model with the one-dimensional unsaturated soil water movement control equation, incorporating the Van Genuchten model in terms of specific water capacity, C, and the soil moisture diffusion coefficient, D. This yields Equation (8):

$$\begin{cases} C(\theta) = \frac{\partial h}{\partial \theta} = \frac{1}{nm\alpha} \cdot \left(Se^{-\frac{1}{m}} - 1\right)^{\frac{1}{n}-1} \cdot Se^{-\frac{1}{m}-1} \cdot \frac{1}{\theta_r - \theta_s} \\ D(\theta) = \frac{K}{C(\theta)} = \frac{K}{\frac{\partial \theta}{\partial h}} = K\frac{\partial h}{\partial \theta} = \frac{K}{nm\alpha} \cdot \left(Se^{-\frac{1}{m}} - 1\right)^{\frac{1}{n}-1} \cdot Se^{-\frac{1}{m}-1} \cdot \frac{1}{\theta_r - \theta_s} \end{cases} \tag{8}$$

The method predicts the instantaneous soil moisture of deep soil layers. For instantaneous soil moisture, which remains constant, $\frac{\partial \theta}{\partial t} = 0$, the governing equation for one-dimensional unsaturated soil water transport can be transformed into Equation (9):

$$0 = D(\theta)\frac{\partial^2 \theta}{\partial z^2} - V\frac{\partial \theta}{\partial z} \tag{9}$$

Assuming the diffusion coefficient is constant, $D = D(\theta_0) = \frac{K(\theta_0)}{nm\alpha} \cdot \left(Se^{-\frac{1}{m}} - 1\right)^{\frac{1}{n}-1} \cdot Se^{-\frac{1}{m}-1} \cdot \frac{1}{\theta_r - \theta_s}$, the general solution of the one-dimensional unsaturated soil water transport governing equation is given by Equation (10):

$$\theta = c1 + c2e^{\frac{V}{D}z} \tag{10}$$

In the equation, $c1$ and $c2$ are undetermined coefficients that can be determined through the boundary conditions at the upper and lower boundaries of the soil layer.

If the surface soil moisture θ_{surface} retrieved from the GNSS signal inversion is known, and the soil moisture θ_0 at a depth of 1 m is obtained through a soil moisture sensor, the undetermined coefficients $c1$ and $c2$ can be determined through Equation (11):

$$\begin{cases} c1 + c2 = \theta_{\text{surface}} \\ c1 + c2 \cdot e^{\frac{V}{D}} = \theta_0 \end{cases} \tag{11}$$

Further simplification leads to Equation (12):

$$\begin{cases} c1 = \frac{\theta_{\text{surface}} \cdot e^{\frac{V}{D}} - \theta_0}{e^{\frac{V}{D}} - 1} \\ c2 = \frac{\theta_{\text{surface}} - \theta_0}{1 - e^{\frac{V}{D}}} \end{cases} \tag{12}$$

Therefore, the particular solution of the one-dimensional unsaturated soil–water movement control equation, representing the relationship between the soil moisture and depth, is given by Equation (13), where the diffusion coefficients D and V are functions of soil–water characteristic parameters α, n, and m.

$$\theta = \frac{\theta_{\text{surface}} \cdot e^{\frac{V(m)}{D(\alpha,m,n)}} - \theta_0}{e^{\frac{V(m)}{D(\alpha,m,n)}} - 1} + \frac{\theta_{\text{surface}} - \theta_0}{1 - e^{\frac{V(m)}{D(\alpha,m,n)}}} e^{\frac{V(m)}{D(\alpha,m,n)}z} \tag{13}$$

After constructing the deep soil moisture prediction model, this study fits the soil–water characteristic parameters using soil moisture data from different depths. Employing the deep soil moisture prediction model, soil moisture data from two or more different depths obtained from sensors are nonlinearly fitted to obtain the soil–water characteristic parameters α, n, and m, completing the construction of the deep soil moisture prediction model.

After completing the construction of the deep soil moisture prediction model, inputting the surface soil moisture data obtained from GNSS-R signal inversion into the deep soil moisture prediction model allows the relationship between the predicted soil moisture θ and the soil depth z for depths ranging from 0 to 1 m to be determined, as described with Equation (13). This enables the real-time and dynamic monitoring of deep soil moisture over a wide range.

2.3. Obtaining Soil Moisture Data

2.3.1. Soil Moisture Data from Sensors

Using soil moisture sensors to obtain soil moisture at different depths is convenient and direct in field operations, with high data accuracy. The working principles of soil moisture sensors are also diverse. Common soil moisture sensors are based on time-domain reflectometry (TDR), frequency-domain reflectometry (FDR), the standing wave ratio (SWR) method, the capacitance method, the resistance method, or the tensiometer method.

The soil moisture at a depth of 1 m from the surface tends to be relatively stable. Therefore, at a representative location within the monitoring area, the soil moisture at a depth of 1 m was chosen as the lower boundary condition for the deep soil moisture prediction model. Additionally, data from two or more sets of soil moisture at different depths are required from this location to fit the soil–water characteristic model and obtain soil–water characteristic parameters, thus establishing the soil–water characteristic model. Apart from surface soil moisture data obtained through GNSS-R signal inversion, other soil moisture data are acquired using soil moisture sensors. In this study, the soil moisture sensor provided soil moisture θ_0 at a depth of 1 m and several sets of soil moisture θ between 0 and 1 m depth, sourced from the International Soil Moisture Network. The International Soil Moisture Network is a work of international cooperation to establish and maintain a global in situ soil moisture database. This database is an essential means of validating and improving global satellite products and land surface, climate, and hydrological models.

2.3.2. Soil Moisture Data from GNSS-R Inversion

This method requires surface soil moisture data to be obtained from GNSS-R inversion in the monitoring area as input variables for the deep soil moisture prediction model. Due to the high accuracy of surface soil moisture data obtained through GNSS-R signal inversion based on land-based or unmanned aerial vehicle receivers, this study utilized soil moisture sensor measurements sourced from the International Soil Moisture Network to substitute for surface soil moisture obtained from GNSS-R signal inversion.

2.4. Model Verification

2.4.1. Research Area

Due to the real-time prediction method for deep soil moisture relying on GNSS-R detection data, ensuring the accuracy of surface soil moisture retrieved from GNSS-R inversion requires the selection of regions where environmental factors minimally affect GNSS-R signal inversion results. Urban areas are prone to significant interference from various signals and light pollution, which can greatly disrupt GNSS-R signal inversion. Therefore, for our research area, we selected a field in Goodwell, Texas County, Oklahoma, USA (latitude: 36.60° N; longitude: 101.64° W), located far from urban areas, to minimize interference from urban environments. This area is situated in the Great Plains region of the United States, characterized by extensive land distribution with minimal interference factors affecting GNSS-R signals.

2.4.2. Model Verification Process Design

To validate the accuracy of the real-time prediction approach to deep soil moisture combining GNSS-R data and a water movement model in unsaturated soil, we compared the soil moisture data obtained from soil moisture sensors at a specific location in the study area with the soil moisture derived from the proposed method in this study. This comparison aims to determine whether the soil moisture data obtained from the real-time prediction method for deep soil moisture align well with the data obtained from soil moisture sensors. If the agreement is satisfactory, it would demonstrate the feasibility of this real-time prediction approach to deep soil moisture, combining GNSS-R data and a water movement model in unsaturated soil.

The first step involves selecting two random locations, labeled A and B, within the study area as monitoring points. We compared the data obtained from soil moisture sensors at these monitoring points with the soil moisture derived from the novel method proposed in this study, thereby validating the method. Before measuring the soil moisture at the monitoring points, it is essential to establish the soil–water characteristic model and the deep soil moisture prediction model for the study area using soil moisture data collected on a specific day.

We obtained surface soil moisture data, as well as soil moisture data, at depths of 0.1 m, 0.2 m, 0.5 m, and 1 m from the International Soil Moisture Network for a specific day at the monitoring points. Utilizing the surface soil moisture data and data from depths of 0.1 m, 0.2 m, 0.5 m, and 1 m, we performed nonlinear fitting according to Equation (13), thereby obtaining the soil–water characteristic model and the deep soil moisture prediction model for the study area. Subsequently, after obtaining the surface soil moisture data from GNSS-R signal inversion at subsequent time points for the monitoring points, we input these data into the deep soil moisture prediction model to derive the relationship function between soil moisture and depth at the monitoring points. Based on the relationship function between soil moisture and depth at the monitoring points, we obtained the soil moisture at each depth between 0 and 1 m for both monitoring points, A and B, as depicted in Figure 2.

Figure 2. Position relationship diagram between monitoring point **A** and monitoring point **B**.

Afterwards, soil moisture sensors could be utilized to measure the soil moisture at the surface and at depths of 0.1 m, 0.2 m, 0.5 m, and 1 m at the monitoring points. The measured results could then be compared with the results obtained from the real-time prediction method for deep soil moisture, which integrates GNSS-R detection data and soil–water characteristics. This comparison serves to validate the feasibility of the real-time prediction approach to deep soil moisture combining GNSS-R data and a water movement model in unsaturated soil.

2.4.3. Model Evaluation Index

In this paper, the data obtained from the real-time prediction approach to deep soil moisture combining GNSS-R data and a water movement model in unsaturated soil are considered predicted values, while the data obtained from soil moisture sensors are regarded as true values. Statistical indicators such as relative error (δ) and mean square error (MSE) are employed to measure the deviation of predicted values from true values. Additionally, the coefficient of determination (R^2) is used to assess the degree of agreement between predicted values and true values, thereby determining whether the prediction model can accurately forecast deep soil moisture. This process aims to validate the feasibility of the real-time prediction approach to deep soil moisture combining GNSS-R data and water movement model in unsaturated soil.

We represent the true values using TR (m^3/m^3) and the predicted values using PV (m^3/m^3). The absolute error (Δ) is defined as the difference between the predicted value, PV, and the true value, TR, as shown in Equation (14).

$$\Delta = PV - TR \quad (14)$$

The relative error (δ) is defined as the absolute error (Δ) divided by the true value (TR) multiplied by 100% to obtain a percentage representation, as shown in Equation (15).

$$\delta = \frac{\Delta}{TR} \times 100 \quad (15)$$

The mean square error (MSE) is defined as the expected value of the square of the difference between the predicted value, PV, and the true value, TR, as shown in Equation (16). A mean square error (MSE) closer to 0 indicates the better predictive performance of the prediction model.

$$MSE = \frac{1}{m}\sum_{i=1}^{m}(PV_i - TR_i)^2 \quad (16)$$

The coefficient of determination (R^2), also known as the goodness of fit, is commonly used in predictive models to assess the degree of agreement between predicted values and true values. A coefficient of determination (R^2) closer to 1 indicates the better predictive performance of the model. The coefficient of determination (R^2) is expressed in Equation (17).

$$R^2 = 1 - \frac{\sum_{i=1}^{m}(PV_i - TR_i)^2}{\sum_{i=1}^{m}(\overline{TR_i} - TR_i)^2} \quad (17)$$

in which $\overline{TR_i}$ represents the mean value of the true values.

3. Results

Firstly, soil moisture data at the surface and depths of 0.1 m, 0.2 m, 0.5 m, and 1 m at monitoring points A and B were obtained from the International Soil Moisture Network for 20 October, 25 October, 30 October, 4 November, 9 November, 14 November, and 19 November 2023. The soil moisture data from 25 October, 30 October, 4 November, 9 November, 14 November, and 19 November 2023 at monitoring points A and B were primarily used to compare with the results obtained from the real-time prediction approach for deep soil moisture proposed in this paper. The soil moisture data from 20 October 2023 at monitoring points A and B were mainly used to establish the soil–water characteristic model and the deep soil moisture prediction model for the study area.

Next, using the soil moisture data from monitoring points A and B at the surface and depths of 0.1 m, 0.2 m, 0.5 m, and 1 m on 20 October 2023, we performed the nonlinear fitting of Equation (13) to establish the soil–water characteristic model and the deep soil moisture prediction model for the study area. Subsequently, utilizing the surface soil moisture data from monitoring points A and B on 25 October, 30 October, 4 November, 9 November, 14 November, and 19 November 2023, we obtained the predicted values of soil moisture at depths from 0 to 1 m using the deep soil moisture prediction model.

Finally, the predicted values and true values of soil moisture at the surface and depths of 0.1 m, 0.2 m, 0.5 m, and 1 m for monitoring points A and B on the six dates were compared. The results are shown in Figures 3 and 4. From the figures, it can be qualitatively observed that the predicted values of soil moisture at the monitoring points correspond well with the true values.

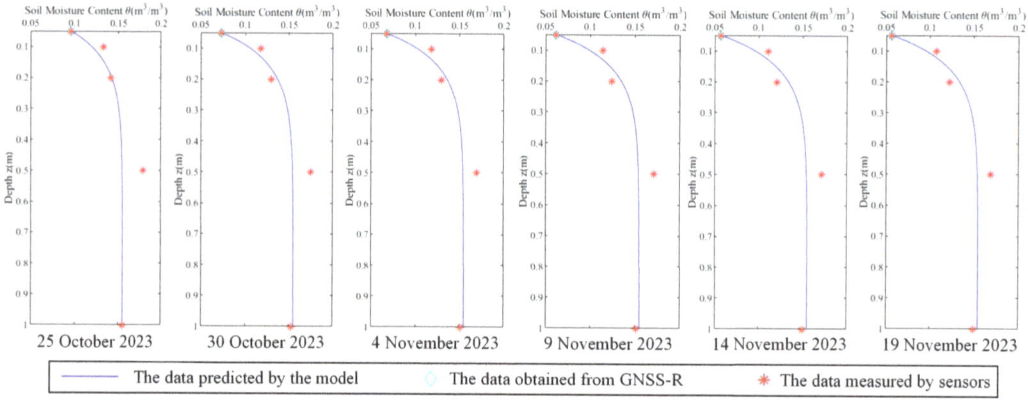

Figure 3. Comparison between predicted and true soil moisture values at monitoring point A over six days.

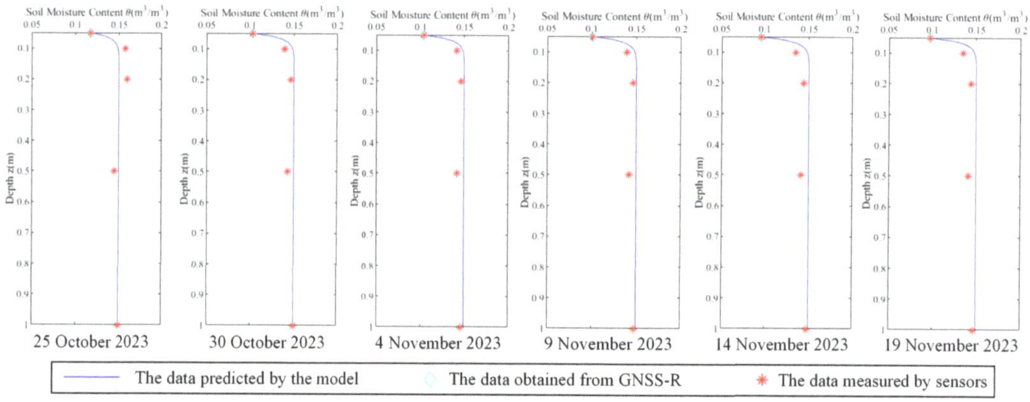

Figure 4. Comparison between predicted and true soil moisture values at monitoring point B over six days.

3.1. Assessment of Predicted Values

3.1.1. Relative Error of the Predicted Values

Subsequently, the relative error (δ) between the predicted and true soil moisture values at different depths for monitoring points A and B over the six days was utilized to quantitatively evaluate the accuracy of the model predictions, as depicted in Figures 5 and 6. On 25 October 2023, the relative error (δ) of the predicted soil moisture values at different depths for monitoring point A ranged from −13.97% to 0.71%, while for monitoring point B, it ranged from −6.25% to 3.45%. Similarly, on 30 October 2023, the relative error (δ) for monitoring point A varied from −12.00% to 5.38%, and for monitoring point B, it ranged from 2.74% to 6.47%. On 4 November 2023, the relative error (δ) for monitoring point A ranged from −12.71% to 5.43%, while for monitoring point B, it ranged from −2.74% to 5.63%. On 9 November 2023, the relative error (δ) for monitoring point A ranged from −13.16% to 8.87%, and for monitoring point B, it ranged from 1.35% to 6.47%. On 14 November 2023, the relative error (δ) for monitoring point A varied from −13.51% to 10.74%, and for monitoring point B, it ranged from 1.35% to 9.63%. Finally, on 19 November 2023, the relative error (δ) for monitoring point A ranged from −10.28% to 9.84%, while for monitoring point B, it ranged from 2.04% to 9.63%. The relative errors (δ) of soil moisture predictions relative to true values at different depths for monitoring points A and B over the six days were all within a small range, indicating minimal deviation between predicted and true soil moisture values. Therefore, this validates the feasibility of the real-time prediction approach to deep soil moisture combining GNSS-R data and a water movement model in unsaturated soil.

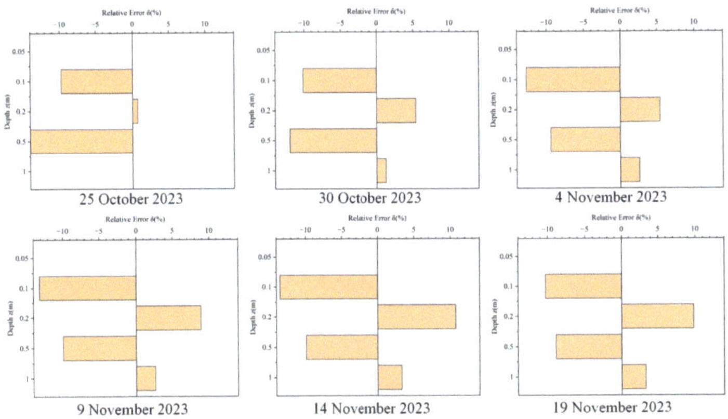

Figure 5. The relative errors of soil moisture predictions at different depths for monitoring point A over six days.

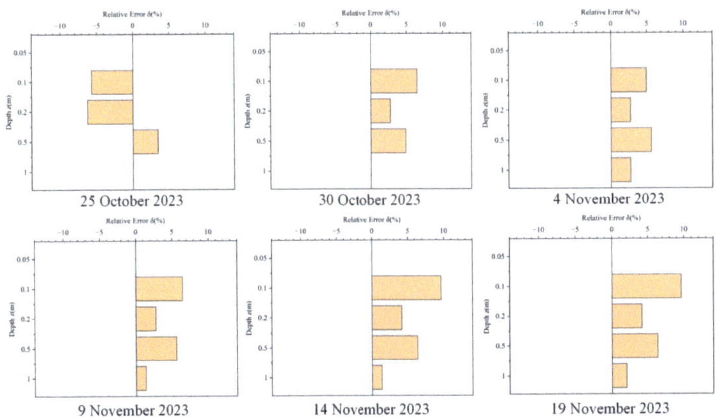

Figure 6. The relative errors of soil moisture predictions at different depths for monitoring point B over six days.

3.1.2. MSE and R^2 of the Predicted Values

Based on the mean square error (MSE) and the coefficient of determination (R^2) relative to the true values of soil moisture predictions over the six days, a more in-depth quantitative analysis of the model's prediction accuracy could be conducted.

According to computations, the mean square error (MSE) of soil moisture predictions for monitoring point A on 25 October 2023 was 1.59×10^{-4}, while for monitoring point B, it was 4.12×10^{-5}. On 30 October 2023, the MSE for monitoring point A was 1.28×10^{-4}, and for monitoring point B, it was 2.92×10^{-5}. For 4 November 2023, the MSE for monitoring point A was 1.09×10^{-4}, and for monitoring point B, it was 2.90×10^{-5}. On 9 November 2023, the MSE for monitoring point A was 1.30×10^{-4}, and for monitoring point B, it was 3.30×10^{-5}. For 14 November 2023, the MSE for monitoring point A was 1.42×10^{-4}, and for monitoring point B, it was 5.80×10^{-5}. Finally, on 19 November 2023, the MSE for monitoring point A was 1.03×10^{-4}, and for monitoring point B, it was 5.90×10^{-5}, as shown in Figures 7 and 8. It is evident that the prediction accuracy from the surface to the deep soil moisture prediction models shows a high degree of fit with the true values. Therefore, it validates the feasibility of the real-time prediction approach to deep soil moisture combining GNSS-R data and a water movement model in unsaturated soil.

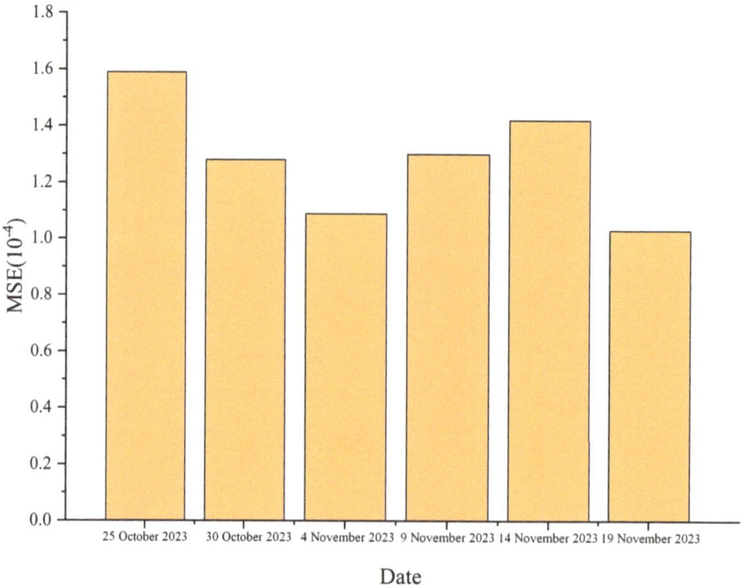

Figure 7. The mean square error (MSE) of monitoring point A over six days.

Through calculations, it was determined that the coefficient of determination R^2 for the soil moisture predictions at monitoring point A on 25 October 2023 was 78.68%, while at monitoring point B, it was 82.80%. On 30 October 2023, the coefficient of determination R^2 for soil moisture predictions at monitoring point A was 89.27%, and at monitoring point B, it was 89.88%. By 4 November 2023, the coefficient of determination R^2 for soil moisture predictions at monitoring point A increased to 90.89%, whereas at monitoring point B, it decreased slightly to 88.71%. On 9 November 2023, the coefficient of determination R^2 for soil moisture predictions at monitoring point A was 90.54%, and at monitoring point B, it was 89.08%. By 14 November 2023, the coefficient of determination R^2 for soil moisture predictions at monitoring point A was 90.60%, and at monitoring point B, it was 83.73%. Finally, on 19 November 2023, the coefficient of determination R^2 for soil moisture predictions at monitoring point A was 93.19%, while at monitoring point B, it was 82.45%, as illustrated in Figures 9 and 10. It is evident that the predictive models for soil moisture content from the surface to deep layers exhibit a high degree of conformity between predicted and actual values. Therefore, the feasibility of the real-time prediction approach to deep soil moisture content, which integrates GNSS-R detection data and soil moisture characteristics, can be validated.

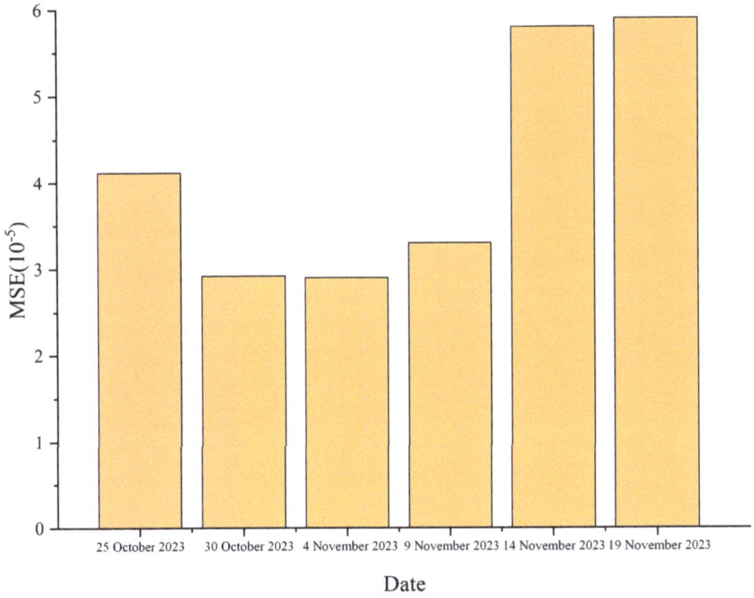

Figure 8. The mean square error (MSE) of monitoring point B over six days.

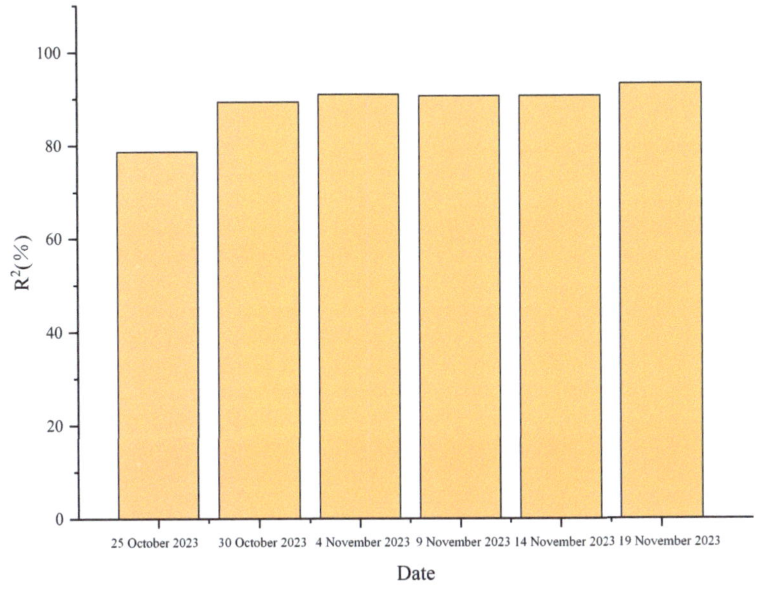

Figure 9. The coefficient of determination (R^2) for monitoring point A's soil moisture prediction over six days.

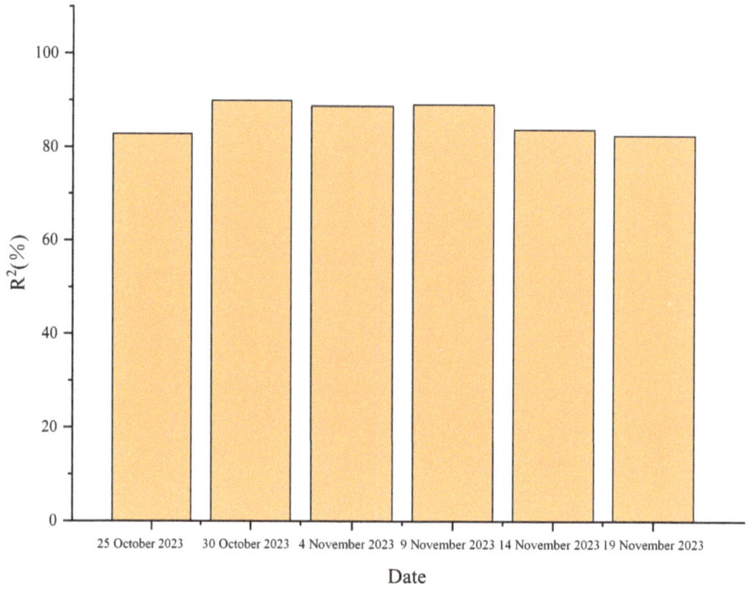

Figure 10. The coefficient of determination (R^2) for monitoring point B's soil moisture prediction over six days.

When combining the mean square errors (MSEs) and coefficients of determination (R^2) relative to the true values for soil moisture predictions at monitoring points A and B at depths from the surface of 0.1 m, 0.2 m, 0.5 m, and 1 m on 25 October, 30 October, 4 November, 9 November, 14 November, and 19 November 2023, it is evident that the deep soil moisture prediction model achieves a high degree of fit between predicted and actual values. Thus, it validates the feasibility of the real-time prediction approach to deep soil moisture combining GNSS-R data and water movement model in unsaturated soil.

3.2. Effects of Soil–Water Characteristic Model on Predicted Values

Using the real-time prediction approach to deep soil moisture combining GNSS-R data and a water movement model in unsaturated soil, soil moisture predictions for monitoring points A and B on 25 October 2023 were obtained, revealing significant discrepancies in the predicted values, as shown in Figure 11. The deep soil moisture prediction model used for monitoring both points, A and B, is consistent, with the only variable being the soil–water characteristic parameters α, n, and m in the deep soil moisture prediction model. However, these parameters, α, n, and m, are related to the soil properties at monitoring points A and B. Therefore, a comparison of the soil–water characteristic models between monitoring points A and B was warranted.

Figure 12 illustrates the permeability coefficient characteristic curves of monitoring points A and B, while Figure 13 depicts the matric suction head characteristic curves of monitoring points A and B. It can be observed that the infiltration coefficient characteristic curves of monitoring points A and B are relatively similar, whereas the matric suction head characteristic curves of monitoring points A and B exhibit significant differences. This discrepancy contributes to the considerable disparity in the predicted soil moisture between monitoring points A and B.

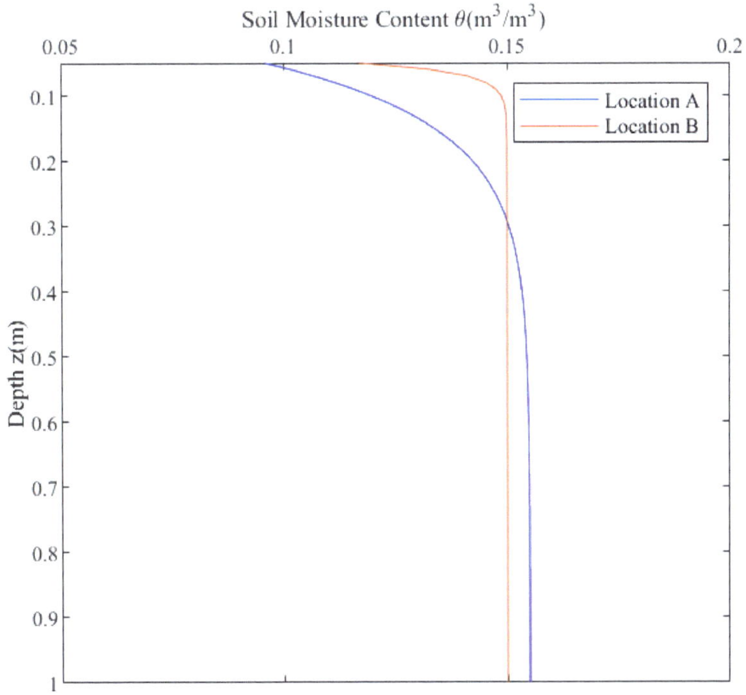

Figure 11. The comparison of predicted values between monitoring points A and B.

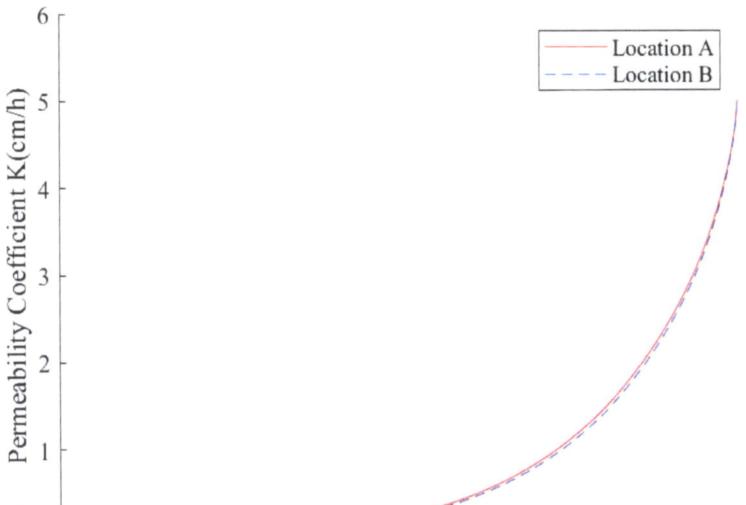

Figure 12. The comparison of the permeability coefficient characteristic curves between monitoring points A and B.

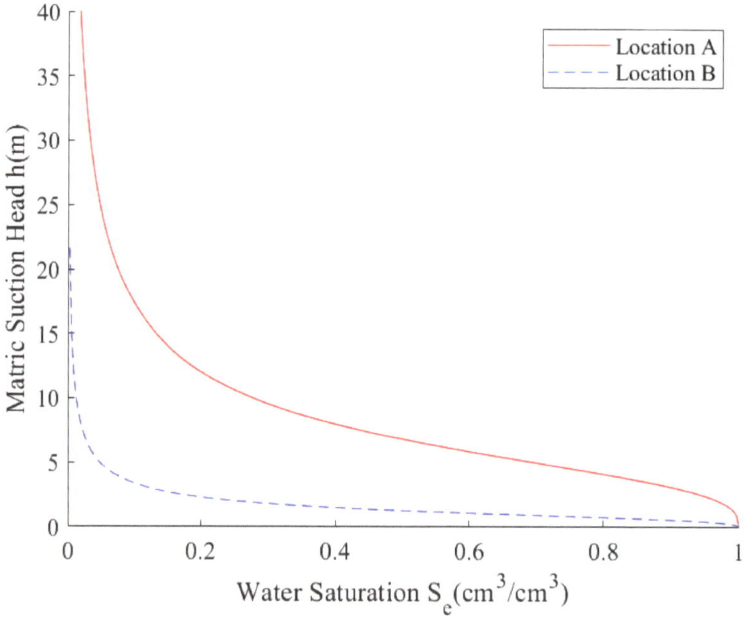

Figure 13. The comparison of matric suction head characteristic curves between monitoring points A and B.

3.3. Effects of Model Lower Boundary Depth on Predicted Values

The real-time prediction approach to deep soil moisture combining GNSS-R data and a water movement model in unsaturated soil proposed in this paper sets the depth of the lower boundary of the moisture movement control equation in unsaturated soil to 1 m. To investigate the influence of the lower boundary depth setting on soil moisture prediction, this paper sets the lower boundary depth to 1 m, 10 m, 20 m, 30 m, 40 m, and 50 m, obtaining the corresponding predicted values for monitoring point A on 25 October 2023 and comparing them, as shown in Figure 14.

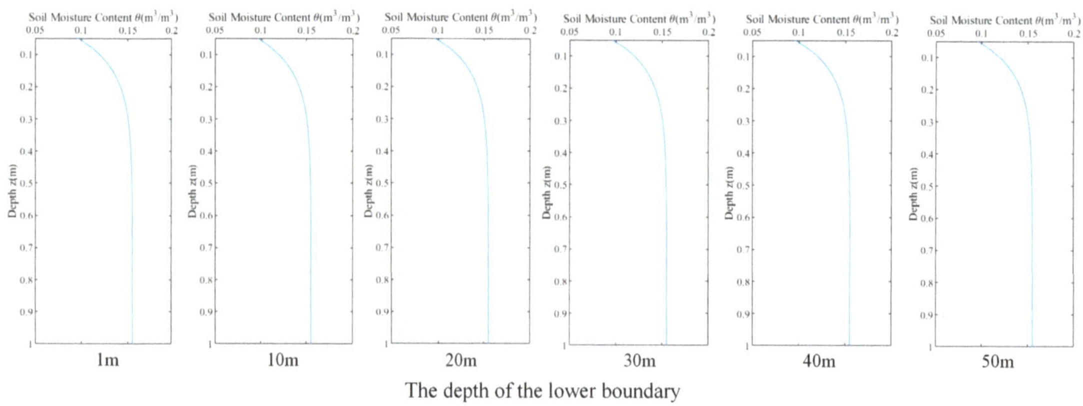

Figure 14. The comparison of the permeability coefficient characteristic curves between monitoring point A and monitoring point B.

From the graph, it can be observed that the lower boundary depth of the unsaturated soil moisture movement control equation has no effect on soil moisture prediction values when it exceeds 1 m. Therefore, choosing 1 m as the lower boundary depth of the unsaturated soil moisture movement control equation is reasonable.

4. Discussion

Due to the close correlation between soil moisture and variables such as soil properties and climatic conditions, the deep soil moisture prediction model proposed in this paper incorporates variables such as soil properties and climatic conditions during the construction process, aiming to achieve applicability under different soil properties and climatic conditions. As our model integrates the soil–water characteristic model, the deep soil moisture prediction model we obtained is parameterized. Through the initial data fitting process, the parameters in the model are solved, reflecting the characteristics of soil types. Since the soil composition in many regions within the depth range of 0–1 m is relatively homogeneous, resembling a homogeneous porous medium, our model is applicable in regions where the soil composition is uniform.

In cases of heterogeneous soil, it is possible to conduct stratified research on the soil in the study area. By combining the corresponding soil–water characteristic parameters of each soil layer, our model can be applied to conduct research effectively. Due to the close correlation between surface soil moisture data and climatic conditions, incorporating GNSS-R-retrieved surface soil moisture data into the process of predicting deep soil moisture through modeling implies that the model takes climate condition variables into consideration. In summary, the real-time prediction approach to deep soil moisture combining GNSS-R data and a water movement model in unsaturated soil proposed in this paper exhibits strong applicability and holds potential for future deployment in various regions. However, further optimization is necessary to enhance the method's universality.

5. Conclusions

The real-time monitoring of deep soil moisture has significant implications for controlling the quality of compacted soil construction, geological disaster monitoring and forecasting, the precise management of agricultural production, and other areas. However, traditional methods of soil moisture monitoring suffer from inefficiency and high costs when applied to large-scale areas. Additionally, methods that rely on GNSS-R signal inversion to obtain soil moisture data are limited due to their shallow monitoring depth. To address the limitations of current soil moisture monitoring methods, particularly their inability to dynamically monitor deep soil moisture over large areas in real time, this study innovatively proposed a real-time prediction approach to deep soil moisture combining GNSS-R data and a water movement model in unsaturated soil. This approach utilizes surface soil moisture data obtained through GNSS-R signal inversion and integrates it with soil–water characteristics and soil moisture data at a depth of 1 m. By employing a deep soil moisture prediction model, it is possible to obtain predicted soil moisture values for depths ranging from 0 to 1 m, enabling the large-scale, real-time monitoring of deep soil moisture. The proposed method was validated using data provided by the International Soil Moisture Network, focusing on a field in Goodwell, Texas County, OK, USA, as the study area.

By calculating the relative error δ of the predicted soil moisture values relative to the true values at monitoring points A and B in the study area at the surface and depths of 0.1 m, 0.2 m, 0.5 m, and 1 m on six days (25 October, 30 October, 4 November, 9 November, 14 November, and 19 November 2023), it was found that the relative error δ remained within a relatively low range. The relative error δ at monitoring point A was controlled within $\pm 14\%$, while at monitoring point B, it was controlled within $\pm 10\%$. Therefore, it is evident that the deep soil moisture prediction model fits the true values quite well. Furthermore, by computing the mean square error (MSE) and coefficient of determination (R^2) of the soil moisture predictions relative to the true values at monitoring points A and

B in the study area at the surface and depths of 0.1 m, 0.2 m, 0.5 m, and 1 m on the same six days, it was observed that the MSE at monitoring point A ranged from 1.03×10^{-4} to 1.59×10^{-4}, and at monitoring point B, it ranged from 2.90×10^{-5} to 5.90×10^{-5}, indicating a relatively low overall control level. The R^2 at monitoring point A ranged from 78.68% to 93.19%, while at monitoring point B, it ranged from 82.45% to 89.88%, indicating a relatively high overall control level. Therefore, it can be concluded that the deep soil moisture prediction model achieves a high level of fitting with the true values. Consequently, the soil moisture data predicted using this method align well with actual soil moisture data, validating the feasibility of the real-time prediction approach to deep soil moisture combining GNSS-R data and a water movement model in unsaturated soil. The soil–water characteristic model influences soil moisture prediction values, rendering the method applicable across different geographic features. Moreover, the depth of the lower boundary in the unsaturated soil moisture transport control equation does not affect soil moisture prediction values when it exceeds 1 m, justifying the selection of 1 m as a reasonable depth for the lower boundary in the unsaturated soil moisture transport control equation.

The real-time prediction approach to deep soil moisture combining GNSS-R data and a water movement model in unsaturated soil achieves the real-time and dynamic monitoring of deep soil moisture over a wide range, which can play a significant role in various fields, such as agricultural production, geological disaster management, engineering construction, and heritage preservation.

Author Contributions: Conceptualization, H.S.; methodology, H.S. and X.L.; software, C.Y.; validation, Y.S., W.B. and C.Y.; formal analysis, W.L.; investigation, X.L.; resources, Y.S., W.B. and W.L.; data curation, W.B. and C.Y.; writing—original draft preparation, X.L.; writing—review and editing, H.S.; supervision, Y.S., W.B. and W.L.; project administration, Y.S., H.S. and W.L. All authors have read and agreed to the published version of the manuscript.

Funding: This work is supported by the Youth Cross Team Scientific Research Project of the Chinese Academy of Sciences (JCTD-2021-10) and in part by National Natural Science Foundation of China 42074042 and 42104032.

Data Availability Statement: The original contributions presented in the study are included in the article, further inquiries can be directed to the corresponding author.

Conflicts of Interest: The authors declare no conflicts of interest.

References

1. Craig, R.F. *Craig's Soil Mechanics*; CRC Press: Boca Raton, FL, USA, 2004.
2. Das, B.M. *Principles of Geotechnical Engineering*; Cengage Learning: Boston, MA, USA, 2021.
3. Gardner, W. The capillary potential and its relation to soilmoisture constants. *Soil Sci.* **1920**, *10*, 357–360. [CrossRef]
4. Gardner, W.; Widtsoe, J.A. The movement of soil moisture. *Soil Sci.* **1921**, *11*, 215–232. [CrossRef]
5. Haines, W.B. The volume-changes associated with variations of water content in soil. *J. Agric. Sci.* **1923**, *13*, 296–310. [CrossRef]
6. Pham, H.Q.; Fredlund, D.G.; Barbour, S.L. A study of hysteresis models for soil-water characteristic curves. *Can. Geotech. J.* **2005**, *42*, 1548–1568. [CrossRef]
7. Richards, L.A. Capillary conduction of liquids through porous mediums. *Physics* **1931**, *1*, 318–333. [CrossRef]
8. Brooks, R.H. *Hydraulic Properties of Porous Media*; Colorado State University: Fort Collins, CO, USA, 1965.
9. Van Genuchten, M.T. A closed-form equation for predicting the hydraulic conductivity of unsaturated soils. *Soil Sci. Soc. Am. J.* **1980**, *44*, 892–898. [CrossRef]
10. Baver, L.D. *Soil Physics*; John Wiley and Sons, Inc.: New York, NY, USA, 1956; Volume 81.
11. Fang, H.Y.; Daniels, J.L. *Introductory Geotechnical Engineering: An Environmental Perspective*; CRC Press: Boca Raton, FL, USA, 2017.
12. Kaplan, E.D.; Hegarty, C. *Understanding GPS/GNSS: Principles and Applications*; Artech House: Boston, MA, USA, 2017.
13. Wigneron, J.P.; Jackson, T.J.; O'Neill, P.; De Lannoy, G.; de Rosnay, P.; Walker, J.P.; Ferrazzoli, P.; Mironov, V.; Bircher, S.; Grant, J.P. Modelling the passive microwave signature from land surfaces: A review of recent results and application to the L-band SMOS and SMAP soil moisture retrieval algorithms. *Remote Sens. Environ.* **2017**, *192*, 238–262. [CrossRef]

14. Masters, D.; Axelrad, P.; Katzberg, S. Initial results of land-reflected GPS bistatic radar measurements in SMEX02. *Remote Sens. Environ.* **2004**, *92*, 507–520. [CrossRef]
15. Egido, A.; Ruffini, G.; Caparrini, M.; Martin, C.; Farres, E.; Banque, X. Soil moisture monitorization using GNSS reflected signals. *arXiv* **2008**, arXiv:0805.1881.
16. Kebiao, M.; Mengyang, Z.; Jianming, W.; Huajun, T.; Qingbo, Z. The study of soil moisture retrieval algorithm from gnss-r. In Proceedings of the 2008 International Workshop on Education Technology and Training & 2008 International Workshop on Geoscience and Remote Sensing, Shanghai, China, 21–22 December 2008; Volume 1, pp. 438–442.
17. Sreedeep, S.; Reshma, A.C.; Singh, D.N. Measuring soil electrical resistivity using a resistivity box and a resistivity probe. *Geotech. Test. J.* **2004**, *27*, 411–415. [CrossRef]
18. Whalley, W.R.; Dean, T.J.; Izzard, P. Evaluation of the capacitance technique as a method for dynamically measuring soil water content. *J. Agric. Eng. Res.* **1992**, *52*, 147–155. [CrossRef]
19. Schwartz, B.F.; Schreiber, M.E.; Yan, T. Quantifying field-scale soil moisture using electrical resistivity imaging. *J. Hydrol.* **2008**, *362*, 234–246. [CrossRef]
20. Robinson, D.A.; Campbell, C.S.; Hopmans, J.W.; Hornbuckle, B.K.; Jones, S.B.; Knight, R.; Ogden, F.; Selker, J.; Wendroth, O. Soil moisture measurement for ecological and hydrological watershed-scale observatories: A review. *Vadose Zone J.* **2008**, *7*, 358–389. [CrossRef]
21. ASTM Committee D-18 on Soil and Rock. *Standard Test Methods for Laboratory Determination of Water (Moisture) Content of Soil and Rock by Mass*; ASTM: West Conshohocken, PA, USA, 2005.
22. Hillel, D. *Introduction to Soil Physics*; Academic Press: Cambridge, MA, USA, 2013.
23. Zazueta, F.S.; Xin, J. Soil moisture sensors. *Soil Sci.* **1994**, *73*, 391–401.
24. Selig, E.T.; Mansukhani, S. Relationship of soil moisture to the dielectric property. *J. Geotech. Eng. Div.* **1975**, *101*, 755–770. [CrossRef]
25. Topp, G.C.; Davis, J.L.; Annan, A.P. Electromagnetic determination of soil water content using TDR: I. Applications to wetting fronts and steep gradients. *Soil Sci. Soc. Am. J.* **1982**, *46*, 672–678. [CrossRef]
26. Campbell, J.E. Dielectric properties and influence of conductivity in soils at one to fifty megahertz. *Soil Sci. Soc. Am. J.* **1990**, *54*, 332–341. [CrossRef]
27. Arulanandan, K. Dielectric method for prediction of porosity of saturated soil. *J. Geotech. Eng.* **1991**, *117*, 319–330. [CrossRef]
28. Jacobsen, O.H.; Schjønning, P. A laboratory calibration of time domain reflectometry for soil water measurement including effects of bulk density and texture. *J. Hydrol.* **1993**, *151*, 147–157. [CrossRef]
29. Robinson, M.; Dean, T.J. Measurement of near surface soil water content using a capacitance probe. *Hydrol. Processes* **1993**, *7*, 77–86. [CrossRef]
30. Gaskin, G.J.; Miller, J.D. Measurement of soil water content using a simplified impedance measuring technique. *J. Agric. Eng. Res.* **1996**, *63*, 153–159. [CrossRef]
31. Gardner, C.; Dean, T.J.; Cooper, J.D. Soil water content measurement with a high-frequency capacitance sensor. *J. Agric. Eng. Res.* **1998**, *71*, 395–403. [CrossRef]
32. Rinaldi, V.A.; Francisca, F.M. Impedance analysis of soil dielectric dispersion (1 MHz–1 GHz). *J. Geotech. Geoenvironmental Eng.* **1999**, *125*, 111–121. [CrossRef]
33. Baumhardt, R.L.; Lascano, R.J.; Evett, S.R. Soil material, temperature, and salinity effects on calibration of multisensor capacitance probes. *Soil Sci. Soc. Am. J.* **2000**, *64*, 1940–1946. [CrossRef]
34. Hilhorst, M.A. A pore water conductivity sensor. *Soil Sci. Soc. Am. J.* **2000**, *64*, 1922–1925. [CrossRef]
35. Kelleners, T.J.; Soppe, R.; Ayars, J.E.; Skaggs, T.H. Calibration of capacitance probe sensors in a saline silty clay soil. *Soil Sci. Soc. Am. J.* **2004**, *68*, 770–778. [CrossRef]
36. Rohini, K.; Singh, D.N. Methodology for determination of electrical properties of soils. *J. Test. Eval.* **2004**, *32*, 62–68. [CrossRef]
37. Rao, B.H.; Singh, D.N. Moisture content determination by TDR and capacitance techniques: A comparative study. *Int. J. Earth Sci. Eng* **2011**, *4*, 132–137.
38. Mittelbach, H.; Lehner, I.; Seneviratne, S.I. Comparison of four soil moisture sensor types under field conditions in Switzerland. *J. Hydrol.* **2012**, *430*, 39–49. [CrossRef]
39. Bowers, S.A. *Reflection of Radiant Energy from Soils*; Kansas State University: Manhattan, KS, USA, 1971.
40. Li, Z.L.; Leng, P.; Zhou, C.; Chen, K.S.; Zhou, F.C.; Shang, G.F. Soil moisture retrieval from remote sensing measurements: Current knowledge and directions for the future. *Earth-Sci. Rev.* **2021**, *218*, 103673. [CrossRef]
41. Edokossi, K.; Calabia, A.; Jin, S.; Molina, I. GNSS-reflectometry and remote sensing of soil moisture: A review of measurement techniques, methods, and applications. *Remote Sens.* **2020**, *12*, 614. [CrossRef]
42. Shen, X.; Walker, J.P.; Ye, N.; Wu, X.; Boopathi, N.; Yeo, I.Y.; Zhang, L.; Zhu, L. Soil moisture retrieval depth of P-and L-band radiometry: Predictions and observations. *IEEE Trans. Geosci. Remote Sens.* **2020**, *59*, 6814–6822. [CrossRef]
43. Shi, Y.; Ren, C.; Yan, Z.; Lai, J. High spatial-temporal resolution estimation of ground-based global navigation satellite system interferometric reflectometry (GNSS-IR) soil moisture using the genetic algorithm back propagation (GA-BP) neural network. *ISPRS Int. J. Geo-Inf.* **2021**, *10*, 623. [CrossRef]

44. Liang, Y.j.; Ren, C.; Wang, H.Y.; Huang, Y.B.; Zheng, Z.T. Research on soil moisture inversion method based on GA-BP neural network model. *Int. J. Remote Sens.* **2019**, *40*, 2087–2103. [CrossRef]
45. Shi, Y.; Ren, C.; Yan, Z.; Lai, J. Improving soil moisture retrieval from GNSS-interferometric reflectometry: Parameters optimization and data fusion via neural network. *Int. J. Remote Sens.* **2021**, *42*, 9085–9108. [CrossRef]

water

MDPI

Article

Geometry, Extent, and Chemistry of Fermentative Hot Spots in Municipal Waste Souk Sebt Landfill, Ouled Nemma, Beni Mellal, Morocco

Yousra El Mouine [1,2], Amal El Hamdi [1,2], Abderrahim Bousouis [3], Youssouf El Jarjini [2], Meryem Touzani [4], Vincent Valles [1], Laurent Barbiero [5,*] and Moad Morarech [2]

1 Laboratory Mediterranean Environment and Agro-Hydrosystem Modelling (UMR 1114 EMMAH), Hydrogeology Department, Avignon University, 84000 Avignon, France; yousra.elmouine@gmail.com (Y.E.M.); amalgoaea@gmail.com (A.E.H.); vincent.valles@outlook.fr (V.V.)

2 Laboratory in Applied and Marine Geosciences, Geotechnics and Georisk (LR3G), Faculty of Science Tetouan, Abdelmalek Essaâdi University, Tetouan 93002, Morocco; youssouf.eljarjini@etu.uae.ac.ma (Y.E.J.); mmorarech@uae.ac.ma (M.M.)

3 Laboratoire de Géosciences, Faculté des Sciences, Université Ibn Tofaïl, BP 133, Kénitra 14000, Morocco; ab-derrahim.bousouis@uit.ac.ma

4 National Institute of Agronomic Research, Rabat 10060, Morocco; meryem.touzani@inra.ma

5 Institut de Recherche pour le Développement, Géoscience Environnement Toulouse, CNRS, University of Toulouse, Observatoire Midi-Pyrénées, UMR 5563, 14 Avenue Edouard Belin, 31400 Toulouse, France

* Correspondence: laurent.barbiero@get.omp.eu

Abstract: The presence of fermentative hotspots in municipal waste dumps has been reported for several decades, but no study has focused on their size and shape. The uncontrolled landfill of Soub Sekt, covering an area of about 8 hectares in the Tadla plain in Morocco, is the source of a permanent pollution plume in the groundwater, detected by self-potential (SP) measurements. The study aims to detect and characterize these hotspots as well as the leachates that form within them. These hotspots are typically circular and smaller than 3 m in size, and they are concentrated within recent waste deposits. Intense electron transfer activities, particularly during redox reactions leading to metal solubilization, result in very low SP values (down to −60 mV), facilitating their detection. Several successive field campaigns suggest that they are active for 2–3 weeks. Due to the low permeability of the soils, highly mineralized leachates (average Electrical Conductivity 45 mS cm^{-1}) rich in organic ions accumulate on the soil surface at the base of the waste windrows. There, they evolve by concentration due to evaporation and oxidation due to slow diffusion of atmospheric O_2. Despite the small size of the hotspots generating the leachates, the accumulation of leachates in ponds and the low soil permeability limits the percolation rate, resulting in moderate but permanent groundwater pollution.

Keywords: self-potential; redox potential; leachate plume; landfill; Tadla; Morocco

1. Introduction

Municipal waste landfills, true biological reactors, are the scene of fermentative processes closely linked to temperature, the nature of the waste, as well as the level of organic matter and moisture, the latter varying due to numerous factors such as climate, cultural and dietary norms, and the possible presence of waste selective sorting [1]. Municipal waste landfills also stand out due to a strong heterogeneity of deposited materials, varying according to their nature, water content, and the proportion of fermentable compounds [2]. The succession of deposits over time maintains this heterogeneity in the intensity of fermentative processes, depending on whether the deposit is recent and rich in fermentable compounds or old and has already undergone these processes. Furthermore, waste decomposition leads to the production of gas and leachate, as well as heat generation, due to continuous aerobic and anaerobic processes [3]. Although a wild landfill can generally be considered an anaerobic reactor, where organic matter degradation occurs, the composition of the leachates generated, which can contaminate groundwater resources due to their infiltration, varies considerably depending on the location [4–6].

Geophysical methods employing non-invasive approaches are highly adept at mapping the scope of fermentative activities. Among various geophysical methodologies, self-potential (SP) measurements have been applied for approximately three decades in research concerning landfill pollution [7–9]. In recent years, there has been a surge in research focusing on contamination, attributed to the remarkable sensitivity of self-potential measurements to redox conditions in shallow aquifers [10–13]. Notably, the detectable self-potential observed at the surface of the soil is responsive to spatial fluctuations in charge flux within the matrix, variations in redox gradients and temperature, signals generated by microorganisms, or other activities associated with the migration of contaminants [7,12,14–17]. Mapping of fermentative activity in a European landfill using the self-potential (SP) technique has highlighted the existence of "hot spots", areas subject to intense fermentative processes under the control of microbial activities and geochemical reactions [9,18]. These phenomena, often studied in soil, intensify with the abundance of organic matter and aeration conditions. Fermentative processes there are intense, generating long-term heat, notably through gas escape, such as methane, which can spontaneously ignite [19]. Few studies have focused on these hot spots, with most merely mentioning their existence [18]. Are they present in all operating municipal waste landfills and distributed randomly or concentrated within recent waste? Analysis of these specific areas could provide crucial information to monitor landfill activities and estimate the location of unauthorized or unlisted landfills. The self-potential method has also been used outside landfills [20–23] to map pollution plumes in aquifers resulting from the arrival of leachates from landfill fermentations. These studies have also confirmed the existence of hot spots through very negative values, explained by the intensity of biogeochemical processes.

This study aims to verify the presence and locate fermentative hot spots within a landfill based on the type and approximate age of the waste. It also aims to characterize the shape, size, and spatial variability of fermentative "hot spots" that can cause aquifer contamination, as well as the chemical composition of the leachates they produce, with a focus on the specific conditions of hot and arid zones in North Africa.

2. Materials and Methods

2.1. Study Area

The study was conducted at the Souk-Sebt Ouled Nemma landfill, an uncontrolled and unsupervised landfill for which no measures have been taken to limit its impact on the environment. It covers approximately 8 hectares in the Tadla plain, the main agricultural production region of Morocco (Figure 1). The site is located on the left bank of the Oum Er Rbia river, which traverses the plain from east to west. The groundwater level, situated approximately 5–6 m beneath a hardened limestone crust, flows in a regional direction towards the north-northwest [24]. The climate in this area is semi-arid to arid continental, characterized by a dry season spanning from April to October and a wet season from November to March. The average yearly temperature hovers around 20 °C, with an annual precipitation average of about 430 mm [25]. The Tadla plain receives water from the Atlas Mountains to the south. Apart from this natural water source, an extensive irrigation system was established in the late 1990s, drawing water from the Bin El Ouidane dam, situated roughly 100 km upstream from the region. This irrigation network is aging; it has numerous leaks that contribute locally to groundwater recharge, causing fluctuations in groundwater levels during irrigation periods. Surface soils in the region have fine textures and moderate permeability. There are many unauthorized landfills within the plain [26]. The Souk-Sebt landfill is surrounded by agricultural plots generally irrigated from late April to late June. The soil in the landfill, compacted by the weight of trucks dumping waste, is rather impermeable, and the leachate produced infiltrates very slowly. During winter, although rainfall is not abundant, it is sufficient to generate leachate pools within the landfill, each reaching up to 1 m in depth and extending over several tens of square meters [22].

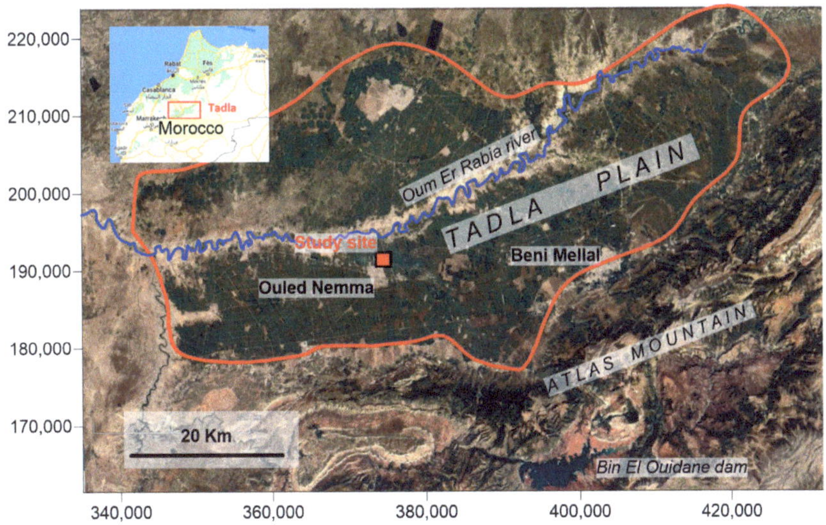

Figure 1. Location of the study site on the left bank of the Oum Er Rabia river, Tadla plain, Morocco.

2.2. Soub Sekt Landfill

The Soub Sekt landfill, located north of the town of Ouled Nemma, primarily receives household and agricultural waste, but more recently, organic residues from a confectionery factory have been regularly dumped there. Active since the 2000s, the landfill receives regular inputs of materials deposited in windrows. The oldest waste is typically at the base of the windrows (Figure 2a). However, after several years, the waste covers almost the entire surface. Remodeling of old waste is therefore regularly carried out to free up space for new waste and maintain access for dump trucks. This results in areas of accumulation of old materials and areas of recent deposits. Notably, there is a sector of old waste mixed with construction materials, a sector of old deposits in overlapping windrows, and a sector of recent waste deposition. This morphology changes little, but the quantity of added deposits varies over time. Thus, the northeast of the landfill has proven to be a very active sector in terms of fermentative processes. This landfill is the source of a pollution plume that remains stable throughout the season, flowing towards the north-northwest, consistent with our understanding of the regional groundwater flow. The plume, continuously fed by leachates infiltrating beneath the landfill, was detected through PS measurements and sampling during a previous study [22] (Figure 2b). The rapid flow of the groundwater suggests pollution over a long distance, although it has not been mapped in detail. The concerning aspect of the groundwater quality's deterioration is heightened by the fact that rural communities in this area rely directly on aquifer pumping for their water needs, given the isolation of their residences.

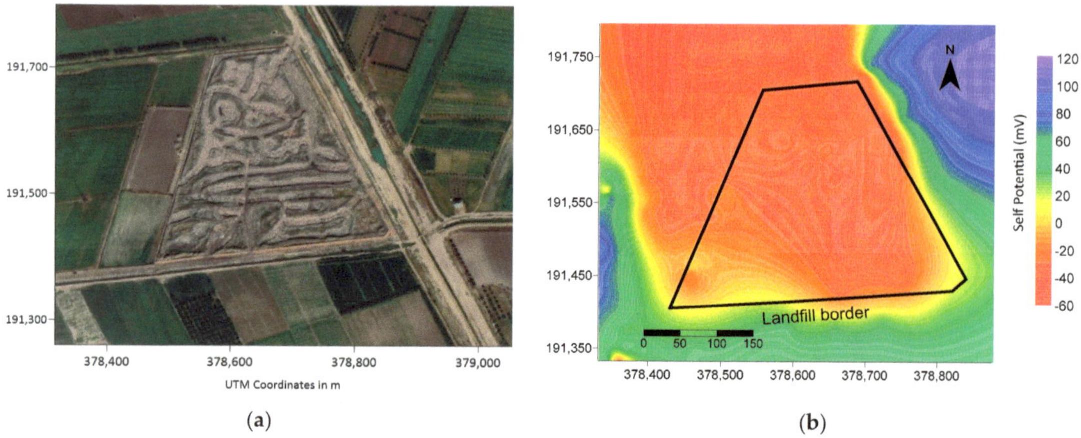

Figure 2. (**a**) Aerial view of the Soub Sekt landfill site showing windrow deposits of waste; (**b**) pollution plume generated by the Soub Sekt landfill detected from self-potential measurements (adapted from El Mouine et al. [22]).

2.3. Self-Potential Measurements

The self-potential measurements were carried out using impolarizable electrodes of type PMS 9000: Pb-$PbCl_2$ NaCl from SDEC France. These probes have an internal resistance of 450 Ohms. They are sealed with porous wood, and an electrolytic solution of NaCl ensures electrical continuity between the inside of the electrode and the soil. The connections are made with multi-stranded copper cable with a section of 0.75 mm^2 surrounded by plastic insulation. The voltmeter used (Widewing Multimeter UNIT-T 71-C, Petiau type, SDEC, Rousset, France) has a high input impedance (40 MΩ) for reliable measurement. An electrode was installed at a fixed point outside the landfill, defining the baseline potential. The second electrode was moved over the landfill. For each measurement of potential difference, the position of the mobile electrode was measured by GPS and noted. The surface of the landfill and its immediate external environment were swept to promote contact with the ground. The measurement is noted when the measured potential is stabilized, i.e., when the voltage fluctuations do not exceed 2 mV. Due to the irregularity of the terrain, the measurements were not carried out according to a strict grid but rather to cover the area substantially based on obstacles, with particular attention to the north of the landfill identified as more reactive from a fermentative standpoint. Curved lines of measurements were followed with a measurement taken every 5 m along these lines. Measurements were thus sometimes taken between two windrows, sometimes at the top or on the sides of the windrows. Measurements were also taken around the perimeter of the landfill. When a hot spot was detected by a low value of electrical potential, the mobile electrode was moved to detect its center associated with the minimum value. From this position, a series of measurements was taken in a star pattern in 8 directions (North, South, East, West, Northeast, Southeast, Southwest, Northwest) and at intervals of 0.5 m (0.5, 1, 1.5, and 2 m). Each hot spot was thus characterized by 33 SP measurements (Figure 3). Some hotspots, which may be close to each other with overlapping zones of influence on PS measurements, were not studied; only isolated hotspots were considered.

2.4. SP Data Treatment

The variograms were calculated and fitted with a model including the nugget effect and a spherical adjustment. Raw and directional variograms were examined to (1) verify if the measurement density is sufficient for a good cartographic representation of SP values on the landfill and (2) detect any potential anisotropy or oriented structure of the values. Maps were generated using kriging (Surfer19, Golden Software, www.goldensoftware.com).

For the study of size and shape, 8 hot spots were selected. The center of each was placed at coordinate (0, 0) and assigned a zero potential value (0 mV) to standardize, allowing for the overlay and comparison of measurements taken in a star pattern. The deviations from the center thus enable the calculation of means and standard deviations in different directions. Therefore, 264 measurements were used for geostatistical processing of data regarding the size and shape of the hot spots.

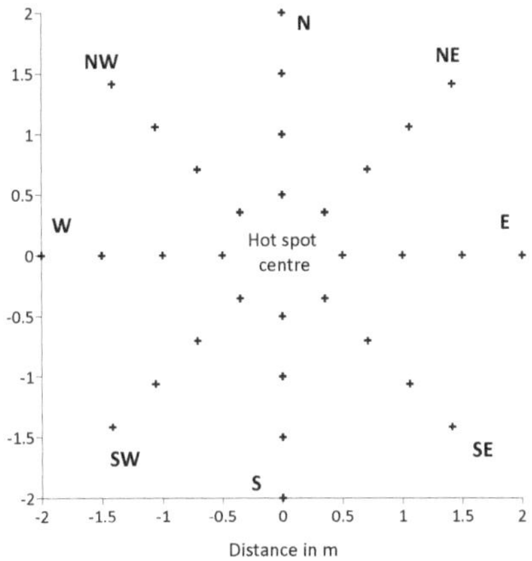

Figure 3. Star SP measurements around the center of fermentative hotspots.

2.5. Leachates Sampling and Analysis

Seven leachate samples were taken near hotspots in the area of recent deposits, as well as towards the center of the landfill and in the sector occupied by older deposits. Leachates are thick, typically odorous liquids, usually black in color but sometimes red or white. In the deepest ponds (0.7 to 1.1 m), a gradient of physico-chemical characteristics between the surface and depth has been observed. We also noted the presence of a layer of floating plastic waste that significantly reduces the exchange surface between the liquid leachate and the atmosphere. Samples were collected at the surface and at depths of around 0.7 m for the deepest ponds. Electrical conductivity, pH, redox potential (Eh), and temperature were measured on-site. The equipment used is a Hanna Instrument HI98150 pH-meter (Hanna Instrument, Lingolsheim, France). The redox electrode is a platinum electrode with a half-cell $KCl/AgCl_2$. The potential of the half-cell, which depends on temperature, was added to the voltmeter reading to obtain the redox potential Eh. These field-acquired data were used to calculate the partial pressure of O_2 according to the formula:

$$2H_2O \rightleftharpoons 4H^+ + 4e^- + O_2\,(g), \quad (1)$$

This reaction, having slow kinetics, results in thermodynamic imbalance [27]. Calculating the equilibrating partial pressure of O_2 offers the advantage of combining pH and Eh measurements to estimate an overall parameter of anoxia, i.e., the logarithm of O_2 patial pressure (pO_2).

Samples were collected in 250 mL HDPE bottles with double closure and without air bubbles. Due to the substantial stock of organic matter in the bottle, microbial activity consuming oxidants is maintained between sampling and laboratory analysis. The samples were transferred to the laboratory (Emmah, Avignon University, France) for analysis of total organic carbon (TOC) by combustion, then, after dilution and filtration to 0.45 μm (cellulose acetate syringe filters), analysis of major anions and cations by ion chromatography and metals by atomic adsorption. Stable isotope analysis of water was conducted on three remaining samples on a Los Gatos Isotope Ratio Infrared Spectrometer (IRIS) (LGR DLT-100, Los Gatos Research Inc., Los Gatos, CA, USA) at the University of Avignon (Accuracy

±0.2‰ vs. V-SMOW for $\delta^{18}O$ and ±1‰ vs. V-SMOW for $\delta^{2}H$). These samples were then sent to ENSCM Laboratory (National High School of Chemistry, Montpellier, France) for analysis of trace elements by ICP-MS. Although these 3 samples are not representative of all the processes occurring in the landfill, principal component analysis was conducted to highlight the main evolutions of leachates within the landfill (XLStat software, addinsoft, https://www.xlstat.com).

3. Results

3.1. Self-Potential Survey of the Landfill Site

The locations of PS measurement points are depicted in Figure 4a. The variogram obtained over the entire landfill is shown in Figure 4b. The spatial structure exhibited a very strong nugget effect, approximately half of the sill. This indicates high variability at the local scale due to the presence of very small hotspots with highly negative values, much smaller than the 5 m distance maintained between successive measurements. Consequently, the distribution map of values (Figure 4a) only reflects the collected data but does not adequately represent the structure of PS values on the study area surface. Based on this result, the exploitation of directional variograms was not meaningful. The map only reflects general trends, with values mostly positive (from −5 to +15 mV) in the extreme southeast of the landfill, composed of older deposits mixed with construction rubble. The southwest third of the landfill was characterized by weakly negative values (from −5 to −30 mV), although some hotspots may be detected there. This sector consists of older waste onto which some recent deposits (around one to two weeks before the measurement campaign) have been accumulated. The northern part of the landfill, as well as a strip along the truck access in the southern part, showed more strongly negative values (from −30 to −60 mV) and are mainly characterized by the concentration of recent deposits. The map highlights the presence of numerous hotspots throughout the landfill, sometimes very close to each other in the northern sector. Several campaigns conducted a few months apart showed that this zoning of the landfill is relatively constant, but the location of hotspots constantly changes depending on the deposits.

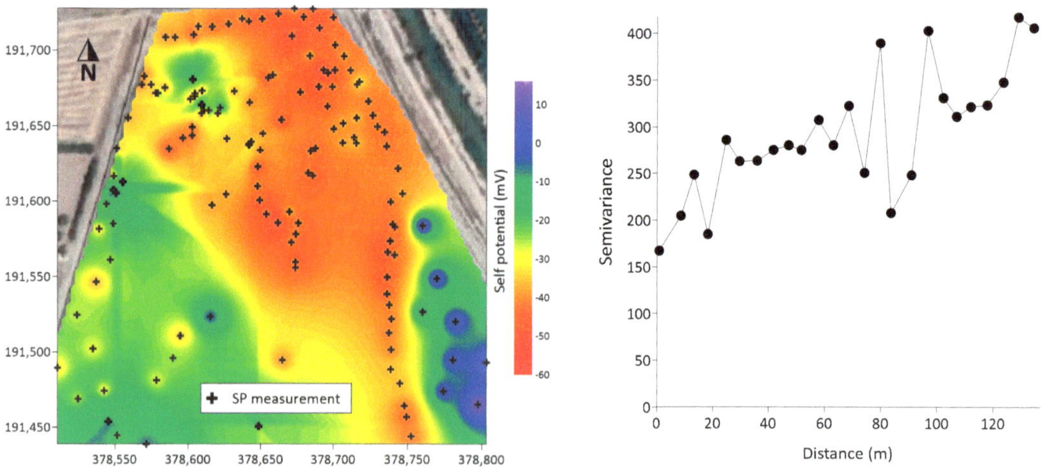

(a) (b)

Figure 4. Distribution map (**a**) and variogram (**b**) of PS values within the landfill.

3.2. Size and Shape of Fermentative Hotspots

In Figure 5a, the color scale has been adapted to highlight abrupt variations in PS around hotspots in accordance with the nugget effect observed on the variogram (Figure 4b).

Hotspots were characterized by very negative PS values (from −40 to −60 mV) but were sporadic. The means and standard deviations of PS values in the measured directions around hotspots are presented in Figure 6, along with the corresponding variogram in Figure 5b. There was an absence of nugget effect with low semivariance for distances less than 0.4 m. Semivariance increased very rapidly for distances from around 1 to 1.6 m, then decreased abruptly for distances greater than 2.5 m, indicating a strong structuring of the analyzed environment, with a structure close to 2.5 m. The results confirmed the small size of these hotspots, around 2.5–3 m in diameter, in accordance with the significant nugget effect observed in Figure 4b at the scale of the landfill. Beyond a distance of 1.5 m from the center, the hotspot is no longer detectable. The same aspect was observed in all directions, with slight variations in terms of potential differences from the center (Figure 6).

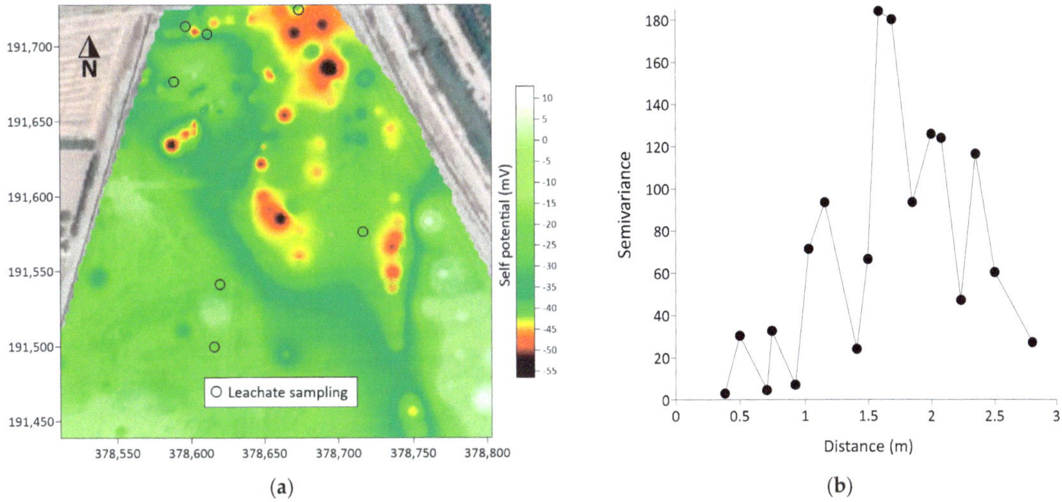

Figure 5. (**a**) Distribution of fermentative hotspots mainly in the northern part of the landfill and leachate sampling; (**b**) variogram of self-potential measurements around hotspots (average over 8 hotspots).

A mean shape can be established from the 8 cases studied (Figure 7), confirming that the hotspots are nearly circular without any preferential orientation. This shape reflects the projection of a sphere onto the surface of the stacked deposits. The fermentative mass is thus punctual or spherical, without elongation in any preferred direction.

3.3. Leachates Characteristics

The leachates were all highly mineralized with an average electrical conductivity of 45 mS cm^{-1}. There was a noted disparity between the sum of major ions and electrical conductivity, which is typically around 1 mmol$_c$ per 50 µS cm^{-1}, for waters in natural environments, whereas, in the case of leachates, it averaged 1 mmol$_c$ per 200 µS cm^{-1}. The pH was generally neutral, ranging from 6.02 (surface black leachate) to 8.35 (white leachate). The calculated partial pressure of O_2 ranged from 10^{-63} to 10^{-50} atm. The levels of organic carbon, major ions, and trace elements were very high (Table S1). PCA conducted on all parameters revealed two major processes: mineralization (Principal Component F1) and redox processes (Principal Component F2, Table 1). The coefficients of the parameters electrical conductivity (EC) and redox potential (Eh) showed a negative correlation, with the most concentrated leachates being generally the least reducing.

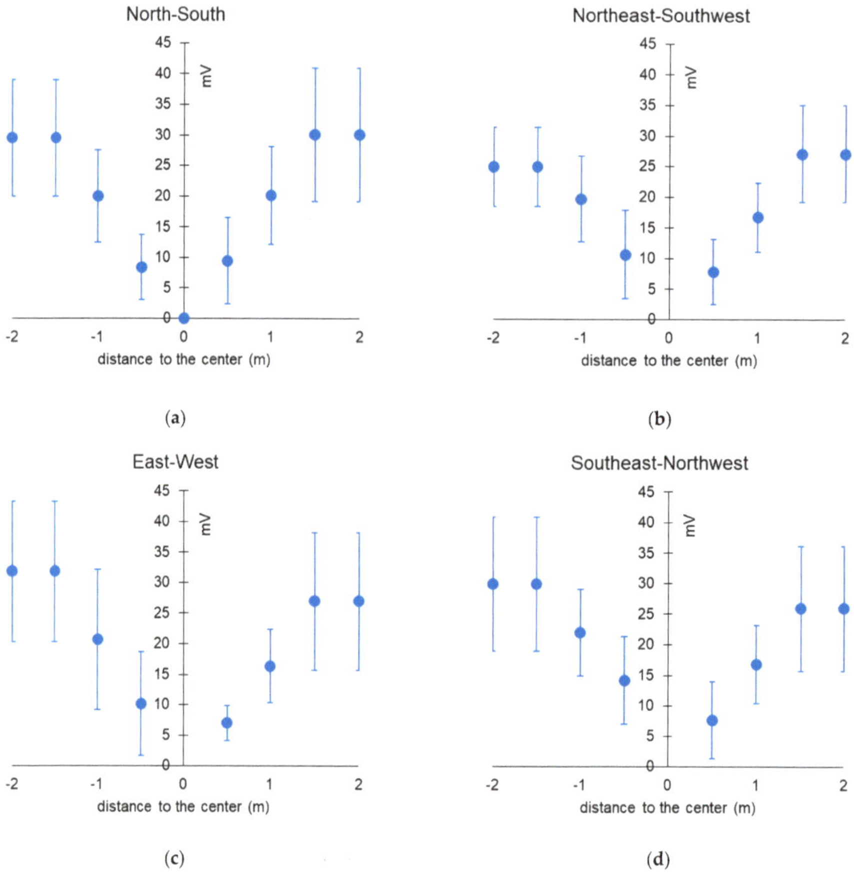

Figure 6. Averages and standard deviations of PS variations for hot spots in the (**a**) north-south, (**b**) north-east-south-west, (**c**) east-west, and (**d**) south-east-north-west directions around the center.

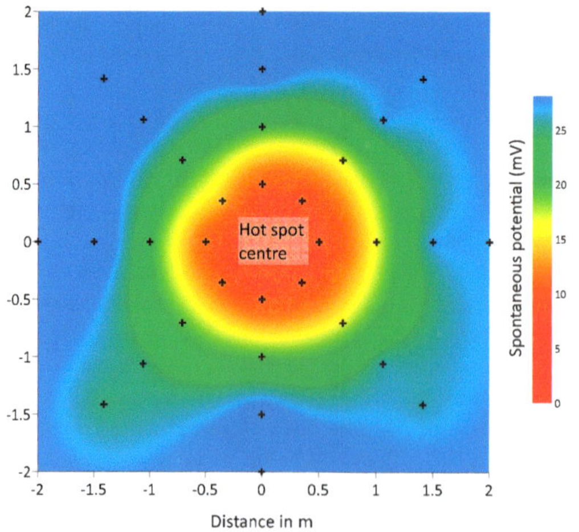

Figure 7. Size and shape based on an average of 8 hotspots.

Table 1. Coordinates of the various parameters on the first two axes of PCA. Note the coordinates of EC on axis F1 and of Eh on axis F2 (values in bold).

Parameter	F1	F2	Parameter	F1	F2	Parameter	F1	F2
Li	1.00	0.01	Zn	1.00	0.06	F^-	1.00	−0.03
B	1.00	−0.01	As	0.94	0.33	Cl^-	0.99	0.16
Na	1.00	0.05	Rb	1.00	0.04	NO_2^-	0.99	−0.14
Mg	1.00	−0.06	Sr	0.99	−0.10	Br^-	0.96	0.27
Al	0.53	0.85	Y	0.99	0.11	NO_3^-	0.85	0.53
S	1.00	−0.07	Mo	−0.10	−0.99	SO_4^{2-}	1.00	−0.09
K	1.00	0.05	Cd	0.81	−0.58	Na^+	0.99	0.15
Ca	0.99	−0.10	Sb	1.00	−0.02	NH_4^+	1.00	0.05
Ti	−0.02	1.00	Cs	1.00	0.06	K^+	0.99	0.13
V	0.99	0.16	Ba	1.00	−0.05	Mg^{2+}	1.00	−0.07
Cr	0.52	0.85	Tl	−0.99	0.17	Ca^{2+}	0.99	−0.16
Mn	0.99	−0.16	Pb	0.95	0.32	SiO_2	−0.04	−1.00
Fe	0.99	−0.16	U	0.89	−0.46	pH	−0.61	0.79
Co	1.00	0.05	EC	0.90	0.43	T °C	0.85	−0.52
Ni	1.00	0.05	HCO_3^-	1.00	−0.04	Eh	0.32	−0.95
Cu	1.00	0.10	TOC	0.85	−0.52			

All samples exhibited highly reducing characteristics. Yet, the observed classification of leachates seems to align with varying levels of reduction, as red or white leachates exhibit lower reducing values compared to black leachates. The latter, which are the most common in the field, exhibited marked stratification of Eh-pH parameters within the ponds. Pourbaix diagrams (Eh-pH) were plotted for nitrogen, iron, sulfur, and carbon (Figure 8).

Deep leachates were close to the domain of methane production, consistent with the observed spontaneous combustion fumaroles on-site. All samples were in the domain of ammoniacal nitrogen, explaining the presence of NH_4^+ in the laboratory-analyzed samples, unlike NO_3^-, which appeared only exceptionally, probably due to slight nitrification during the transportation of some samples. All samples were also in the domain of soluble ferrous iron, explaining the abundance of Fe observed in the leachate analysis. The leachates were close to the iso-activity limit of sulfate-sulfide, a theoretical equilibrium difficult to achieve because sulfides are insoluble and precipitate as they form. This transformation from sulfate to sulfide always occurs above the theoretical line of activity equality, as observed for deep black leachates.

Regarding the isotopic signature of the leachates, two out of the three analyzed samples were close to the local meteoric water line [28], while the third deviated significantly, indicating an evaporation process (Figure 9). This sample also showed partial oxidation on the Pourbaix diagrams.

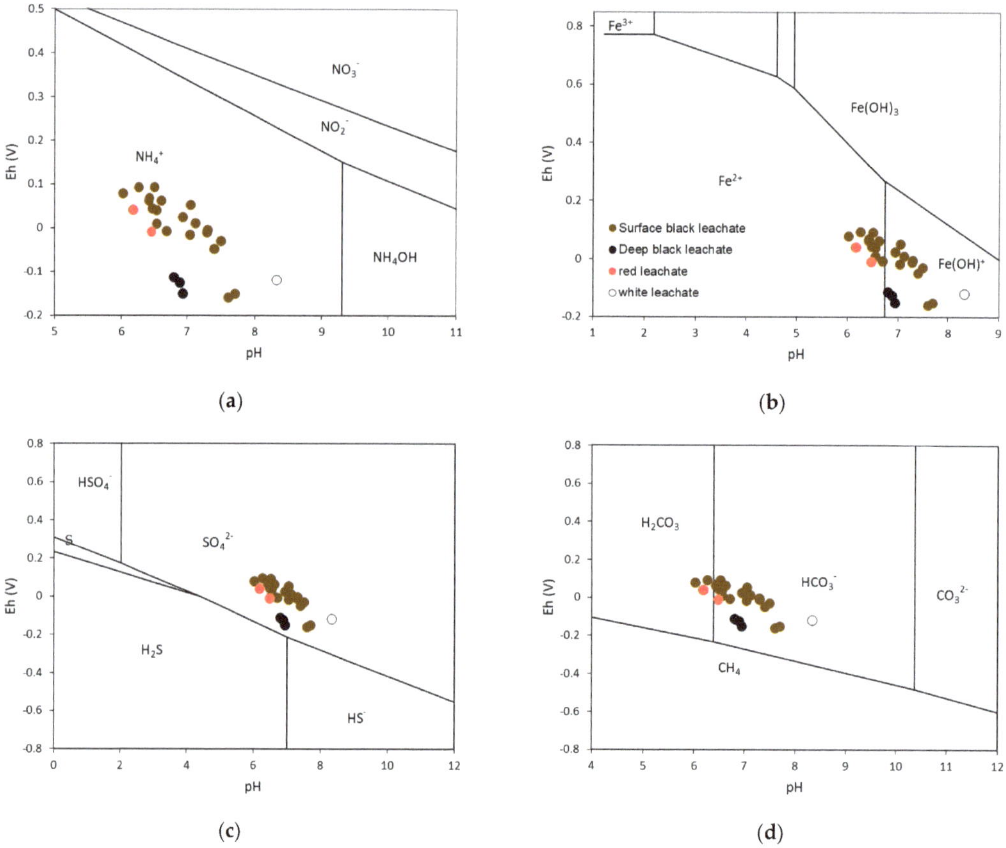

Figure 8. Pourbaix diagram showing the species stability domains of leachates for (**a**) N, (**b**) Fe, (**c**) S, and (**d**) C.

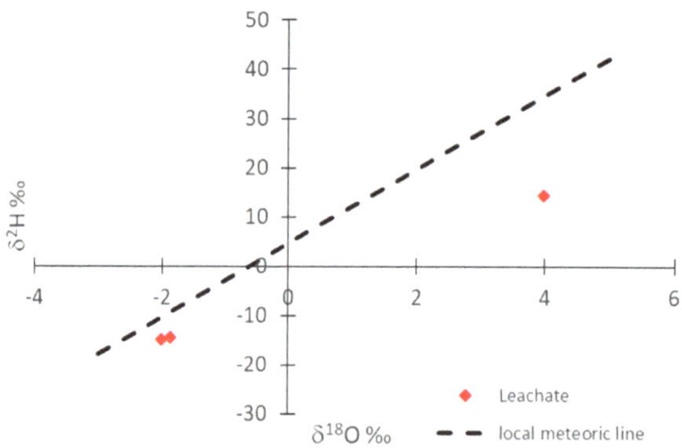

Figure 9. Stable isotopic signature of the leachate and its location with respect to the local meteoric line [28].

4. Discussion

4.1. Intermittent and Fleeting Fermentation Processes

The previous study conducted on the detection and characterization of pollution plumes originating from the Soub Sekt landfill highlighted relatively moderate contamination, mainly due to dilution by leaks from the irrigation network, but persistent throughout the year [22]. However, leachates are produced at small-scale fermentation hotspots, primarily within recent waste. The presence of fermentation hotspots, meaning the very localized nature of fermentation points dispersed within the waste, has been mentioned by some authors [29] but has not been the subject of specific studies, apart from methane gas production and emission in landfills [30,31]. Because of the discreet and non-general nature of these fermentation points, each one is likely to exhibit different characteristics. This is reflected notably in distinct colors in the produced leachate. As a result, the few samples collected probably do not represent all the processes occurring within the landfill. The diversity of leachate is a direct consequence of the localized nature of fermentation. The operational duration of fermentation hotspots is limited to a few weeks, but the continuous deposition of newer waste sustains continuous production at the landfill scale. Leachate pools that form at the base of the waste windrows, typically ranging from 10 to 100 m^2 and sometimes reaching 1 m in depth, serve as intermediate storage for leachates. The low permeability of soils limits leachate flow towards the water table, all contributing to limited but continuous contamination towards the water table. The isotopic signature of leachates [32] shows that rainwaters are the source of leachate percolation but also that they undergo an evaporation process after reaching the soil surface at the base of the waste windrows.

4.2. Intense Fermentation Processes

Leachate samples exhibit high levels of nearly all elements and very high electrical conductivity. From this point of view, leachates are saline. The disparity between the content of inorganic major ions and EC values, the latter being approximately four times higher in relation to the corresponding sum of major ions, confirms that a significant proportion of the leachate ionic load is organic. These compounds have a strong complexing power and promote metal solubilization, which can be facilitated by attack from organic acids on metallic waste [33]. The highly reducing nature of the hotspots favors the transition of metals into their reduced form, which is generally more soluble (with a few exceptions). The very low redox potential values at the core of the hotspots reflect intense electron exchange activity, particularly during redox processes leading to metal solubilization, following the principle of a geobattery [12,18,29,34]. Once metals are solubilized, organo-metallic complexation protects them from subsequent precipitation by oxidation, which is itself delayed by slow oxygen diffusion and the preferential oxidation of easily biodegradable carboxylic acids [35]. Thus, oxidation through atmospheric O_2 diffusion to the liquid phase within the pools primarily oxidizes the most labile forms of organic matter before oxidizing other components. The high organic content acts as a sort of redox buffer, maintaining highly reducing conditions in the leachates over the long term, even after migration from hotspots to pools.

5. Conclusions

Our study confirms the presence of fermentation hotspots in the landfill under investigation. These hotspots, small in size and rich in fermentable organic matter, are the site of intense fermentation processes, accompanied by electron exchanges that allow their detection through self-potential measurements. Low potential (SP) values are detectable on the surface of the waste within a maximum radius of 1.5 m from the center of the hotspot. Fermentation processes liquefy organic materials and fractionate organic molecules into simpler and more soluble organic acids. These organic acids massively solubilize metals, especially those with affinities for organic matter (such as copper, lead, etc.). The solubilization of compounds occurs through reduction and organometallic complexation. Rare rains promote the migration of leachates from hotspots to the soil surface at the base of the waste windrows, diluting the fermentation products somewhat. These rains imprint their isotopic signature on the leachates. Once accumulated in the pools, leachates are subjected to two processes: evaporation due to the aridity of the climate and slow oxidation by oxygen diffusion from the atmosphere to the leachate. This latter process is responsible for vertical variability in redox conditions, with the deepest parts of the leachate pools being more reducing than the interface with the atmosphere. The slow oxidation of leachates by atmospheric O_2, due to the high stock of organic carbon inducing a "redox buffer" effect, maintains the leachate pools under highly reducing conditions. Evaporation leads to an increase in the concentrations of various components. The results obtained from the Souk Sebt landfill cannot be fully extrapolated to other landfills. The low permeability of the soil, compacted by trucks depositing waste, leads to soil sealing and the accumulation of leachates in large pools, which may not necessarily be observed in all landfills. This situation has allowed for the study of leachates and their evolution as they accumulate in pools, a context that, as mentioned in previous work, results in the presence of a permanent pollution plume oriented northwest, in accordance with the regional flow of the aquifer.

Supplementary Materials: The following supporting information can be downloaded at: https://www.mdpi.com/article/10.3390/w16060795/s1, Table S1: Leachates chemistry (ICPMS and Ion chromatography) and physico-chemical field measurements.

Author Contributions: Conceptualization, V.V. and M.M.; methodology, V.V., Y.E.M. and M.M.; software, A.E.H., A.B., Y.E.J. and M.T.; validation, Y.E.M. and A.E.H.; formal analysis, Y.E.M. and A.E.H.; investigation, Y.E.M. and A.E.H.; resources, V.V., M.M. and L.B.; data curation, V.V. and L.B.; writing—original draft preparation, V.V., M.T. and Y.E.M.; writing—review and editing, L.B.; visualization, V.V. and L.B.; supervision, V.V. and M.M.; project administration, M.M. and V.V.; funding acquisition, L.B. and M.M. All authors have read and agreed to the published version of the manuscript.

Funding: The first two authors benefited from a national doctoral scholarship from the University of Tetouan and support for international mobility of doctoral students from the University of Avignon (Perdiguier Program). This research did not receive any other funding.

Data Availability Statement: The data presented in this study are available in Supplementary Materials.

Acknowledgments: The authors thank Avignon and Abdelmalek Essaadi universities and the Dean of the National Institute of Agricultural Research. The authors are also grateful to the staff of the Hydrogeology Laboratory of Avignon for their help and support.

Conflicts of Interest: The authors declare no conflicts of interest. The funders had no role in the design of the study; in the collection, analyses, or interpretation of data; in the writing of the manuscript; or in the decision to publish the results.

References

1. Belevi, H.; Baccini, P. Long-term behavior of municipal solid waste landfills. *Waste Manag. Res.* **1989**, *7*, 43–56. [CrossRef]
2. Han, Z.; Ma, H.; Shi, G.; He, L.; Wei, L.; Shi, Q. A review of groundwater contamination near municipal solid waste landfill sites in China. *Sci. Total Environ.* **2016**, *569–570*, 1255–1264. [CrossRef] [PubMed]
3. Naudet, V.; Revil, A.; Rizzo, E.; Bottero, J.-Y.; Bégassat, P. Groundwater redox conditions and conductivity in a contaminant plume from geoelectrical investigations. *Hydrol. Earth Syst. Sci.* **2004**, *8*, 8–22. [CrossRef]
4. Jhamnani, B.; Singh, S.K. Groundwater Contamination due to Bhalaswa Landfill Site in New Delhi. *Int. J. Civ. Environ. Eng.* **2009**, *1*, 121–125.
5. Nazli, Y.; Hanson, J.L.; Wei-Lien, L. Heat Generation in Municipal Solid Waste Landfills. *J. Geotech. Geoenvironmental Eng.* **2005**, *131*, 1330–1344.
6. Hanson, J.L.; Yeşiller, N.; Oettle, N.K. Spatial and Temporal Temperature Distributions in Municipal Solid Waste Landfills. *J. Environ. Eng.* **2010**, *136*, 804–814. [CrossRef]
7. Martínez-Pagán, P.; Jardani, A.; Revil, A.; Haas, A. Self-potential monitoring of a salt plume. *Geophysics* **2010**, *75*, WA17–WA25. [CrossRef]
8. Revil, A.; Naudet, V.; Nouzaret, J.; Pessel, M. Principles of electrography applied to self-potential electrokinetic sources and hydrogeological applications. *Water Resour. Res.* **2003**, *39*, 1–15. [CrossRef]

9. Revil, A.; Titov, K.; Doussan, C.; Lapenna, V. Applications of The Self-Potential Method to Hydrological Problems. In *Proceedings of the Applied Hydrogeophysics*; Vereecken, H., Binley, A., Cassiani, G., Revil, A., Titov, K., Eds.; Springer: Dordrecht, The Netherlands, 2006; pp. 255–292.
10. Arisalwadi, M. Rahmania Mapping leachate distribution based on the self-potential method in Manggar Landfill, Balikpapan Indonesia. *J. Phys. Conf. Ser.* **2021**, *1763*, 12013. [CrossRef]
11. Stanly, R.; Yasala, S.; Oliver, D.H.; Nair, N.C.; Emperumal, K.; Subash, A. Hydrochemical appraisal of groundwater quality for drinking and irrigation: A case study in parts of southwest coast of Tamil Nadu, India. *Appl. Water Sci.* **2021**, *11*, 53. [CrossRef]
12. Arora, T.; Linde, N.; Revil, A.; Castermant, J. Non-intrusive characterization of the redox potential of landfill leachate plumes from self-potential data. *J. Contam. Hydrol.* **2007**, *92*, 274–292. [CrossRef]
13. Rani, P.; Piegari, E.; Di Maio, R.; Vitagliano, E.; Soupios, P.; Milano, L. Monitoring time evolution of self-potential anomaly sources by a new global optimization approach. Application to organic contaminant transport. *J. Hydrol.* **2019**, *575*, 955–964. [CrossRef]
14. Gallas, J.D.F.; Taioli, F.; Malagutti Filho, W. Induced polarization, resistivity, and self-potential: A case history of contamination evaluation due to landfill leakage. *Environ. Earth Sci.* **2011**, *63*, 251–261. [CrossRef]
15. Naudet, V.; Revil, A.; Bottero, J.-Y.; Bégassat, P. Relationship between self-potential (SP) signals and redox conditions in contaminated groundwater. *Geophys. Res. Lett.* **2003**, *30*, 2091. [CrossRef]
16. Naudet, V.; Gourry, J.C.; Girard, F.; Mathieu, F.; Saada, A. 3D electrical resistivity tomography to locate DNAPL contamination around a housing estate. *Near Surf. Geophys.* **2014**, *12*, 351–360. [CrossRef]
17. Cui, Y.A.; Liu, L.; Zhu, X. Unscented Kalman filter assimilation of time-lapse self-potential data for monitoring solute transport. *J. Geophys. Eng.* **2017**, *14*, 920–929. [CrossRef]
18. Revil, A.; Mendonça, C.A.; Atekwana, E.A.; Kulessa, B.; Hubbard, S.S.; Bohlen, K.J. Understanding biogeobatteries: Where geophysics meets microbiology. *J. Geophys. Res. Biogeosciences* **2010**, *115*, G1. [CrossRef]
19. Artíñano, B.; Gómez-Moreno, F.J.; Díaz, E.; Amato, F.; Pandolfi, M.; Alonso-Blanco, E.; Coz, E.; García-Alonso, S.; Becerril-Valle, M.; Querol, X.; et al. Outdoor and indoor particle characterization from a large and uncontrolled combustion of a tire landfill. *Sci. Total Environ.* **2017**, *593–594*, 543–551. [CrossRef]
20. Barry, A.A.; Yameogo, S.; Ayach, M.; Jabrane, M.; Tiouiouine, A.; Nakolendousse, S.; Lazar, H.; Filki, A.; Touzani, M.; Mohsine, I. Mapping Contaminant Plume at a Landfill in a Crystalline Basement Terrain in Ouagadougou, Burkina Faso, Using Self-Potential Geophysical Technique. *Water* **2021**, *13*, 1212. [CrossRef]
21. Touzani, M.; Kacimi, I.; Kassou, N.; Morarech, M.; Bahaj, T.; Valles, V.; Barbiero, L.; Yameogo, S. El impacto del vertedero de Oum Azza sobre la calidad de las aguas subterráneas de Rabat (Marruecos). *Cuad. Geográficos* **2019**, *58*, 68–82. [CrossRef]
22. El Mouine, Y.; El Hamdi, A.; Morarech, M.; Kacimi, I.; Touzani, M.; Mohsine, I.; Tiouiouine, A.; Ouardi, J.; Zouahri, A.; Yachou, H.; et al. Landfill Pollution Plume Survey in the Moroccan Tadla Using Spontaneous Potential. *Water* **2021**, *13*, 910. [CrossRef]
23. Touzani, M.; Mohsine, I.; Ouardi, J.; Kacimi, I.; Morarech, M.; El Bahajji, M.H.; Bouramtane, T.; Tiouiouine, A.; Yameogo, S.; El Mahrad, B. Mapping the Pollution Plume Using the Self-Potential Geophysical Method: Case of Oum Azza Landfill, Rabat, Morocco. *Water* **2021**, *13*, 961. [CrossRef]
24. Massoni, C.; Missante, G.; Beaudet, G.; Combes, M.; Etienne, H.P.; Ionesco, T. La plaine du Tadla. *Cah. Rech. Agron.* **1967**, *24*, 163–194.
25. ABHOER. *Etude D'actualisation du Plan Directeur D'aménagement Intègre des Ressources en eau (PDAIRE) de la zone D'action de L'agence du bassin Hydraulique de l'Oum Er Rbia: Données Génerales sur la Plaine du Tadla*; ABHOER: Rabat, Morocco, 2008.
26. Merzouki, H.; Hanine, H.; Lekhlif, B.; Latrache, L.; Mandi, L.; Sinan, M. Physicochemical Characterization of Leachate Discharge Fkih Ben Salah from Morocco. *J. Mater. Environ. Sci.* **2015**, *6*, 1354–1363.
27. Van Loon, G.W.; Duffy, S.J. *Environmental Chemistry: A Global Perspective*, 4th ed.; Oxford University Press: Oxford, UK, 2017.
28. El Ouali, A.; Roubil, A.; Lahrach, A.; Mudry, J.; El Ghali, T.; Qurtobi, M.; El Hafyani, M.; Alitane, A.; El Hmaidi, A.; Essahlaoui, A.; et al. Isotopic Characterization of Rainwater for the Development of a Local Meteoric Water Line in an Arid Climate: The Case of the Wadi Ziz Watershed (South-Eastern Morocco). *Water* **2022**, *14*, 779. [CrossRef]
29. Naudet, V.; Revil, A. A sandbox experiment to investigate bacteria-mediated redox processes on self-potential signals. *Geophys. Res. Lett.* **2005**, *32*, L11405. [CrossRef]
30. Gonzalez-Valencia, R.; Magana-Rodriguez, F.; Cristóbal, J.; Thalasso, F. Hotspot detection and spatial distribution of methane emissions from landfills by a surface probe method. *Waste Manag.* **2016**, *55*, 299–305. [CrossRef] [PubMed]
31. Karimi, N.; Ng, K.T.W.; Richter, A. Prediction of fugitive landfill gas hotspots using a random forest algorithm and Sentinel-2 data. *Sustain. Cities Soc.* **2021**, *73*, 103097. [CrossRef]
32. Wimmer, B.; Hrad, M.; Huber-Humer, M.; Watzinger, A.; Wyhlidal, S.; Reichenauer, T.G. Stable isotope signatures for characterising the biological stability of landfilled municipal solid waste. *Waste Manag.* **2013**, *33*, 2083–2090. [CrossRef]
33. Xie, S.; Ma, Y.; Strong, P.J.; Clarke, W.P. Fluctuation of dissolved heavy metal concentrations in the leachate from anaerobic digestion of municipal solid waste in commercial scale landfill bioreactors: The effect of pH and associated mechanisms. *J. Hazard. Mater.* **2015**, *299*, 577–583. [CrossRef]

34. Rittgers, J.B.; Revil, A.; Karaoulis, M.; Mooney, M.A.; Slater, L.D.; Atekwana, E.A. Self-potential signals generated by the corrosion of buried metallic objects with application to contaminant plumes. *Geophysics* **2013**, *78*, EN65–EN82. [CrossRef]
35. Ilgen, G.; Glindemann, D.; Herrmann, R.; Hertel, F.; Huang, J.-H. Organometals of tin, lead and mercury compounds in landfill gases and leachates from Bavaria, Germany. *Waste Manag.* **2008**, *28*, 1518–1527. [CrossRef] [PubMed]

water

MDPI

Article

An Integrated Approach for Saturation Modeling Using Hydraulic Flow Units: Examples from the Upper Messinian Reservoir

Nader H. El-Gendy [1], Walid M. Mabrouk [2], Mohamed A. Waziry [3], Thomas J. Dodd [4], Fathy A. Abdalla [5], Dimitrios E. Alexakis [6,*] and Moataz Khairy Barakat [1]

1 Geology Department, Faculty of Science, Tanta University, Tanta 31527, Egypt; naderelgendy52@outlook.com (N.H.E.-G.); moatazbarakat@science.tanta.edu.eg (M.K.B.)
2 Department of Geophysics, Faculty of Science, Cairo University, Giza 12613, Egypt; wmabrouk@sci.cu.edu.eg
3 Dana Gas Company, Cairo 11774, Egypt; wazery_m@yahoo.com
4 British Geological Survey, Lyell Center, Research Avenue South, Edinburgh EH21 6ER, UK; tdodd@bgs.ac.uk
5 College of Science, King Saud University, Riyadh 11451, Saudi Arabia; fabdalla@ksu.edu.sa
6 Laboratory of Geoenvironmental Science and Environmental Quality Assurance, Department of Civil Engineering, School of Engineering, University of West Attica, 250 Thivon & P. Ralli Str., GR 12241 Athens, Greece
* Correspondence: d.alexakis@uniwa.gr

Abstract: The Upper Messinian reservoirs located in the Salma Field of the Nile Delta area contain variable facies. The key reservoir interval of the Abu Madi Formation was deposited in fluvial to deltaic environments. These fine-grained facies form significant reservoir heterogeneity within the reservoir intervals. The main challenges in this study are reservoir characterizing and predicting the change in reservoir water saturation (SW) with time, while reservoir production life based on the change in reservoir capillary pressure (Pc). This work applies petrophysical analysis to enable the definition and calculation of the hydrocarbon reserves within the key reservoir units. Mapping of SW away from the wellbores within geo-models represents a significant challenge. The rock types and flow unit analysis indicate that the reservoir is dominated by four hydraulic flow units. HFU#1 represents the highest flow zone indicator (FZI) value. Core analysis has been completed to better understand the relationship between SW and the reservoir capillary pressure above the fluid contact and free water level (FWL), which is used to perform saturation height function (SHF) analysis. The calculated SW values that are obtained from logs are affected by formation water resistivity (Rw) and log true resistivity (RT), which are influenced by the volume of clay content and mud salinity. This study introduces an integrated approach, including evaluation of core measurements, well log analysis covering cored and non-cored intervals, neural analysis techniques (K-mode algorithm), and permeability prediction in non-cored intervals. The empirical formula was predicted for direct calculation of dynamic SW profiles and predicted within the reservoir above the FWL based on the change in reservoir pressure.

Keywords: flow unit; saturation height; J-function; Messinian reservoir; Nile Delta

1. Introduction

The reservoir evaluation is necessary to identify reservoir units and to better understand their relevant reservoir properties [1]. Calculating porosity, permeability, and the study of dynamic flow is necessary to obtain more accurate estimates of reservoir storage volumes and gain an improved understanding of flow performance. Characterizing water saturation (SW) within reservoirs is a key challenge of the hydrocarbon reserves estimation, which strongly influences the creation of static and dynamic reservoir models [2–6].

Reservoir classification techniques that include combined core analysis and well log data can also be used to characterize reservoir flow unit parameters [7–9]. Advanced rock typing techniques, which are based on flow zone indicators (FZI), can be used to calculate flow unit identification and characterization [10–12]. Reservoir parameters such as porosity and permeability have been used to define reservoir hydraulic flow units during the construction of reservoir models [13,14]. Additionally, applying artificial intelligence techniques can potentially deliver solutions for predicting petrophysical parameters in the non-core areas of the reservoir.

The Salma Field is a significant hydrocarbon province in the Nile Delta region [15–18]. Within the Salma Field, the clastic late Messinian Abu Madi Formation is considered as the primary gas-producing interval (Figure 1). Previous studies include comprehensive reservoir analysis, property modeling of petrophysical parameters of the reservoir, and identifying various facies of the flow [19–23]. Calculating water saturation variability within the model based on reservoir flow units can reduce the uncertainty in reservoir flow performance and volumetric evaluation in the Salma Field.

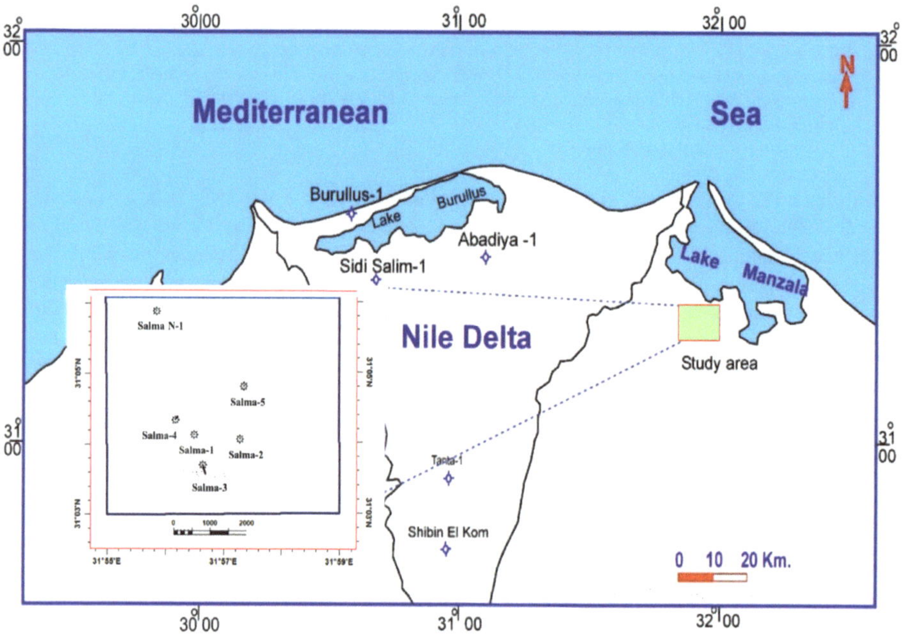

Figure 1. Location map of Salma field.

This study analyzes critical data from the Salma Field and provides an innovative method that links reservoir flow units with reservoir water saturation. In order to achieve this, this study (a) provides permeability calculations that are defined by a flow unit-dependent porosity–permeability correlation; (b) establishes multiple saturation height function (SHF) by using key reservoir parameters; and (c) considers the variation in rock quality in relation to the reservoir flow performance.

This method for calculating water saturation is successful and more precise in the example of the Salma Field, mainly as it decreases the uncertainty in the SW results calculated from resistivity tool readings, which are influenced by shale distribution and/or variation in reservoir water salinity. Fundamentally, this study and proposed methodology are more practical within fields worldwide, mainly when saturation modeling is conducted within heterogeneous reservoirs containing abundant clastic mudstone facies.

2. Geological Setting

This study examines a suite of Messinian-aged sedimentary strata from the Delta region of Egypt (Figure 2). The Messinian section, which comprises the Abu Madi and Qawasim formations, hosts the most potential reservoirs in the Delta area [24]. The Messenian section in the Nile Delta comprises a complexly layered incised valley filled with various facies forming the reservoir and sealing lithological units [24]. These sedimentary successions of the Abu Madi Formation have been interpreted as fluvial to the coastal marine in origin, deposited in a subsiding basin undergoing transgression. The Abu Madi Baltim trend in West El Manzala is defined as a sequence of back-stepping fluvial channels [16,25,26]. The basin follows the same strike as the overall trend of the adjacent gas-producing system of Abu Madi to the west [27].

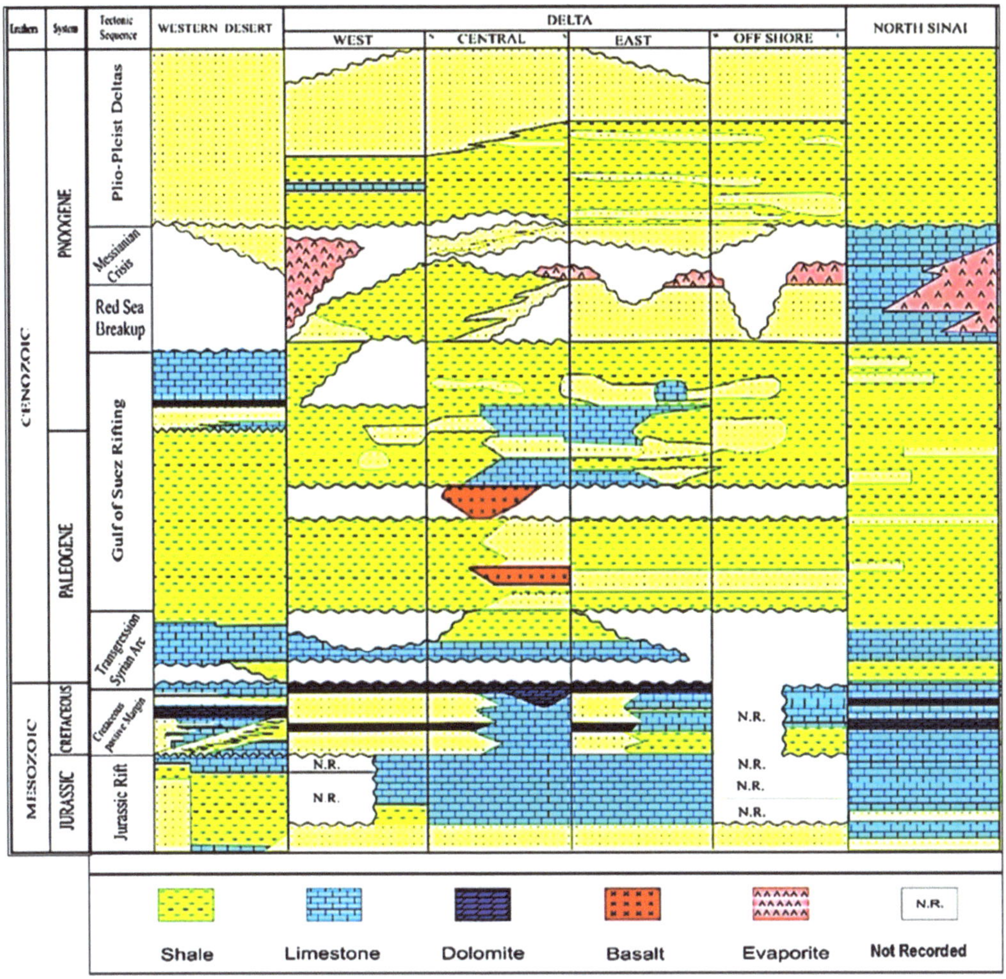

Figure 2. The stratigraphic chart of Egypt includes subsurface sediments and tectonic sequences [16].

The Abu Madi Formation is a fluvial–estuarine sedimentary rock unit with erosive-based channel sandstones at the base of the sequence. In the Salma Field, the Miocene cycle is characterized by the Messinian sandstones, mudstones (shales), and sandy mudstones (shales) of the Abu Madi Formation, which overlay the marine sedimentary rocks of the

Sidi Salim Formation. It varies in composition and comprises different lithologies, but mainly is made up of siltstones and sandstones with variable sedimentary structures. The formation ranges from fluvial to deltaic environments with various depositional conditions [28,29]. The lithofacies associations indicate subfeldspathic arenite-wacke, sub-lithic arenite, and lithic arenites distributed across the reservoir; they also indicate reservoir heterogeneity [30].

3. Materials and Methods

A complete set of well log data from six available wells are used in this study. Fifty-eight core plugs of the Salma-2 well and ninety core plugs of the Salma-4 well were analyzed at the Corex Laboratories in Egypt. Obtained measurements from the core plugs include porosity, grain density, and permeability. Additionally, special core analysis (SCAL) was performed on 32 core plugs of the Salma-2 well and 56 plugs of the Salma-4 well.

Pore throat analysis and capillary pressure measurements were performed on eight samples obtained by covered mercury injection by Corex, from which data was made available for the Salma-2 and Salma-4 wells. Porosity and permeability data were corrected to reservoir net overburden pressure to replicate in situ values [31]. The sedimentology of the core data was investigated (described and interpreted) (Figure 3), and the reservoir interval was divided and characterized based on sedimentary facies and depositional environments (Figure 4). Applying combined and integrated datasets during reservoir characterization provides a robust methodology for static and dynamic reservoir properties attribution and modeling [32–34].

3.1. Reservoir Hydraulic Flow Units (HFUs)

Reservoir facies were classified into rock units based on their dynamic behavior [35]. Facies are categorized and defined based on their reservoir quality index (RQI; [36]) and value of flow zone indicator (FZI) [10,37,38]. The RQI formula is based on the theory that a package of capillary tubes can represent a flow within a porous medium with an average radius. The Kozeny–Carman realistic porous media theory was modified by [39]. The technique to characterize the reservoir quality index (RQI) and determine the FZI was developed by [10] and can be expressed as follows:

$$K = \Phi^3/(1-\Phi)^2 \times FZI^2 \quad (1)$$

$$\sqrt{(K/\Phi)} = [\Phi/(1-\Phi)] \times FZI \quad (2)$$

By defining RQI in μm

$$RQI\ (\mu m) = 0.0314\sqrt{(K/\varnothing)} \quad (3)$$

$$FZI\ (\mu m) = RQI/\Phi z \quad (4)$$

$$\Phi z = \Phi/(1-\Phi) \quad (5)$$

where

Øz = the normalized porosity in a fraction, and
Φ = porosity in a fraction, K = permeability in milli Darcy (mD).

The hydraulic flow concept is used to divide a reservoir into distinct units with unique FZI values [40]. The hydraulic flow unit (HFU) is defined as a representative reservoir volume with consistent petrophysical and fluid properties [10].

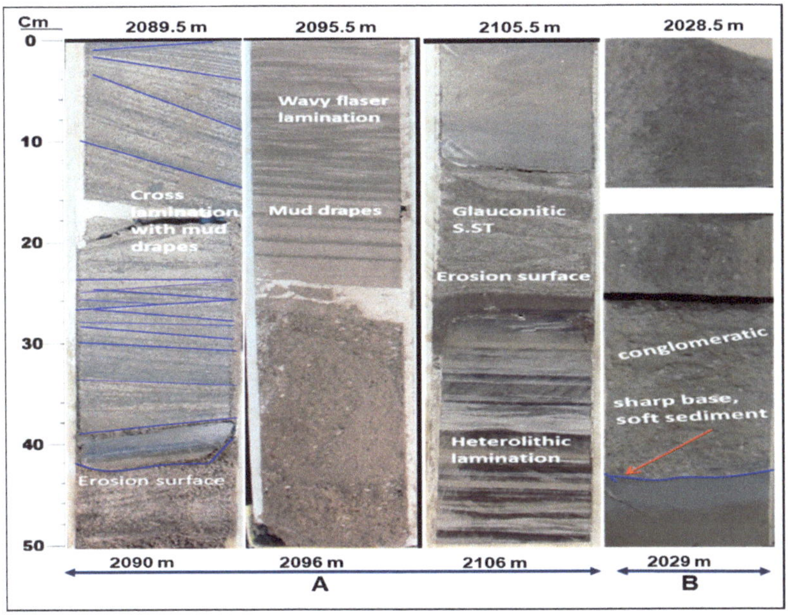

Figure 3. Core photos show Abu Madi Formation's sedimentary facies (**A**) Salma-4 and (**B**) Salma-2 wells.

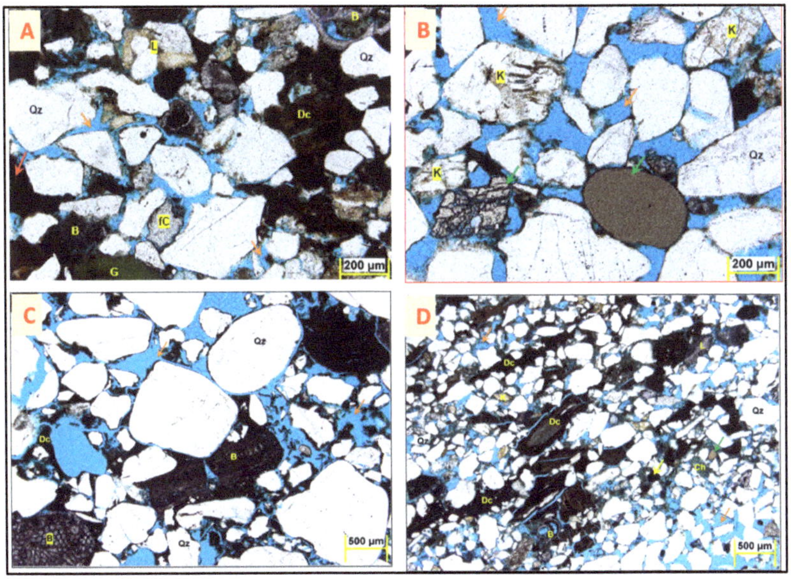

Figure 4. Thin section microphotographs illustrating different sandstones microfacies of Abu Madi Formation: (**A**) Feldspathic arenites and wacke stones, (**B**) Fine- to very fine-grained sandstones, (**C**) poorly cemented, conglomeratic kaolinitic pebbly sandstones, (**D**) subfeldspathic arenites and wacke stone. Mineral symbols: Anhydrite, An; Quartz, Qz; K-feldspars, K; Glauconite, G; Bioclasts, B; Plagioclase feldspars, Ps; Lithic fragments, L; Detrital clays, Dc; Porosity (Orange Arrows); Heavy Minerals (Green Arrows); Residual Hydrocarbons (Red Arrows).

3.2. Well Log Analysis

Petrophysical analysis (i.e., lithology, clay content, and reservoir fluid saturations) was carried out for six wells from the Salma Field. A quantitative reservoir evaluation requires accurately determining shale volume (VSH). Based on mudstone (shale) distribution, mud-rich sandstones possess different properties under different conditions and constraints [41]. Effective porosity (PHIE) was calculated using a neutron-density end point matrix cross-plot [42], which was corrected for VSH and gas effects. The lithology and grain density, which was determined from the core, was used in the evaluation. For reservoir water saturation, the 'Indonesian Model' was applied [43–45]. The calculations were corrected to clay content, as this can reduce resistivity and increase irreducible water saturation values [46].

3.3. Neural Log Analysis

An artificial neural network determined the FZI in non-cored intervals and wells. This was performed using FZIs that were calculated using core analysis and well log prediction [31]. Reservoir flow unit classification and identification were performed for all reservoir intervals, and geological models were produced using the Techlog^{TM} software (Version 2015). Statistical data for reservoir parameters was obtained from the petrophysical analysis of the well logs [47,48]. FZI measurements were obtained from the core and used to predict the FZI curve logistically within non-cored intervals.

3.4. Free Water Level (FWL) and Fluid Contacts

The FWL in a water-wet rock is defined as the lower contact level of fluid at which capillary pressure (CP) is zero [49]. A FWL was used as a reference in modeling the upper SWH functions, where capillary pressure exceeds zero and, therefore, water can be displaced by hydrocarbons. The FWL was defined by plotting formation pressure data against true vertical depth to define different fluid gradients. This is the point at which formation water pressure gradient lines intersect the hydrocarbon pressure gradient (Figure 5).

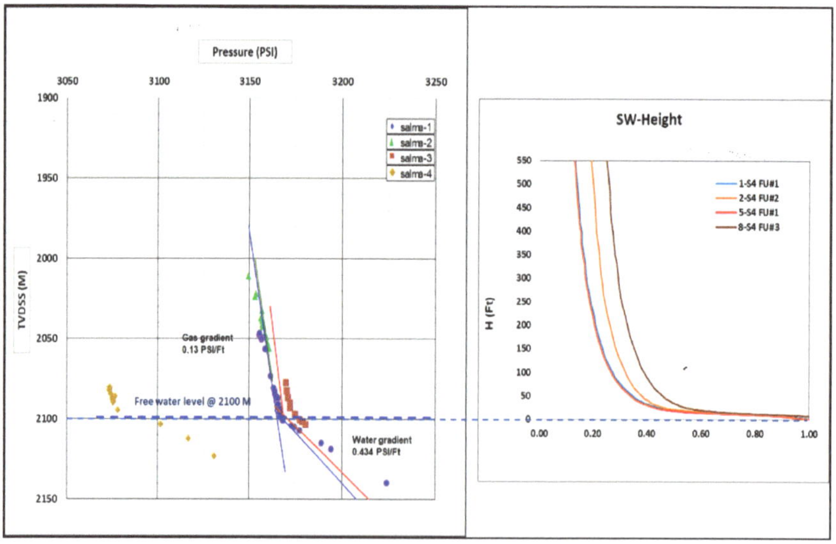

Figure 5. Fluid contact (gas/water contact) and free water level for Salma field.

The CP–SW curve can be predicted and converted into water saturation against height above the free water contact H–SW curve [50,51]. This point represents a fluid contact level, where the capillary pressure was more significant than zero, with the height above FWL. The various contacts' positions may differ from those of the FWL due to rock pore throat sizes, typically where tiny pore throat sizes formed fluid contacts that were marginally above the FWL.

3.5. Mercury Injection Capillary Pressure (MICP)

The mercury injection capillary pressure (MICP) technique was effectively used to determine pore throat size distribution [52]. This method uses mercury as a non-wetting liquid with solids. By applying a pressure up to 2000 psi, mercury can penetrate the pores spaces of the studied samples. The pore volume distribution was established as a function of pore throat radius. The interfacial forces were the source of the fluid rise to what is known as capillary pressure when rock pore volume is occupied with two immiscible fluids. The relationship between pore size and a given pressure was derived by [53], as

$$\mathrm{Pc} = \frac{2\sigma\cos\theta}{\mathrm{r}} \tag{6}$$

where

r = pore radius, σ = the interfacial tension, θ = contact angle,
Pc = capillary pressure (absolute applied pressure).

Capillary pressure represents the interaction of rock and fluid and is controlled by the pore size, interfacial tension, and wettability [54]. The free water level (FWL) from water saturation within the transition zone (the height relation; [49]) is also assumed, as follows:

$$\mathrm{Pc} = \frac{2\sigma\cos\theta}{\mathrm{r}} = (\rho\mathrm{w} - \rho\mathrm{o})\mathrm{gh} \tag{7}$$

$$\mathrm{h\ (ft.)} = \mathrm{Pc}/0.434\ (\rho\mathrm{b} - \rho\mathrm{c}) \tag{8}$$

where

Pc = capillary pressure (absolute applied pressure), σ = the interfacial tension,
ρ = the density of water and hydrocarbon (gas or oil), g = the gravitational acceleration,
h = the height above FWL, ρb = specific gravity of brine,
ρc = specific gravity of hydrocarbons, 0.434 psi/ft = gradient of water.

MICP and mercury saturation analysis were performed and plotted for samples assigned to different flow units for the Salma-2 (Figure 6A) and the Salma-4 wells (Figure 6B).

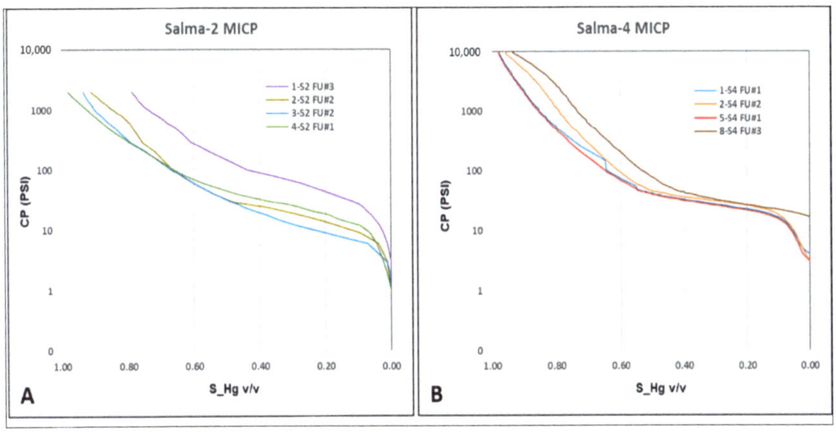

Figure 6. Mercury capillary pressure and saturation test (**A**) Salma-2 well, (**B**) Salma-4 well.

3.6. Reservoir Capillary Pressure and Saturation Height Function

Several studies on saturation height were developed and provided different methods to calculate SW through saturation height (SWH) modeling [51,54]. Different rock types are linked with different saturation height relationships [55]. A flow unit was defined for each sample by applying Equations (3)–(5), with laboratory-measured capillary pressure using mercury injection, where data was corrected from the laboratory fluid system to the reservoir fluid system using Equation (6) and data in Table 1 to convert Pc from lab to reservoir condition. Pc data for the reservoir fluid system were converted to a height above free water level (HAFWL) using Equation (8) and applying a field-free water level at 3100 m TVDSS (using mean sea level as the mean datum). Finally, samples from two wells were used to demonstrate the saturation height function. Data for SW-HAFWL were plotted for the Salma-2 well (Figure 7A) and for the Salma-4 well (Figure 7B).

Table 1. MICP interfacial tension values, contact angles, and descriptions used for laboratory and reservoir conditions.

Parameter	Contact Angles (°)	Parameter Description
σ Res	50	interfacial tension in the reservoir (gas–water)
θ Res	0	contact angle in the reservoir (gas–water)
σ Lab	70	interfacial tension in lab (air–water)
θ Lab	0	contact angle in lab (air–water)
σ Lab	485	interfacial tension in the lab (mercury–air)
θ Lab	140	contact angle in the lab (mercury–air)

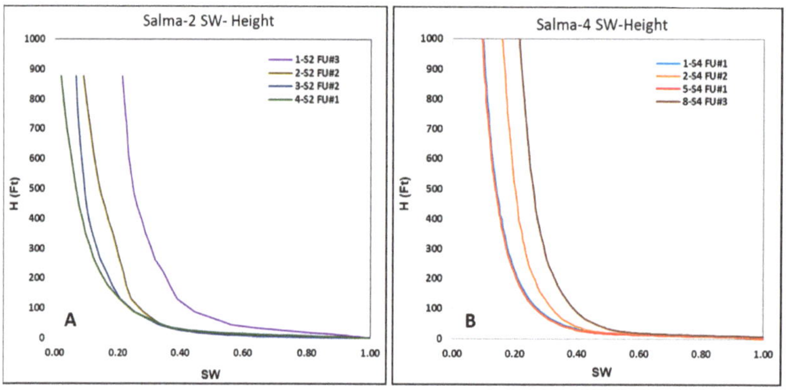

Figure 7. Water saturation (SW) and height above free water level (H) X-Y plot, (**A**) Salma-2 and (**B**) Salma-4.

3.7. Leverett J-Function

The capillary pressure measurements of the core samples represent a limited and small interval of the overall reservoir of the Salma Field. Therefore, additional capillary data must be collected and combined with the saturation curves to represent different reservoir facies and units to create a general equation that can define different reservoir units. Leverett (1941) [36] created a dimensionless capillary pressure–saturation function that they termed the "J-function". This can be used to develop a general equation that represents all typical capillary pressure curves and their dependent factors, including porosity, interfacial tension, and average pore radius. This can be expressed as follows:

$$J = 0.2166 \times Pc/(\sigma \times \cos\theta) \times \sqrt{(K/\Phi)} \quad (9)$$

where

J = Leverett J-function, Pc = capillary pressure, σ = the interfacial tension, θ = contact angle, Φ = porosity in a fraction, K = permeability in mile Darcie's (mD).

In this study, the Leverett J-function was used to convert all capillary-pressure data to a universal curve for the same formation and remove the variances in Pc–SW curves, whilst considering the variations in porosity and permeability for reservoir units. However, J-function SW correlations cannot obtain different formations with a single universal curve and are unable to represent all the reservoir units, so each flow unit should have its own independent J-function.

Finally, the Pc data was converted to a J-function using Equation (9) to normalize Pc within the reservoir system. This included a gas/water fluid system, gas gradient of 0.13 PSI/Ft, and a water gradient 0.434 PSI/ft, which were obtained from the results of pressure data evaluation. J–SW relationships were plotted for the Salma-2 well (Figure 8A) and the Salma-4 well (Figure 8B), which represent the variation related to different HFU.

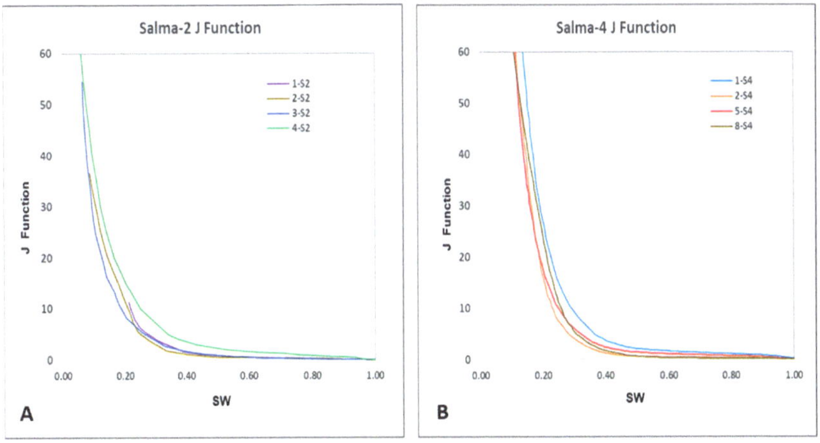

Figure 8. J-function and saturation plot, (**A**) Salma-2 well and (**B**) Salma-4 well.

4. Results

4.1. Facies Evaluation

Analysis of the core samples and integration with the well log data indicated that four key depositional environments represent the Abu Madi Formation facies and environment. These environments are defined on the basis of multiple sedimentary facies and/or facies associations that collectively represent that particular environment.

4.1.1. Flood Plain Environment

Description: Very fine-grained sandstone with symmetrical and asymmetrical ripple cross-lamination. Heterolithic lamination, tidal mud drapes, and reactivation surfaces are common (Figure 3A). These facies comprise poorly cemented, moderately compacted sub-feldspathic arenites and wacke stones (Figure 4A; 2105.7–2106 m). They contain abundant monocrystalline quartz grains and small amounts of K-feldspar and display moderate to good pore interconnectivity.

Interpretation: The heterolithic lamination, tidal mud drapes, and reactivation surfaces indicate the deposition from alternating high and low flow energies and/or changes in flow direction/regime [56]. The recognition of tidal influence on sedimentation indicates a transition from fluvial to marine conditions.

4.1.2. Tidally Influenced Fluvial Channel Environment

Description: Fine- to very fine-grained sandstones with wave ripple cross lamination, wavy bedding, flaser lamination, and abundant reactivation surfaces with mud drapes (Figure 3A), (2095.5–2096 m). These sediments are characterized by fining-upwards successions with erosive bases. These facies are composed of silt to poorly sorted and sub-angular to rounded granule grade sediments (Figure 4B). The sandstones are poorly cemented and moderately compacted, with common pore-filling and grain-coating detrital clays.

Interpretation: The presence of fining-upwards sequences with erosive bases suggests deposition by a sudden event that decelerated rapidly and was erosional at the front and/or base of the flow. The presence of wave ripples cross lamination, wavy bedding, falser lamination, and abundant reactivation surfaces suggest a tidal influence on sedimentation. This suggests that these sediments were deposited in a tidally influenced fluvial channel setting [57,58].

4.1.3. Fluvial Channel Environment

Description: This environment is composed of massively bedded conglomeratic kaolinitic pebbly sandstones with sharp bases that generally lack interbedded mudstones (Figure 3B). The sandstones are poorly cemented and moderately compacted, with common pore-filling and grain-coating detrital clays (Figure 4C). The sandstones are characterized by moderate to good pore interconnectivity. There are some minor instances where light brown parallel laminated mudstones (shales), without bioturbation or trace fossils, exist within the succession.

Interpretation: The massively bedded conglomerates may have been deposited by a high sediment load fluvial current [56,59]. The common scour surfaces and absence of mudstone interbeds between the channel-fill deposits indicate a stacked channel element formed from multiple channel incision and infill stages [60]. The instances where light brown horizontal laminates shales are present, and, in particular, the absence of bioturbation or trace fossils, potentially suggest a continental freshwater depositional setting.

4.1.4. Tidal Channel Environment

Description: The sandstones are very fine-grained and glauconitic and display ripple cross-lamination and trough cross-bedding (Figure 3A; 2089.5–2090 m). The sandstones display fining-upwards sequences and mud drapes along the sedimentary fore sets (for ripple cross-lamination and trough cross-bedding); the presence of mud drapes increases upwards within these intervals. The sandstones are moderately to poorly sorted, subrounded to sub-angular, poorly cemented, and poorly compacted subfeldspathic arenites and wacke stones (Figure 4D). These facies display good to moderate pore interconnectivity.

Interpretation: The presence and abundance of bioturbation, an upwards increase of mud drapes, and glauconite, indicate that these sediments were formed in a shallow water setting, with significant tidal influence [57,58]. The presence of trough-cross bedding and ripple cross-lamination suggests that deposition possibly occurred in a channelized setting, with varying flow rates. Overall, these deposits are deposited in a tidal channel environment, with glauconite suggesting the marginal marine nature of this facies.

4.2. Hydraulic Flow Units (HFU)

The FZI was calculated using RQI and normalized porosity, using the normal distribution of the FZI values and the cumulative curve of FZI. HFUs were defined where the change in the slope of the cumulative curve was interpreted as a change in flow unit bound by the inflexion point. Measured core data for Salma-2 and Salma-4 wells (Figure 9) show four main HFUs controlling reservoir performance in the Abu Madi Formation reservoir. The defined HFUs, reservoir facies (Figure 10), and associated data are summarized in Table 2.

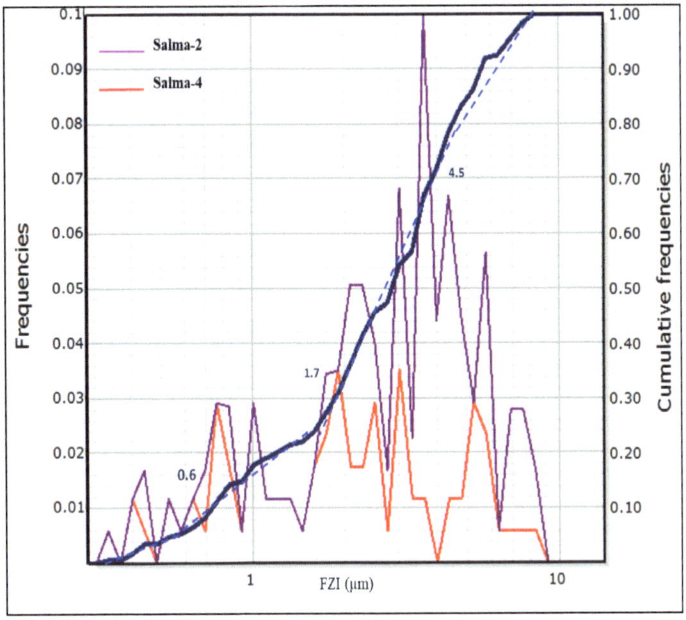

Figure 9. Core FZI distribution and cumulative curves for Abu Madi in Salma-2 and Salma-4 wells. (cumulative curve is bold blue line, slop line is dot blue line).

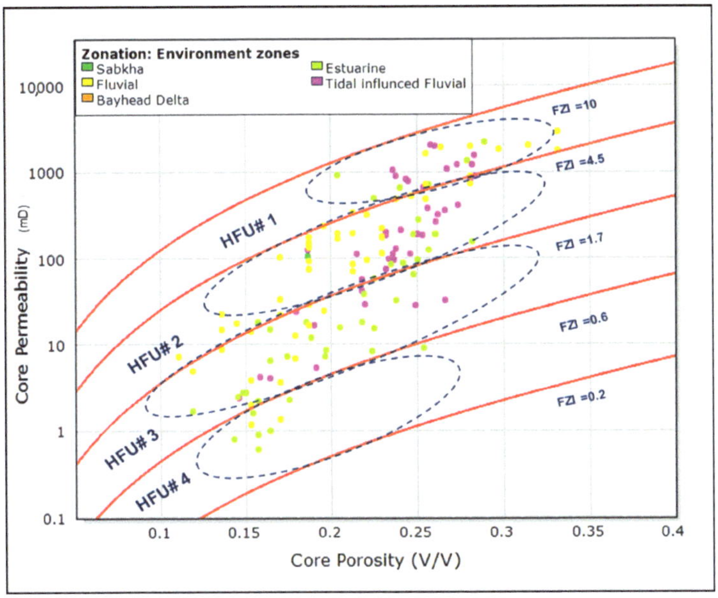

Figure 10. Porosity vs. horizontal permeability X-Y plot for (HFU) of Abu Madi Formation.

Table 2. HFU data for the Abu Madi Formation.

Hydraulic Flow Unit	FZI (μm)	Porosity (%)	Permeability (mD)	Reservoir Quality
HFU# 1	4.5 to 10	25–33	>900	Excellent
HFU# 2	1.7 to 4.5	17–33	70–1000	Good–Very Good
HFU# 3	0.6 to 1.7	12–33	4–100	Moderate–Good
HFU# 4	0.2 to 0.6	15–30	0.6–8	Low

HFU#1: The FZI average value (4.5 to 10 μm) indicates an excellent-quality sandstone reservoir with a porosity range of 25–33% and a permeability of >900 mD. It suggests the occurrence of fluvial channel and tidally influenced fluvial channel facies within the reservoir.

HFU#2: The FZI average value (1.7 to 4.5 μm) is a very good- to good-quality sandstone reservoir, with a porosity range of 17–33% and a permeability of 70–900 mD. This implies the presence of fluvial channel and tidally influenced fluvial channel facies within the reservoir.

HFU#3: The FZI average value (0.6 to 1.7 μm) indicates moderate-quality sandstone reservoirs, with a porosity range of 12–33% and a permeability of 4–100 mD. It is related to the occurrence of tidal channel and floodplain deposits.

HFU#4: The FZI average value (0.2 to 0.6 μm) represents a low-quality sandstone reservoir, with a porosity range of 15–30% and a permeability of 0.6–8 mD. This is likely an indicative of tidal channel and floodplain facies.

4.3. Formation Evaluation

Graphical and computational methods were used to determine the petrophysical properties of the Abu Madi Formation reservoir. Techlog^{TM} software (Version 2015) was used to identify various reservoir parameters, including shale volume, lithology, effective porosity, and water saturation. The raw data of neutron-density cross-plots for the Salma-2 and Salama-4 wells show that many of the data points lie on or close to the sandstone line (Figure 11).

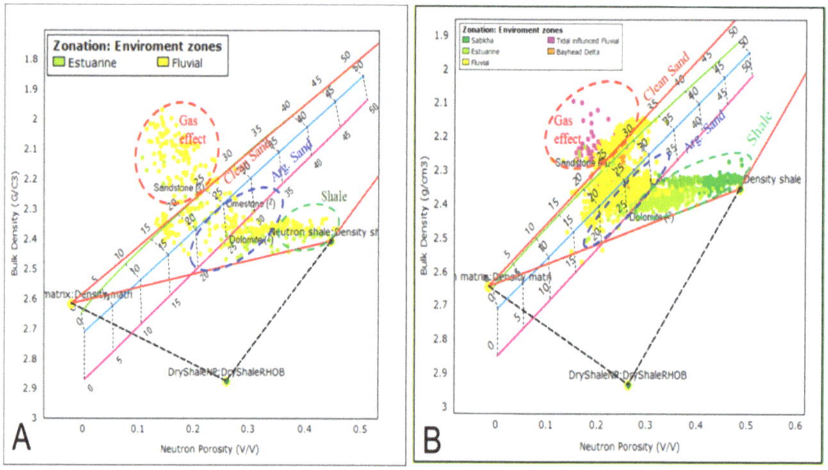

Figure 11. Neutron-density X-Y plot, (**A**) Salma-2 well, (**B**) Salma-4 well. (Red tringle lines represent the area of clay line to clean sand line).

Points plotted near the limestone lines suggest the presence of calcareous cements. Other points lie below the dolomite line as they are composed of 100% shale. Due to the gas effect, some neutron-density points plot away from the sandstone line [28,61,62].

The core data and well log analysis indicate that the reservoirs are composed mainly of sandstone and mudstone (shale) intercalations. Sandstone intervals are characterized by excellent reservoir quality in the Salma-2 well, which is composed of the coarser-grained sandstones of the fluvial channel (Figure 12A) that have an average porosity of 22%, low clay content (average shale volume of 18%), and water saturation of 30–42%. The Abu Madi Formation reservoirs in the Salma-4 well (Figure 12B) vary from argillaceous to clean sandstone intervals, which were deposited in estuarine, tidal, and fluvial environments. These display excellent reservoir parameters within the pay zone, with an average porosity of 24%, an average shale volume of 21%, and water saturations of 35–43%.

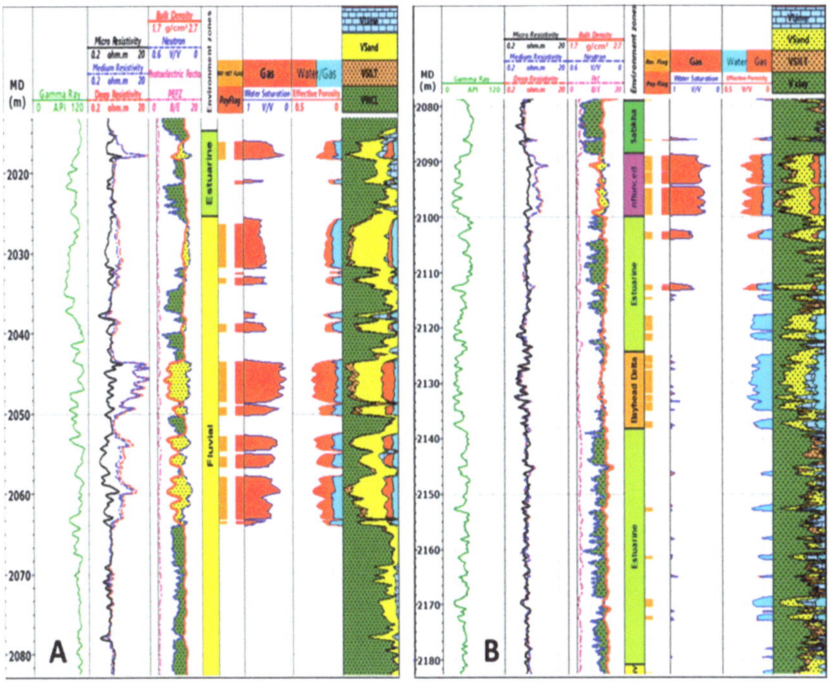

Figure 12. Composite logs showing porosity, water saturation, lithology, and depositional environment for Abu Madi Formation: (**A**) Salma-2 well, (**B**) Salma-4 well. (track-1: Depth, track-2: GR, track-3: resistivity, track 4: density-neutron curves, track-7: SW, track-8: Φ in a fraction).

4.4. Neural Log FZI and Permeability Prediction

The neural log application (K-mode) was originally a statistical technique, but its results are showing to be geologically consistent [28]. Neural log techniques were applied to data from 148 core samples as input data (PHIE and FZI from the data of core analysis), to extrapolate and predict FZI values and define HFUs in the non-cored intervals within the wells using log data (PHIE, Pef, and VSH). Using log and core data within the cored intervals, as well as applying the neural analysis method to FZI on the log, the values were then predicted within non-cored intervals [31]. The results of the FZI curve correlated with the curves produced from the cored intervals after a reasonable number of iterations were conducted (5 runs; each run included 100 iterations in the internal process), until it reached a minimum constant accepted error. Permeability (K) values were calculated in the non-cored intervals as a function of porosity (Φ) and FZI. Re-write Equation (2) in new format as follows [10]:

$$K = 1014\ \text{FZI}^2 \times \Phi^3/(1-\Phi)^2 \tag{10}$$

The same procedure was applied to the other wells (Salma-1, Salma-3, Salma-5, and Salma N-1), which only have well logs that predict FZI in the non-cored intervals (i.e., those 'other wells' are non-cored throughout). The result of predicted FZI on log bases and permeability calculations (Figures 13 and 14) are summarized in Table 3. Although the HFUs are defined by different ranges of FZI, each flow unit may display a wide range of porosity and permeability. Reservoir property model distribution within the entire reservoir is provided by the distributions of FZI values and is defined by the flow unit model. The flow units are used as the basis for the distribution of the porosity and permeability within the range of each unit.

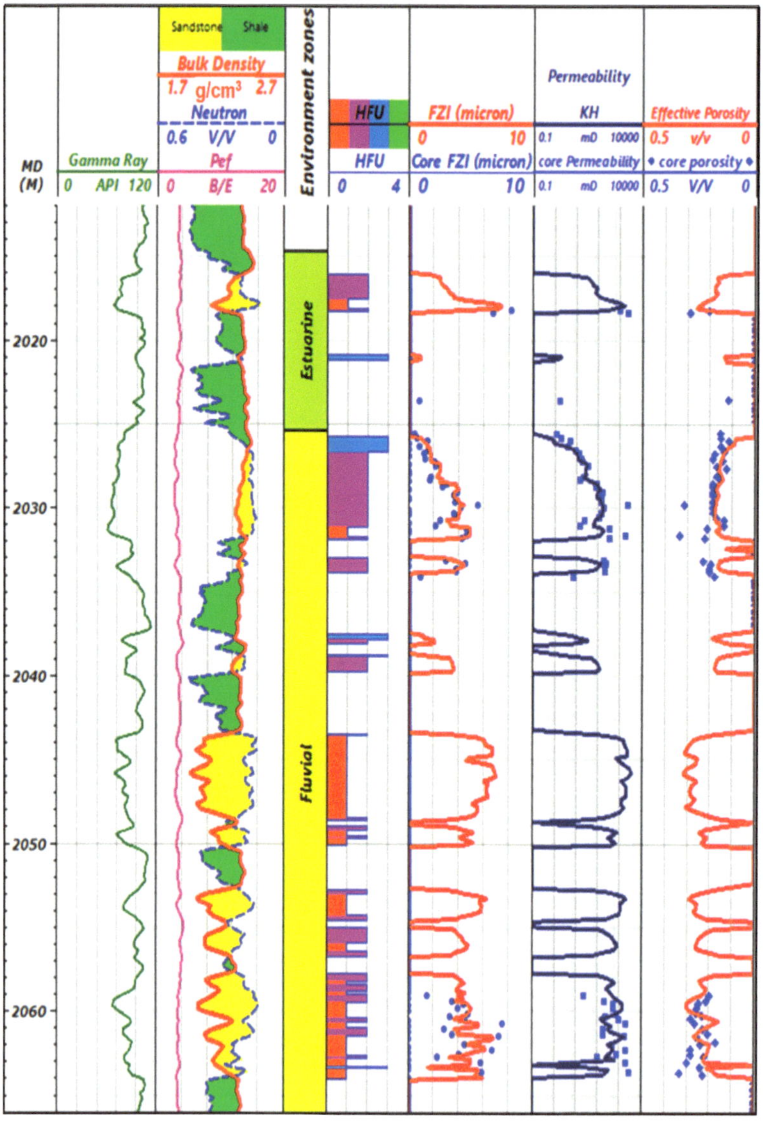

Figure 13. Composite log from neural log showing porosity, permeability, FZI, HFU, and depositional environment for Abu Madi Formation in Salma-2 well. (track-1: Depth, track-2: GR, track-3: density-neutron curves, track-5: HFU, track-6: FZI, track-7: KH, track-8: Φ in a fraction, dot for core data).

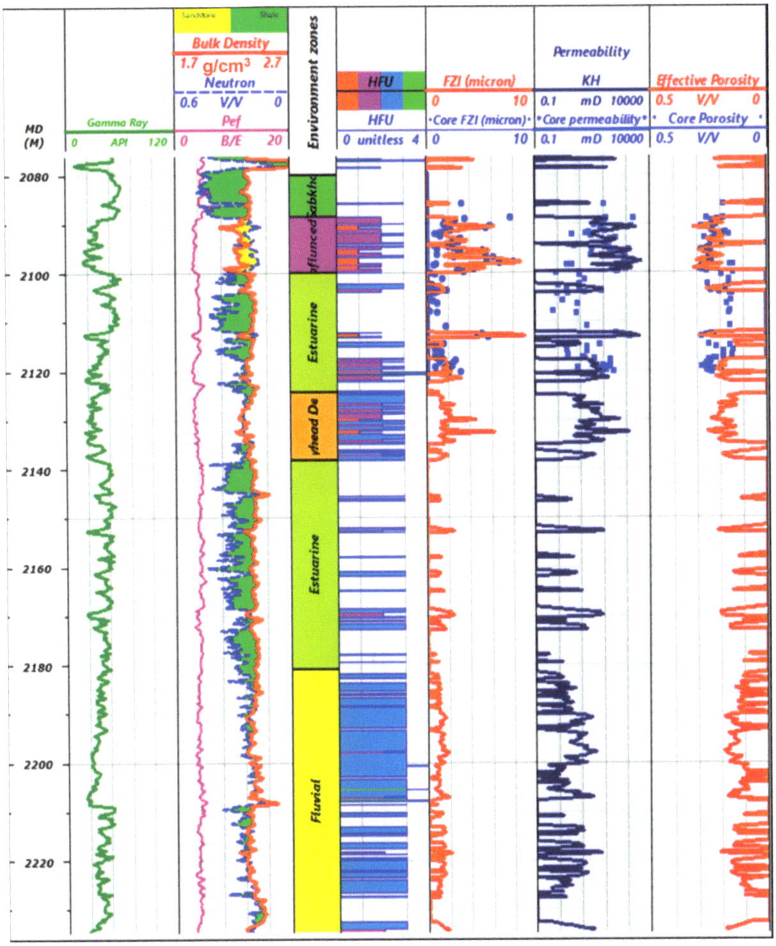

Figure 14. Composite log from neural log showing porosity, permeability, FZI, HFU, and depositional environment for Abu Madi Formation in Salma-4 well. (track-1: Depth, track-2: GR, track-3: density-neutron curves, track-5: HFU, track-6: FZI, track-7: KH, track-8: Φ in a fraction, dot for core data).

Table 3. Petrophysical and neural analysis of Abu Madi Formation.

Well	Zone	Top (m)	Bottom (m)	Gross (m)	Net (m)	Shale (%)	PHIE (%)	SW (%)	FZI (μm)	KH (mD)
Salma-2	Estuarine	2014	2025	11	2.3	24.5	18.9	35.9	3.9	357
	Fluvial	2025	2088	63	23.3	18.9	21.5	43.0	4.5	636
Salma-4	Tidally influenced Fluvial	2088	2100	12	9.8	21.4	23.9	38.9	3.7	677
	Estuarine	2100	2181	67	10.8	25.2	18.8	89.9	2.2	171
	Bayhead Delta	2124	2138	14	10.2	24.4	18.4	99.4	1.9	84
	Fluvial	2181	2291	110	64.9	21.0	19.7	99.8	3.5	684

4.5. Saturation Height Model

The saturation height function was defined on the flow unit's bases to predict water saturation at different reservoir points based on its position above the free water level.

The goal was to create a saturation model that was more relevant to the change in facies and reservoir parameters. Firstly, a single equation was created for each flow unit by registration, to represent the relationship between SW and height above FWL. Secondly, data were plotted, and the resulting curves were compared to the data for each flow unit (Figure 15). The red line represents the best linear fit to the data trend. Finally, this study developed an equation for a water saturation calculation with a direct relation for HAFWL without detection for capillary pressure and incorporated this for each flow unit as follows:

$$\mathrm{SW} = \mathrm{a} \times \mathrm{h}^{\mathrm{b}} \tag{11}$$

a = 1.4863, b = (−0.432) for HFU#1
a = 1.3103, b = (−0.391) for HFU#2
a = 1.2406, b = (−0.275) for HFU#3 and 4
SW = water saturation (v/v); h = height above free water level (ft).

On this basis, the water saturation (SW) was determined for all wells in the Salma Field.

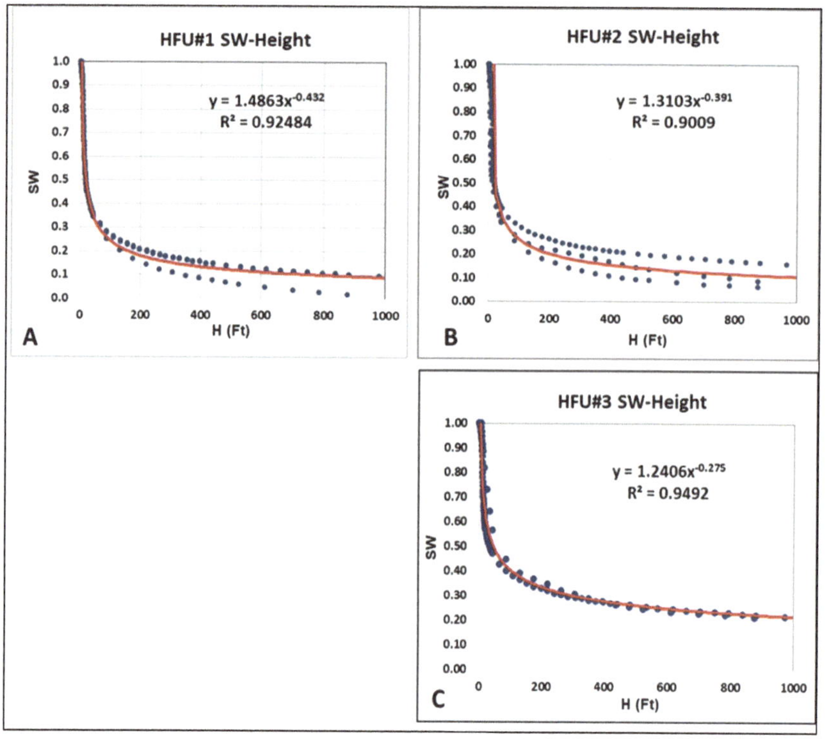

Figure 15. SW–H-function curves for Abu Madi Formation: (**A**) HFU#1, (**B**) HFU#2, (**C**) HFU#3. (Dot blue line = core data, solid red line = predicted best fit line).

4.6. Water Saturation Using J-Function

The Leverett J-function was calculated by considering the change of reservoir porosity and permeability values within the flow unit and Pc. The values of J–SW data for different flow units are plotted (Figure 16) and a regression was performed to fit the normalized data using a single equation to fit the J–SW data for each flow unit (HFU#1 to HFU#3), with different parameters (Figure 16A–C). The best fit is represented by the red line.

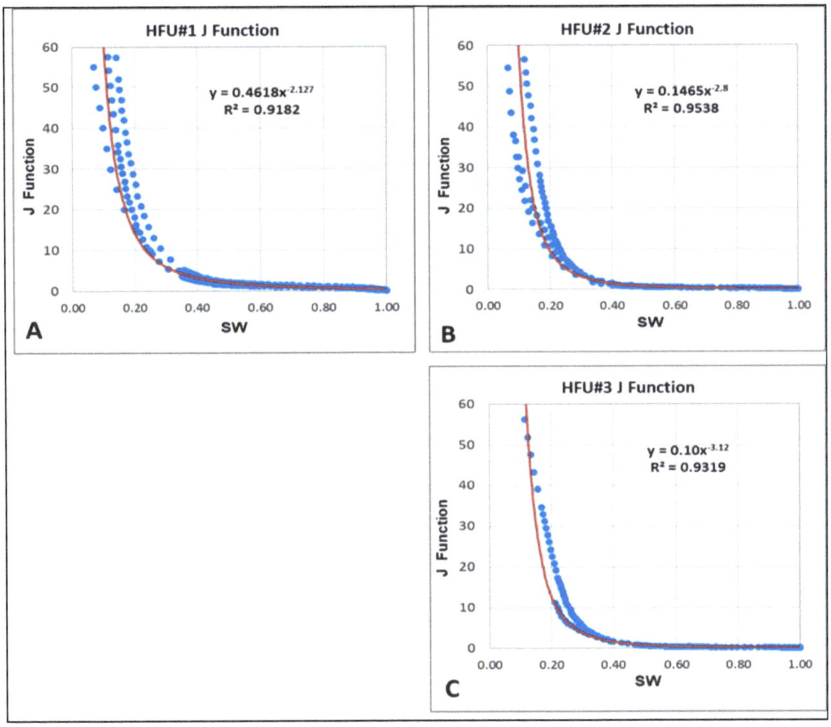

Figure 16. SW–J-function curves for Abu Madi Formation: (**A**) HFU#1, (**B**) HFU#2, (**C**) HFU#3. (Dot blue line= core data, solid red line= predicted best fit line).

Finally, this study developed an equation with the different parameter for each flow unit, which is as follows:

$$J = a \times W^{b} \qquad \text{or } SW = (J/a)^{1/b} \tag{12}$$

a = 0.4618, b = (−2.127) for HFU#1
a = 0.1465, b = (−2.8) for HFU#2
a = 0.100, b = (−3.12) for HFU#3 and 4
where SW = water saturation (v/v); J = Leverett J-function.

The water saturation (SW_JF) was computed for all wells in the Salma Field and was based on predefined flow units, with calculated Pc, and J-function above free water level. Calculated SW_JF was plotted and correlated with previous SW calculated from resistivity (Figure 17). When we compared the water saturation predicted by the saturation height function and the J-function, the J-function was interpreted to be more reliable, largely as the values were variable depending on the porosity and permeability of the reservoir. In comparison, the saturation height function varied in response to changes in porosity and permeability.

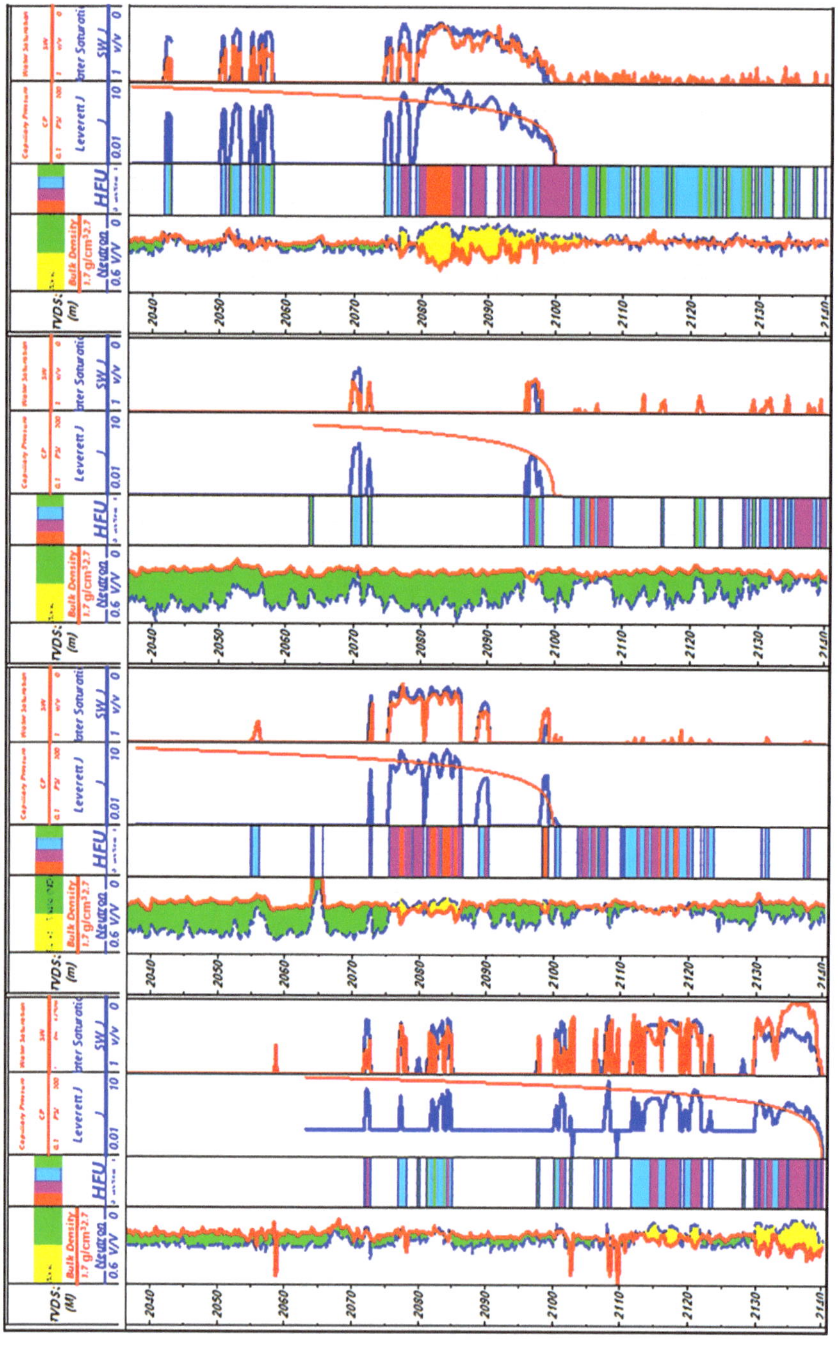

Figure 17. SW–H, and SW–J-function for Salma Field wells. (track-1: Depth, track-2: density-neutron curves, track-3: HFU, track-4: J-Function, track-5: SW of J-Function, for each well).

5. Discussion

5.1. Depositional Environments, Flow Units, and Control upon Reservoir Quality

A wide range of facies and different flow units characterize the Abu Madi Formation reservoir. The variation in the depositional environments principally controls the reservoir parameters of these flow units. The fluvial channel and tidal channel deposits form the highest-quality reservoir facies, in which grain sizes range from silt- to granule-grade sandstones that are occasionally conglomeratic. This variability in grain sizes (poor-sorting) is typically formed at the base of these facies, with beds becoming moderately sorted in the upper parts. The sandstone of the fluvial channel facies has very good pore interconnectivity. The high FZI values indicate very good-quality reservoirs with an effective pore system dominated by HFU#1 and HFU#2.

However, when the reservoir's depositional environment transitions into more estuarine conditions, the sedimentary deposits are dominated by siltstone, mudstone (shale), and some mud-rich sandstones. Despite the higher mud content, these types of deposits have moderate FZI values (HFU-#3). Abundant in mudstone (shale) and highly argillaceous sandstone intervals present heterolithic facies of the estuarine environment, with poor reservoir quality. The reservoir is characterized by low FZI values, with an ineffective pore system dominated by HFU#4.

5.2. Flow Unit Identification and Validation of Irreducible Water Saturation and SW Estimation

Two methods were used to build the saturation model: the saturation height function and the J-function. Based on different flow units, three models for each method were applied. Data from the two methods show that small pores retain a fluid volume regardless of any existing pressure (irreducible water saturation). Data from the SW–H model (Figure 15) show that HFU#1 has a minimum irreducible water saturation (10%) and is lower in SW versus height above free water level. HFU#2 has a minimum irreducible water saturation (11%) and low to moderate SW values versus height above free water level. HFU#3 has a minimum irreducible water saturation (20%), with high SW values versus height above free water level. J–SW curves with different parameters related to different flow units (Figure 16) and water saturation (SW_JF) were computed for all wells in the Salma Field based on predefined flow units. Generally, there is a good match between SW predicted by the two methods and SW calculated based upon the resistivity log data. The Salma-2 and the Salma-4 well data show good agreement between modelled and calculated SW.

The predicted SW_J show reasonable matching in the deeper and middle zones above the free water level, where most of the reservoir is blocky and clean sandstones of HFU#1 and HFU#2 are present. SW_J is almost lower than SW in the upper pay zone based on the resistivity log (Figure 17). The difference in observed value is interpreted to be from an overestimation of SW from resistivity, largely due to the low-resistivity values of thin beds, resulting from the shoulder effect of adjacent shale layers, or low resistivity for interbedded sandstone and mudstones (heterolithic).

6. Conclusions

Rock typing classification, reservoir quality assessment, and petrophysical characterization aid in the subdivision of the reservoirs into fluid flow units and rock types.

Based on the results of this study, it is concluded that four HFUs control the reservoirs of the Salma Field in Egypt. These are HFU#1 (excellent reservoir facies), HFU#2 (very good to good reservoir quality), HFU#3 (moderate quality), and HFU#4 (low quality). The neural log technique (K-mode) has succeeded in predicting FZI, permeability, and petrophysical parameters in the un-core interval in the study wells. The capillary pressure analysis and output of a water saturation curve independent of a resistivity log measurement provided a more consistent method than the conventional log-based analysis in low-resistivity zones, the latter having issues around the undefined effect of clay content and shale distribution within the reservoir.

Understanding and prediction of the current fluid contacts was achieved by applying the saturation height and J-function models, and the technique has been shown to be a successful method for extrapolating water saturation for reservoir zones away from the well and can predict throw reservoir production life and be used as a base of reservoir dynamic model.

Author Contributions: Conceptualization, N.H.E.-G., M.A.W. and M.K.B.; methodology, N.H.E.-G., W.M.M., M.A.W. and M.K.B.; software, M.A.W. and M.K.B.; investigation writing—original draft preparation, N.H.E.-G., M.A.W., W.M.M., M.K.B. and F.A.A.; reviewing and editing, M.A.W., M.K.B., F.A.A., T.J.D. and D.E.A.; supervision and visualization, N.H.E.-G., W.M.M., M.K.B., F.A.A. and D.E.A. All authors have read and agreed to the published version of the manuscript.

Funding: This research was funded by the Researchers Supporting Project number (RSPD2023R1120), King Saud University, Riyadh, Saudi Arabia.

Data Availability Statement: Restrictions apply to the access of the data, which were utilized under permission for the current study and are therefore not publicly available. The data supporting the findings of this investigation can be obtained from [the EGPC]. However, data are accessible from the authors upon justifiable request and with [the EGPC's] consent.

Acknowledgments: The authors extend their appreciation to King Saud University for funding through Researchers Supporting Project number (RSPD2023R1120), King Saud University, Riyadh, Saudi Arabia. Thanks are due to EGPC, (DANA GAS) Company, and El-Wastani Petroleum Company for providing the required data.

Conflicts of Interest: The authors confirm that there is no conflict of interest between all authors even Mohamed who is affiliated with the Dana Gas Company and that there is no commercial or financial relationships in the present work.

References

1. Leila, M.; Kora, M.A.; Ahmed, M.A.; Ghanem, A. Sedimentology and reservoir characterization of the upper Miocene Qawasim Formation, El-Tamad oil field onshore Nile Delta, Egypt. *Arab. J. Geosci.* **2016**, *9*, 17. [CrossRef]
2. Novak, K.; Malvik, T.; Velic, J.; Simon, K. Increased hydrocarbon recovery and CO_2 storage in Neogene sandstones, a Croatian example: Part II. *Environ. Earth Sci.* **2014**, *71*, 3641–3653. [CrossRef]
3. Abdel Aal, A.; Price, R.; Vital, J.; Sharallow, J. Tectonic evolution of the Nile delta, its impact on sedimentation and hydrocarbon potential. In Proceedings of the 12th EGPC Exploration and Production Conference, Cairo, Egypt, 12–15 November 1994; pp. 19–34.
4. Abdullah, E.; Al-Areeq, N.; Elmahdy, M.; Barakat, M.K. A new insight into the structural architecture of Sharyoof field, Say'un–Masilah basin, Yemen. *Arab. J. Geosci.* **2021**, *14*, 1977. [CrossRef]
5. Ali, A.M.; Radwan, A.E.; Abd El-Gawad, E.A.; Abdel-Latief, A.A. 3D Integrated Structural, Facies and Petrophysical Static Modeling Approach for Complex Sandstone Reservoirs: A Case Study from the Coniacian–Santonian Matulla Formation, July Oilfield, Gulf of Suez, Egypt. *Nat. Resour. Res.* **2021**, *31*, 385–413. [CrossRef]
6. Radwan, A.E. Modeling the Depositional Environment of the Sandstone Reservoir in the Middle Miocene Sidri Member, Badri Field, Gulf of Suez Basin, Egypt: Integration of Gamma-Ray Log Patterns and Petrographic Characteristics of Lithology. *Nat. Resour. Res.* **2021**, *30*, 431–449. [CrossRef]
7. Beiranvand, B.; Kamali, M.R. Petrophysical evaluation and determination of rock types in a carbonate reservoir in SW Iran with interpretation of petrography and geophysical well logs. *Iran. Int. J. Sci.* **2004**, *5*, 203–221.
8. Al-Ibadi, H.; Al-Jawad, S.N. Permeability Evaluation of Carbonate Reservoir Using Hydraulic Unit Analyse: Case Study from Middle East Region. In Proceedings of the 82nd EAGE Conference & Exhibition, Amsterdam, The Netherlands, 8–11 December 2020; pp. 1–5.
9. Dolson, J.; Shann, M.; Matbouly, S.; Harwood, C.; Rashed, R.; Hammouda, H. The petroleum potential of Egypt. *AAPG Mem.* **2001**, *74*, 453–482.
10. Amaefule, J.; Altunbay, M.; Tiab, D.; Kersey, D.; Keelan, D. Enhanced reservoir description, using core and log data to identify hydraulic (flow) units and predict permeability in uncored intervals/wells. In Proceedings of the SPE 26436, Annual Technical Conference and Exhibition, Houston, TX, USA, 3–6 October 1993; pp. 3–6.
11. El-Sharawya, M.; Leila, M.; Bakr, A.; Kamela, A. Petrophysical evaluation of the Messinian Abu Madi Formation in Salma delta gas field, northeastern onshore Nile Delta, Egypt. *J. Environ. Sci.* **2020**, *49*, 46–53.
12. El-Adl, H.; Leila, M.; Ahmed, A.; Anan, T.; El-Shahat, A. Integrated sedimentological and petrophysical rock-typing of the Messinian Abu Madi Formation in South Batra gas field, onshore Nile Delta, Egypt. *Mar. Pet. Geol.* **2021**, *124*, 104835. [CrossRef]

13. Mirzaei-Paiaman, A.; Ostadhassan, M.; Rezaee, R.; Saboorian, H.; Chen, Z. A new approach in petrophysical rock typing. *J. Pet. Sci. Eng.* **2018**, *166*, 445–464. [CrossRef]
14. Abdullah, E.; Al-Areeq, N.; Al-Masgari, A.; Barakat, M.K. Petrophysical evaluation of the Upper Qishn clastic reservoir in Sharyoof oil Field, Sayun-Masilah Basin, Yemen. *ARPN J. Eng. Appl. Sci.* **2021**, *16*, 2375–2394.
15. EGPC (Egyptian General Petroleum Corporation). *Nile Delta and North Sinai, Field Discoveries and Hydrocarbon Potentials (A Comprehensive Overview)*; EGPC: Cairo, Egypt, 1994; p. 387.
16. Barakat, M.K. Modern Geophysical Techniques for Constructing a 3D Geological Model on the Nile Delta, Egypt. Ph.D. Dissertation, Technical University of Berlin, Berlin, Germany, 2010. [CrossRef]
17. Hassan, S.; Darwish, M.; Tahoun, S.S.; Radwan, A.E. An integrated high-resolution image log, sequence stratigraphy and palynofacies analysis to reconstruct the Albian–Cenomanian basin depositional setting and cyclicity: Insights from the southern Tethys. *Mar. Pet. Geol.* **2022**, *137*, 105502. [CrossRef]
18. El-Gendy, N.H.; El-Nikhely, A.H.; Zakaria, A.H.; Hellish, H.M.; Nabawy, B.S.; Barakat, M.K. Pre-Stack Imaging for the Messinian Sub Salt Sequences in the Levantine Basin of the East Mediterranean: A Case Study, Offshore Lebanon. *Iraqi Geol. J.* **2023**, *56*, 1–18. [CrossRef]
19. Corbett, P. Petroleum Geoengineering: Integration of static and dynamic models. In *Society of Exploration Geophysicists and European Association of Geoscientists and Engineers*; European Association of Geoscientists & Engineers: Amsterdam, The Netherlands, 2009.
20. Soleimani, M.; Shokri, B.J. 3D static reservoir modeling by geostatistical techniques used for reservoir characterization and data integration. *Environ. Earth Sci.* **2015**, *74*, 1403–1414. [CrossRef]
21. Ghandra, V.; Barnett, A.; Corbett, P.; Geiger, S.; Wright, P.; Steele, R.; Milroy, P. Effective integration of reservoir rock-typing and simulation using near-wellbore upscaling. *Mar. Pet. Geol.* **2015**, *67*, 307–326. [CrossRef]
22. Corbett, P.W.M.; Duarte, G.L.B. Understanding subsurface fluvial architecture from a combination of geological well test models and well test data. *Geol. Soc.* **2019**, *488*, 237–257. [CrossRef]
23. Aadland, A.I.; Hassan, A.A. Hydrocarbon potential of the Abu El Gharadig basin in the Western Desert, Egypt. In Proceedings of the 8th Arab Petroleum Congress, Algiers, Algeria, 2 June 1972; p. 19.
24. Leila, M.; Moscariello, A. Seismic stratigraphy and sedimentary facies analysis of the pre- and syn-Messinian salinity crisis sequences, onshore Nile Delta, Egypt. Implications for reservoir quality prediction. *Mar. Petrol Geol.* **2019**, *101*, 303–321. [CrossRef]
25. Barakat, M.K.; Dominik, W. Seismic studies on the Messinian rocks in the onshore Nile Delta, Egypt. In Proceedings of the 72nd EAGE Conference, and Exhibition, Barcelona, Spain, 14–17 June 2010; pp. 5422–5426.
26. Hamouda, A.Z.; El-Gendy, N.; El-Gharabawy, S.; Salah, M.; Barakat, M.K. Marine geophysical surveys and interpretations on the ancient Eunostos Harbor Area, Mediterranean Coast, Egypt. *Egypt. J. Pet.* **2023**, *32*, 47–55. [CrossRef]
27. Leila, M.; El Sharawy, M.; Bakr, A.; Mohamed, A.K. Controls of facies distribution on reservoir quality in the Messinian incised-valley fill Abu Madi Formation in Salma delta gas field, northeastern onshore Nile Delta, Egypt. *J. Nat. Gas Sci. Eng.* **2022**, *97*, 104360. [CrossRef]
28. El-Gendy, N.; Radwan, A.; Waziry, M.; Dodd, T.; Barakat, M.K. An integrated sedimentological, diagenesis, rock typing, image logs, artificial neural networks, and well logging analysis for reservoir quality assessment of the heterogeneous fluvial-deltaic Messinian Abu Madi reservoirs of the Salma Field, onshore East Nile Delta, Egypt. *Mar. Pet. Geol.* **2022**, *145*, 105910.
29. Abu El-Ata, A.S.; El-Gendy, N.H.; El-Nikhely, A.H.; Raslan, S.; El-Oribi, M.; Barakat, M.K. Seismic characteristics and AVO response for non-uniform Miocene reservoirs in offshore eastern Mediterranean region, Egypt. *Sci. Rep.* **2023**, *13*, 8897. [CrossRef] [PubMed]
30. Gunter, G.; Finneran, H.; Hartmann, D.; Miller, J. Early determination of reservoir flow units using an integrated petrophysical method. In Proceedings of the SPE Annual Technical Conference and Exhibition, San Antonio, TA, USA, 5–8 October 1997.
31. Dubois, M.K.; Byrnes, A.P.; Bhattacharya, S.; Bohling, G.C.; Doveton, J.H.; Barba, R.E. *Hugoton Asset Management Project (HAMP): Hugoton Geo Model Final Report*; KGS Open File Report; 2006. Available online: https://www.kgs.ku.edu/PRS/publication/2007/OFR07_06/index.html (accessed on 28 November 2023).
32. Jones, R.R.; McCaffrey, K.J.W.; Clegg, P.; Wilson, R.W.; Holliman, N.S.; Holdworth, R.E. Integration of regional to outcrop digital data: 3D visualization of multiscale geological models. *Comput. Geosci.* **2009**, *35*, 4–18. [CrossRef]
33. El-Nikhely, A.; El-Gendy, N.H.; Bakr, A.M.; Zawra, M.S.; Ondrak, R.; Barakat, M.K. Decoding of seismic data for complex stratigraphic traps revealing by seismic attributes analogy in Yidma/Alamein concession area Western Desert, Egypt. *J. Petrol Explo. Prod. Technol.* **2022**, *12*, 3325–3338. [CrossRef]
34. Radwan, A.E. Three-dimensional gas property geological modeling and simulation. In *Sustainable Geoscience for Natural Gas Sub-Surface Systems*; Wood, D.A., Cai, J., Eds.; Elsevier: Amsterdam, The Netherlands, 2022; pp. 29–45.
35. Varavur, S.; Shebl, H.; Salman, S.M.; Shibasaki, T.; Dabbouk, C. Reservoir rock type definition in a giant cretaceous carbonate. In Proceedings of the SPE Middle East Oil and Gas Show and Conference, Manama, Bahrain, 12–15 March 2005.
36. Leverett, M.C. Capillary behavior in porous solids. *Trans. AIME* **1941**, *142*, 159–172. [CrossRef]
37. Barakat, M.K.; Nooh, A.Z. Reservoir quality using the routine core analysis data of Abu Roash "C" in Badr El Din-15 oil field, Abu Gharadig basin, North Western Desert, Egypt. *J. Afr. Earth Sci.* **2017**, *129*, 683–691. [CrossRef]
38. Nabawy, B.S.; Shehata, A.M. Integrated petrophysical and geological characterization for the Sidi Salem-Wakar sandstones, off-shore Nile Delta, Egypt. *J. Afr. Earth Sci.* **2015**, *110*, 160–175. [CrossRef]
39. Carman, P.C. Fluid flow through a granular bed. *Trans. Inst. Chem. Eng.* **1937**, *15*, 150–156. [CrossRef]

40. Al-Ajmi, F.A.; Holditch, S.A. Permeability estimation using hydraulic flow units in a central Arabia reservoir. In Proceedings of the SPE Annual Technical Conference and Exhibition. SPE 63254, Dallas, TX, USA, 1–4 October 2000.
41. Hamada, G.M.; AL-Awad, M.N.; Alsughayer, A. Variable Saturation Exponent Effect on the Determination of Hydrocarbon Saturation. In Proceedings of the SPE Asia Pacific Oil and Gas Conference and Exhibition, Melbourne, Australia, 8–10 October 2002. [CrossRef]
42. Bateman, R.M. *Openhole Log Analysis and Formation Evaluation*; Schlumberger Education Services: Houston, TX, USA, 1985.
43. Archie, G. The electric resistivity logs as an aid in determining some reservoir characteristics. *Trans. Am. Int. Mech. Eng.* **1942**, *146*, 54–62.
44. Poupon, A.; Leveaux, J. Evaluation of Water Saturation in Shaly Formations. *Log Anal.* **1971**, *12*, SPWLA-1971-vXIIn4a1.
45. Bhatt, A.; Helle, H.B.; Ursin, B. Application of committee machines in reservoir Characterization while drilling: A novel neural network approach in log analysis. In Proceedings of the 6th Nordic Symposium on Petrophysics, Trondheim, Norway, 15–16 May 2001.
46. Tiab, D.; Donaldson, E.C. *Petrophysics: Theory and Practice of Measuring Reservoir Rock and Fluid Transport Properties*; Gulf Publishing: Houston, TX, USA, 1996; 926p.
47. Gandhi, A.; Torres-Verdín, C.; Voss, B.; Gabulle, J.; Seminario, F. Construction of Reliable Static and Dynamic Multi-Layer Petrophysical Models in Camisea Gas Reservoirs, Peru. In Proceedings of the SPWLA 51st Annual Logging Symposium, Perth, Australia, 19–23 June 2010.
48. Heidari, Z.; Torres-Verdín, C.; Mendoza, A.; Wang, G.L. Assessment of residual hydrocarbon saturation with the combined quantitative interpretation of resistivity and nuclear logs. *Petrophysics* **2011**, *52*, 217–237.
49. Wardlaw, N.; Taylor, R. Mercury Capillary Pressure Curves and the Interpretation of Pore Structure and Capillary Behaviour in Reservoir Rocks. *Bull. Can. Pet. Geol.* **1976**, *24*, 225–262.
50. Skelt, C.; Harrison, B. An Integrated Approach to Saturation Height Analysis. In Proceedings of the SPWLA 36th Annual Logging Symposium, Paris, France, 26–29 June 1995; pp. 1–6.
51. Wiltgen, N.A.; Le Calvez, J.; Owen, K. Methods of Saturation Modeling Using Capillary Pressure Averaging and Pseudos. In Proceedings of the SPWLA 44th Annual Logging Symposium, Galveston, TX, USA, 22–25 June 2003; pp. 22–25.
52. Jennings, R. Unsystematic security price movements, management earnings forecasts, and revisions in consensus analyst earnings forecasts. *J. Account. Res.* **1987**, *25*, 90–110. [CrossRef]
53. Siddique, J.I.; Landis, F.A.; Mohyuddin, M.R. Dynamics of Drainage of Power-Law Liquid into a Deformable Porous Material. *Open J. Fluid Dyn.* **2014**, *4*, 403. [CrossRef]
54. Harrison, B.; Jing, X.D. Saturation height methods and their impact on volumetric hydrocarbon in place estimates. In Proceedings of the SPE Annual Technical Conference and Exhibition, New Orleans, LA, USA, 30 September–3 October 2001. [CrossRef]
55. Lucia, F.J. *Carbonate Reservoir Characterization*; Springer: Berlin/Heidelberg, Germany, 1999.
56. Miall, A. A review of the braided river depositional environment. *Earth Sci.* **1977**, *13*, 1–62. [CrossRef]
57. Terwindt, J. Litho-facies of inshore estuarine and tidal-inlet deposits. *Geol Mijnb.* **1971**, *50*, 515–526.
58. Van den Berg, J.; Boersma, J.; Van Gelder, A. Diagnostic sedimentary structures of the fluvial-tidal transition zone- evidence from deposits of the Rhine and Meuse. *Neth. J. Geosci.* **2007**, *86*, 306–387. [CrossRef]
59. Tucker, M. *Sedimentary Petrology*; Blackwell Scientific: Oxford, UK, 2001; p. 272.
60. Bridge, J. Fluvial facies models, recent Developments. *Facies Model. Revisit.* **2006**, *84*, 85–170. [CrossRef]
61. Vandré, C.; Cramer, B.; Gerling, P.; Winsemann, J. Natural gas formation in the western Nile delta (Eastern Mediterranean): Thermogenic versus microbial. *Org. Geochem.* **2007**, *38*, 523–539. [CrossRef]
62. Nabawy, B.S.; Lashin, A.; Barakat, M.K. Implementation of lithofacies and microfacies types on reservoir quality and heterogeneity of the Late Cretaceous Upper Bahariya Member in the Shurouk Field, Shoushan Basin, North-Western Desert, Egypt. *J. Asian Earth Sci.* **2022**, *224*, 105014. [CrossRef]

water

MDPI

Article

Identification of Breaches in a Regional Confining Unit Using Electrical Resistivity Methods in Southwestern Tennessee, USA

Md Rizwanul Hasan [1], Daniel Larsen [1,*], Scott Schoefernacker [2] and Brian Waldron [3]

1 Department of Earth Sciences, The University of Memphis, Johnson Hall, 488 Patterson St., Memphis, TN 38152, USA; mhasan4@memphis.edu
2 Protect Our Aquifer, 1910 Madison Ave. #130, Memphis, TN 38104, USA; schoefernacker@gmail.com
3 Center for Applied Earth Science and Engineering Research, The University of Memphis, 1000 Wilder Tower, 3675 Alumni Ave., Memphis, TN 38152, USA; bwaldron@memphis.edu
* Correspondence: dlarsen@memphis.edu; Tel.: +1-(901)-870-4355

Abstract: Electrical resistivity and borehole data are applied to delineate lithostratigraphic boundaries and image the geometry of confining-unit breaches in Eocene coastal-plain deposits to evaluate inter-aquifer exchange pathways. Eight dipole–dipole array surveys were carried out, and apparent resistivity was inverted to examine the lateral continuity of lithologic units in different water-saturation and geomorphic settings. In addition, sensitivity analysis of inverted resistivity profiles to electrode spacing was performed. Resistivity profiles from Shelby Farms (SF) highlight the effect of varied electrode spacing (2.5, 5, and 10 m), showing an apparent ~0.63 to 0.75 depth shift in resistivity-layer boundaries when spacing is halved, with the 10 m spacing closely matching borehole stratigraphy. Grays Creek and Presidents Island profiles show clay-rich Eocene Cook Mountain Formation (CMF), with resistivity ranging from 10 to 70 Ω-m, overlying the Eocene Memphis Sand—a prolific water-supply aquifer. Resistivity profiles of SF and Audubon Park reveal sandy Cockfield Formation (CFF) paleochannels inset within and through the CMF, providing hydrogeologic connection between aquifers, and clarifying the sedimentary origin of confining-unit breaches in the region. The results underscore the efficacy of the electrical resistivity method in identifying sand-rich paleochannel discontinuities in a low-resistivity regional confining unit, which may be a common origin of breaches in coastal-plain confining units.

Keywords: aquitard breaches; coastal-plain deposits; electrical resistivity; electrode spacing; paleochannels

1. Introduction

Non-invasive geophysical methods such as electrical resistivity (ER) are emerging as cost-effective tools for geologic and hydrogeologic studies [1], especially to characterize and delineate subsurface hydrogeologic and geologic structures to improve groundwater management [2,3]. ER has been used to identify the aquifers and aquitards in different geological settings [4–6], which is crucial for groundwater management. Defining aquifers and confining-unit boundaries using geophysical methods helps to create better hydrogeologic models for improving groundwater budgets and identifying groundwater contamination flow paths [7–9]. Characterizing and identifying a confining unit, especially the layer above a semi-confined aquifer, is critical to prevent contaminants from flowing into an underlying aquifer. The presence of confining-unit breaches (thinning or absence of confining clay layer) raises the potential for groundwater contamination.

The dynamic nature of the study area's coastal-plain depositional setting has led to the development of a complex stratigraphic framework. Over time, sedimentation processes, erosion, and the influence of relative-sea-level change have shaped subsurface geology, resulting in a heterogeneous distribution of geological formations. Previous studies provide evidence for breaches in the upper Claiborne confining unit (UCCU), which includes the clay-rich Cook Mountain Formation and heterolithic Cockfield Formation, overlying the

Memphis aquifer, mainly comprising the Memphis Sand [10–15]. The Memphis aquifer is a regionally important water-supply aquifer in Memphis, Shelby County, Tennessee, and most of the northern Mississippi embayment region [16,17]. Recent studies suggest that the breaches may result from the paleo-channel incision during Eocene sea-level fall and Quaternary incision by western Tennessee tributaries to the Mississippi River [18,19].

ER surveys of the subsurface primarily depend on lithology and pore-fluid conditions such as salinity and water saturation. The resistivity of the soil decreases with increasing soil water saturation and pore-fluid salinity. However, the salinity of the pore-fluid affects soil resistivity more than the saturation [20–23]. Previous studies have highlighted the significance of considering factors such as soil saturation, pore-fluid salinity, and landscape characteristics when analyzing ER data [24–29]. The lateral and depth resolution of the ER survey data depends on the chosen array type, length, and electrode spacing. The array length is vital for depth of investigation (DOI) and depends on the electrode spacing. Understanding the suitable electrode spacing can help interpret layer boundaries and properties accurately. The effect of different electrode configurations on subsurface investigation and resolution is discussed in Sections 2.2.1 and 2.2.3.

Numerous studies have employed ER to identify fault zones and groundwater contamination pathways [7,30,31]. Whereas previous research has identified breaches in the UCCU using diverse data sources [11,14,32–34], their locations and extents remain poorly constrained. Notably, none have defined the subsurface geometry of breaches or clarified their paleo-channel origin. This study addresses knowledge gaps in understanding the geometry and origins of breaches in the Mississippi embayment region and coastal-plain settings, in general, utilizing ER with borehole data to visualize and conceptualize near-surface hydrogeology and confining-unit breaches. The research employs ER surveys to not only delineate lithostratigraphic units and assess confining-layer continuity but also to confirm breach presence and explore its potential causes. This innovative approach goes beyond traditional applications of ER, such as hydrostratigraphic unit delineation or identifying fault zones and contamination pathways, showcasing the development of an interpretation methodology for breaches in aquitards. The identification of breaches and assessment of their underlying causes underscore the importance of ER in shallow sub-surface hydrogeologic investigation.

2. Materials and Methods

2.1. Geologic and Hydrogeologic Settings

The Mississippi embayment spans eight south-central United States states and consists of unconsolidated aquifers and aquitards [17,35]. Shelby County, Tennessee, situated within the embayment, relies entirely on groundwater for public supply, withdrawing 696,000 m^3/day in 2015 [36]. The geologic setting of the study area comprises alluvial floodplains of modern streams underlain by Pleistocene to Holocene sand and gravel with overlying silty alluvium [14,17] (Figure 1). Uplands are prevalent in the remainder of the county, with Pliocene and Pleistocene fluvial-terrace deposits overlain by Pleistocene loess [10,37].

Shelby County has three primary freshwater aquifers: the shallow, Memphis, and Fort Pillow. The shallow aquifer, with a thickness ranging from 0 to 30 m, comprises alluvial and fluvial deposits extending over the entire county [10,16,38,39]. This aquifer encompasses the Mississippi River Valley Alluvial (MRVA) aquifer on the west side of the bluff line [40]. In the eastern county, the shallow aquifer aligns with the unconfined region of the Memphis aquifer, serving as a crucial recharge zone [11].

The UCCU, a regional confining unit for the Memphis aquifer, lies below the shallow aquifer and comprises the Eocene Cook Mountain and Cockfield Formations (older to younger). The thickness of the UCCU ranges from 0 to 61 m. The Cockfield Formation includes sand, silt, clay, and lignite, commonly in one or more fining-upward sequences of sand grading up to silt, clay, and lignite [10,41]. The Cook Mountain Formation is mostly laminated with thinly bedded silty clay and very-fine-grained sand [14,42]. The UCCU overlies and provides confinement to the Memphis aquifer but is thin or absent in the southern and eastern Shelby County [11,15]. The Memphis aquifer is 152–275 m thick and composed mostly of sand with clay and minor lignite. The Flour Island confining unit, underlies the Memphis aquifer, separating the Memphis and Fort Pillow aquifers, and composed of clay, silt, sand, and lignite.

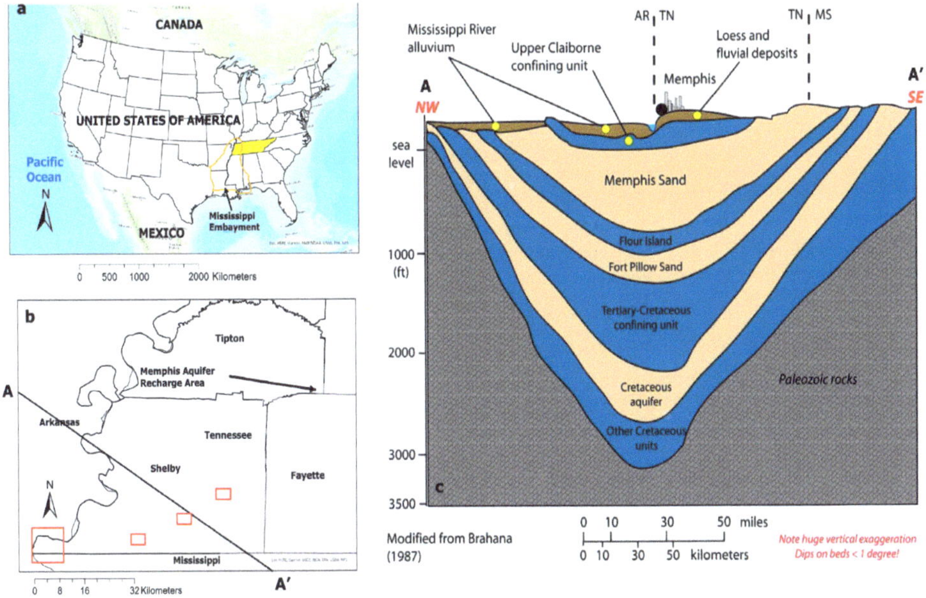

Figure 1. (**a**) Location of the Mississippi embayment. (**b**) The map shows the study area in Shelby County and survey locations (red squares). The cross-section line A–A′ extends northwest in Arkansas and southeast in Mississippi [11]. (**c**) Cross-section of Mississippi embayment stratigraphy along the cross-section line A–A′. The black dot indicates the projected intersection of the cross-section line and the Mississippi River in Figure 1b. (Ref. [43] modified Figure 1c from [44]; current revision shows the direction, study areas on (**b**), and the intersection of the cross-section line on (**c**)).

This research focuses primarily on the UCCU, which locally includes sand-rich breaches that provide little or no confinement to the Memphis aquifer [11,12]. Although most of the recharge to the Memphis aquifer occurs in the region of the subcrop east of Shelby County [16], modeling of focused recharge through breaches contributes as much as 50 percent of the water withdrawn from the Memphis aquifer [13,44], some of which has associated water-quality impacts [11,13].

2.2. Methodology

The ER data collected for this study were from five different locations in Shelby County (Figure 2). Four surveys were conducted at Shelby Farms (SF) and Gray's Creek (GC) to evaluate the impact of electrode spacing in stratigraphic delineation. Data collected from SF and GC examined if 10 m electrode spacing best corresponded to the interpretation of the top of the layers. Data from Audubon Park (AP) and President's Island (PI) further investigated suspected breaches suggested by [11]. In addition, two boreholes were drilled at PI to provide control and verify the interpretation of the inverted resistivity data. Geologic and geophysical logs from nearby boreholes are used to constrain the accuracy of the interpretations.

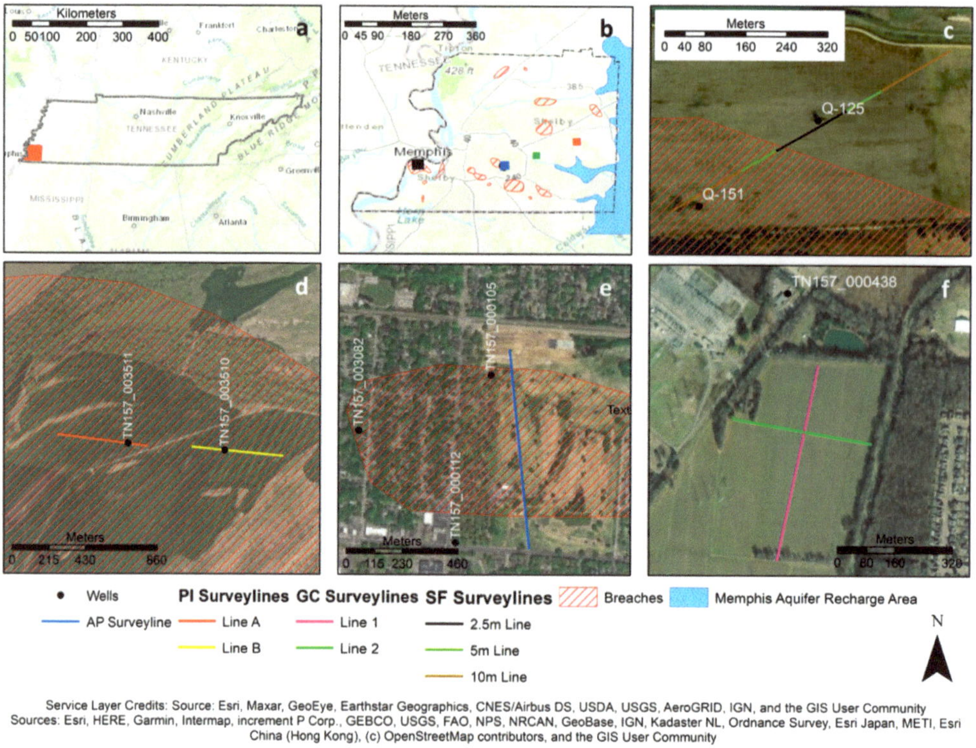

Figure 2. Map showing (**a**) the location of Shelby County within the state of Tennessee; (**b**) map showing the locations of known or suspected breaches in Shelby County (red polygon with hatching) and Memphis aquifer recharge area (blue polygon) [15]. The black, blue, green, and red squares represent PI, AP, SF, and GC, respectively; (**c**) image showing the location of survey lines in SF. The black, leaf green, and brown lines represent 2.5 m, 5 m, and 10 m survey lines, respectively: (**d**) the location of survey lines in PI. The red and yellow lines represent lines A and B, respectively; (**e**) the location of the survey line at AP (blue line); and (**f**) the location of the survey lines in GC. The magenta and green lines represent lines 1 and 2, with 10 and 7 m electrode spacing, respectively.

2.2.1. Subsurface Electrical Resistivity

Soil properties, such as porosity, permeability, rock texture and type, liquid chemicals, and saturation, as well as clay content of the subsurface layer, affect the electrical resistivity values (ERV) [26,27]. Among these parameters, the clay content impacts subsurface conductivity most [27–29] in sedimentary deposits. A resistivity survey applies electrical current to the subsurface through current electrodes, measuring the resulting potential difference at the surface through potential electrodes. Current flows radially from a single-point source in a homogeneous, isotropic subsurface, developing equipotential surfaces perpendicular to the current flow [45].

Electrical resistivity has developed as a powerful geophysical method for investigating sub-surface stratigraphy in fluvial depositional environments. Fluvial systems, characterized by the dynamic interchange of sediment transport, deposition, and erosion, leave discrete signatures in the subsurface resistivity distribution. Ref. [46] studied the potential of geoelectrical methods for characterizing shallow sediments in riverbeds. Combining sensitivity analysis and measurement configurations, ER maximizes shallow riverbed hydrostratigraphy mapping effectiveness. In recent years, research has shown the utility of ER for stratigraphic properties studies [2,8,30].

2.2.2. Data Collection

ER data were collected using a SuperSting R8 Wi-Fi instrument with 56 channels. Eight surveys were conducted, each serving a specific purpose in the investigation. Five surveys were carried out at SF and GC to examine the influence of different electrode spacing on delineating stratigraphic boundaries. The electrode spacing in these surveys was adjusted to 2.5, 5, and 10 m at SF and 7 and 10 m at GC, resulting in survey lengths varying from 137.5 to 540 m. Both of these locations are in agricultural fields. However, the SF field has numerous monitoring wells and pipelines along the margins of the field.

Three more surveys were conducted to investigate the suspected breach locations identified in previous studies. These surveys utilized an electrode spacing of 10 m, which is shown to be the spacing that produces profiles that best conform to borehole-based stratigraphic boundaries and lithology. Two surveys at PI had survey lengths of 540 m, whereas a roll-along study was performed at AP, covering a distance of 840 m. PI is an agricultural field with a buried petroleum pipeline running between the two survey lines (~100 m from the ends of each line). AP is a golf course with <10 m relief, buried irrigation lines, asphalt and concrete cart paths, modified land surface for greens and fairways, and possibly other shallowly buried infrastructure.

Data points collected along survey lines at PI and SF are 762 and 1524 for AP and GC, respectively. Data collected at GC contain more data points than the other survey lines due to extended data coverage, which allows overlapping data collection to improve the data inversion process. The extended data coverage was not used for the other survey lines due to minimal improvement in the inversion process. Data were collected at different times during 2021 through 2022, depending on the degree of soil saturation and access. The water table varied for different locations but remained consistent for a specific area. The number of survey lines does not necessarily represent the whole research area but comprises a sample of geological settings where breaches were identified using previous studies [10,11,33,34] or suspected from recent borehole data. The number of survey lines is also limited due to land-access permission and lack of accessibility due to tree, power, and supply lines.

2.2.3. Array Selection and Electrode Spacing

Array configuration in an ER survey controls the amount of data collected in the field and influences the interpretation of the inverted resistivity data. The dipole–dipole array [47] was chosen for this study as it is sensitive to lateral variations in the subsurface [48]. Before selecting the dipole–dipole array for this study, preliminary surveys were carried out using different array configurations, including the Wenner and Schlumberger arrays. These initial array choices, however, showed limitations in terms of lateral resolution, particularly concerning the targeted objective of distinguishing lateral variations within a clay layer. The study by [7] also confirms the suitability of using the dipole–dipole array in a subsurface-lateral-variation study. The data acquisition using these arrays yielded comparatively fewer data points when compared with the dipole–dipole arrays. The reduced data quantity within the preliminary survey arrays requires increased confidence in interpolation techniques, thereby introducing additional assumptions into the data analysis process. Thus, the dipole–dipole array is more suitable for identifying vertical contacts between high and low-resistive materials within a depth range in the subsurface. In the present study, confining-unit breaches have high-resistive material (sand or gravel) juxtaposed with low-resistive material like clay.

Understanding electrode spacing is essential for DOI. The DOI is approximately 20% of the array length [45,46]. We also compared the obtained data to verify the DOI, which confirms the above statement (Appendix C). As previous studies have lacked investigation of the optimum electrode spacing in delineating stratigraphic boundaries, this study investigated the 2.5 m, 5 m, 7 m, and 10 m electrode spacing coupled with a dipole–dipole array to find the best balance of depth resolution and survey length for near-surface sediments in the northern Mississippi embayment.

2.2.4. Data Processing

Apparent resistivity data collected in the field were analyzed using AGI EarthImager 2D, V. 2.1.5 – a proprietary software. The contact resistance (CR) and noise percentage were checked for bad data and removed if the CR was above 4000 ohms or the noise percentage was greater than 20%. We chose a smooth inversion model for the initial setting, where data were removed if the maximum error was greater than 5%. Terrain files containing elevation data were applied for terrain correction for the surveys conducted in Audubon Park and President's Island, except Shelby Farms and Gray's Creek. Terrain files were not used for Shelby Farms and Grays Creek due to the focus on DOI and electrode-spacing analysis.

Furthermore, terrain variations are <1 m at this location. EarthImager 2D automatically chooses the finite element method for forward modeling when a terrain correction is completed. The default equation solver for the forward is Cholesky Decomposition, which the software manufacturer sets. We selected a maximum of ten iterations and a maximum RMSE of 5% as termination criteria for resistivity inversion settings. The inversion is continued until the RMSE is reduced below 5%. Figure 3 shows the data-processing workflow for this study. The resistivity scale for this study was set to the same scale (10–3500 Ω-m) for a consistent interpretation of the stratigraphic layers.

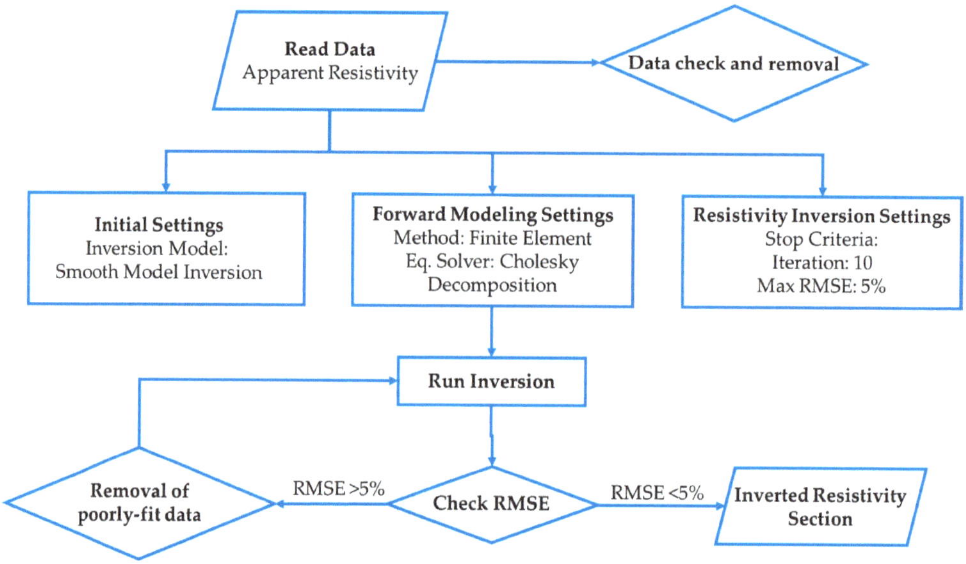

Figure 3. Data-processing workflow utilized along the current study.

A zone of no data was present from the distance of 350 to 510 m of the roll-along survey at AP due to triangulation of the overlapped data points below the depth of 55 m (Appendix B). The absence of data points below 55 m of depth and between the distances mentioned above was interpolated during the inversion process.

2.2.5. Data Interpretation

The resistivity range for glacial sediments suggested by [49] was used to interpret the inverted resistivity sections. The ranges for this study are also chosen by comparing the resistivity values collected from the field with nearby borehole data to ensure accuracy in interpretations. The presence of inorganic silt increases the resistivity value of clay [48]. Thus, in this study, ERV less than 50 Ω-m represents clay; 50–100 Ω-m is silty clay with increasing silt and fine sand percentage toward higher ERV; and greater than 100 Ω-m defines sand and gravel, with increasing ERV reflecting decreased silt and clay and increased gravel.

The water content and salinity decrease the resistivity of a layer [50,51]. For a 1% increase in water content in silty sand, the ERV decreases by 5% when the water content is ≤35% (calculated from the graph of four electrode methods for silty sand presented in [52]). The changes in the ERV are negligible as water content reaches an equilibrium state [52]. The ERV decreases rapidly when specific conductance (SC) is ≥0.02 mS/cm [51]. At SF, SC ranges from 0.065 to 1.33 mS/cm [53], whereas the SC ranges from 0.045 to 0.140 mS/cm near GC [54] at 25 °C for the alluvial and the Memphis aquifer. Near the PI area, the Memphis aquifer SC ranges from 0.150 to 0.393 mS/cm [14]. At the Sheahan well field near AP, SC ranges from 0.102 to 0.151 mS/cm [32] in the Memphis aquifer. The SC of the MRVA aquifer near the PI area is higher than that of the Memphis aquifer and ranges from 0.55 to 0.89 mS/cm. Hence, the ERV in the MRVA aquifer will be lower than in the Memphis aquifer given that all factors (clay fraction, porosity, salinity) are similar for both.

3. Results

3.1. Shelby Farms

The subsurface geology at the SF site is well constrained by geologic borehole data [12,33,34], shallow seismic surveys [18], and previous ER surveys [7], which mainly focused on the impact of specific conductance from landfill leachate. Thus, SF provides a suitable location for assessing the effect of ER electrode spacing on profile correspondence to stratigraphic units designated using borehole and geophysical data. Analysis of ER profiles based on electrode spacings of 2.5, 5, and 10 m are compared to information obtained from geophysical well logs and geologic descriptions from boreholes Q-151 and Q-125, and the latter of which is ~5 m from the survey line.

The top of the high ERV zone is estimated at approximately 16 m, based on the depth of sand-rich deposits in borehole Q-125, inferred from the contrast in ERV shown in Figure 4a at an electrode spacing of 10 m. The inverted resistivity profile (IRP) generated with a 5 m electrode spacing (Figure 4b) exhibits a high ERV boundary at a depth of 12 m—4 m higher than that of the 10 m electrode spacing. The IRP conducted with a 2.5 m electrode spacing (Figure 4c) shows the top of the high ERV boundary at a depth of 7.5 m. Based on these observations, when the electrode spacing is reduced by half, the depth to the top sand-rich zone decreases by a factor of ~0.75–0.63. Q-125 contains a metal protective casing and is ~5 m from the survey lines. The metal casing can act as a conductor, allowing electrical current to bypass the subsurface and travel directly through the casing [55]. This phenomenon can introduce biases or distortions in the survey data near the well, leading to overestimating subsurface resistivity. Although this effect may be significant for the 2.5 m electrode spacing, it is less likely to be a problem for the 5 m electrode spacing. The apparent decrease in the depth of the high ERV zone is not clearly understood. However, it may be related to the processing or inherent characteristics of the ER technique. Comparison to the borehole-based lithology indicates that the 10 m electrode spacing best represents lithological changes, especially those below 5 m depth.

The profiles from the shorter electrode spacings (2.5 and 5 m) have value because they resolve the upper 10–15 m of subsurface in greater detail than the longer spacing. For example, Figure 4c explicitly shows a low-resistivity zone (<50 Ω-m) within the top 4 m that corresponds well to the upper alluvium and an underlying moderate to high ER layer (ranging from 80 to 200 Ω-m) that fits well to the sand and gravel alluvium. Whereas in Figure 4a, the upper ~10 m are an intermediate ER layer ranging from 80 Ω-m to 200 Ω-m with little discernable structure. Furthermore, the intermediate ER layer is underlain by a high ERV zone, 100 Ω-m to greater than 3500 Ω-m, with noticeable lower resistivity discontinuities (Figure 4a,b). This interval is not apparent in Figure 4c, perhaps due to the distortion caused by the steel casing or poor depth resolution. The size and geometry of the discontinuities are consistent with paleo-channel features identified from shallow-shear-wave seismic analysis along approximately the same line [18]. The revised stratigraphic interpretation of the SF by [19] indicates that the paleo-channel features are in the Eocene Cockfield Formation, based on stratigraphic position and texture.

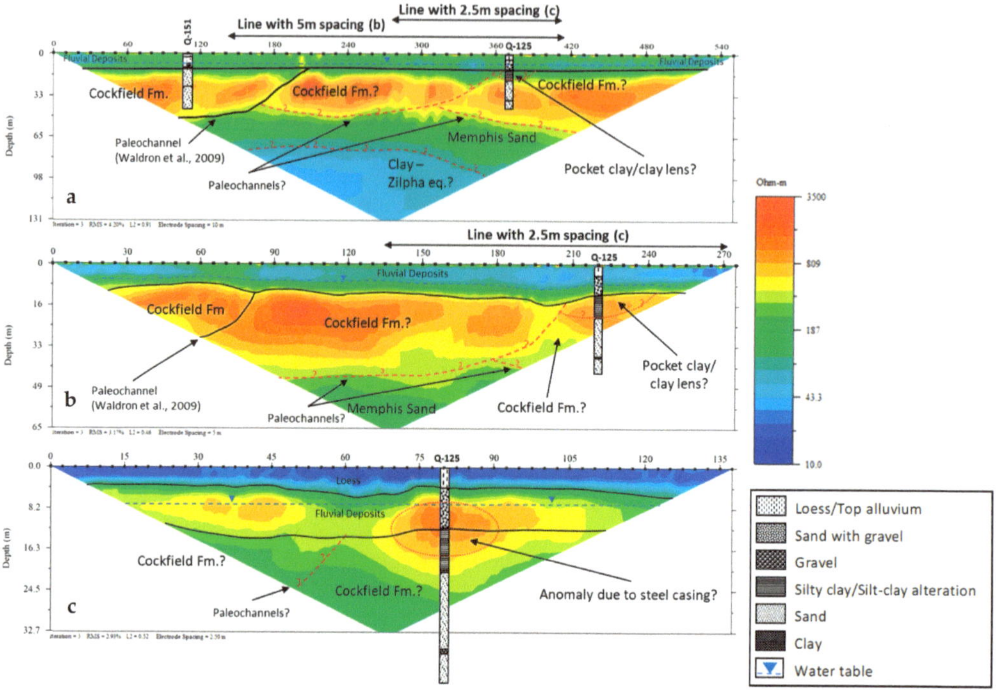

Figure 4. Comparison of interpreted geology at SF with different electrode spacing: (**a**) IRP with 10 m electrode spacing, (**b**) IRP with 5 m electrode spacing, and (**c**) IRP with 2.5 m electrode spacing. The IRPs are complemented with superimposed lithologic logs, providing additional context for interpretation. The blue dashed line with the triangle above is the inferred water table. The black lines on each IRP indicate the inferred depth of the sand-rich, high-resistivity interval based on borehole data in wells Q-125 and Q-151 and the paleo-channel feature based on [18]. Additionally, the red dashed lines represent interpreted Eocene paleo-channel margins, while the red dotted line highlights the anomaly caused by the presence of a steel casing.

The high ERV layer (ranging from 100 to 200 Ω-m) observed at depths between 50 and 80 m in Figure 4a is interpreted as the Memphis Sand, which aligns with [56] and recent stratigraphic revisions by [19]. The low ER layer, below 50 Ω-m, observed below 80–90 m (Figure 4a) corresponds to a silty clay layer locally observed in the middle of the Memphis Sand, which may be equivalent to the Zilpha clay defined in Mississippi [17]. The low-resistivity layer is not visible in Figure 4b,c due to the shallower DOI.

3.2. Grays Creek

Electrical resistivity surveys were conducted at the GC site to investigate the resistivity ranges for the Cook Mountain Formation and Memphis Sand and a potential breach inferred from borehole data. Two orthogonal surveys, a long line with 10 m electrode spacing and a shorter one with 7 m electrode spacing, were used to assess further the influence of electrode spacing on the depth of strong resistivity contrasts (the boundary between the top high-resistivity zone and the low-resistivity zone below it).

In the IRPs from GC (Figure 5a,b), the alluvium is a low-to-moderate ERV layer ranging from 15 to 80 Ω-m and is observed in the top 12 m. Similar to the results at SF, the sand-rich lower alluvium is distinct in the IRP with shorter spacing. The low ERV layer (<50 Ω-m) extending from a depth of ~12 to ~32 m is the Cook Mountain Formation. The high ERV interval, ranging from 100 to 300 Ω-m, was observed in both IRPs below ~32 m and represents the Memphis Sand.

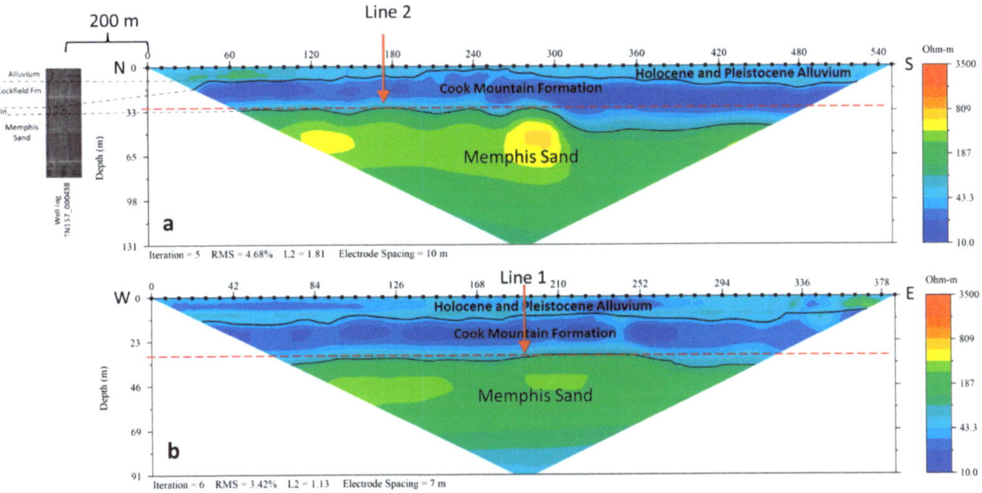

Figure 5. The IRPs of (**a**) line 1 (10 m spacing) and (**b**) line 2 (7 m spacing) carried out along the GC. The solid black lines represent stratigraphic layer boundaries, and the red dashed lines show the average depth of the top of the Memphis Sand. No data for the water table is available at this site.

Figure 5a,b provides insights into the average depths of the layer boundaries between lithologic units. The average depth for the boundary between the Cook Mountain Formation and Memphis Sand is estimated at ~32 m for the 10 m spacing and ~32.5 m for the 7 m spacing survey, suggesting little or no apparent depth shift. The boundary depth between the Cook Mountain Formation and Memphis Sand, determined by the well log (TN157_000438) ~200 m north of GC line 1, is 32.5 m mbgs (meter below ground surface). Similarly, the depth of the base of the alluvium in both IRPs is ~8 m, which is similar to or a little deeper than that observed in the nearby borehole. The absence of a distinct shift may be related to very high resistivities or the absence of steel casing adjacent to the survey lines.

The inferred boundary between the alluvium and Cook Mountain Formation deepens to the north, which is away from the direction of the stream. The deeper contact is interpreted to reflect alluvium underlain by the Cockfield Formation inset into the Cook Mountain Formation, which is evident in the adjacent well log. Although there is no evidence of the presence of a paleochannel, the inferred shape and slope of the Cockfield—Cook Mountain formation contact to the north towards the well is believed to represent a paleo-channel just north of line 1 at GC. Land inaccessibility limited further assessment of the inferred paleo-channel and potential breach.

3.3. President's Island

Both IRPs discussed at PI are oriented west to east on President's Island (Figure 6), a former Mississippi River island now connected to the east bank of the Mississippi River. The primary goal of conducting these two profiles was to investigate a previously proposed breach beneath most of PI [11], recent evidence for which is suggested locally from inverted airborne electromagnetic (AEM) data [29]. Because no lithologic control from borehole data was available at the site, cored boreholes were drilled, and gamma logs (Appendix D) were obtained along each line to provide detailed geologic information (Figure 5).

Line A is positioned ~730 m east of the river. A discontinuous layer with a moderate ERV range of ~50–110 Ω-m is present from 65 to 58 masl (meter above sea level) in Line A (Figure 6a). At 60–20 masl, a high ERV zone (>100 Ω-m) is observed, which connects to the surface at both ends of the survey line. Combining these two intervals forms the MRVA with the low-resistivity upper layer being the fine-grained, silty alluvium. At ~34 to 20 masl, a low ERV layer (<50 Ω-m) underlies the MRVA with ~14 m of relief along the contact. This unit extends to ~23 m masl or more. It is represented by laminated-to-thinly-bedded silty clay and silty, very-fine-grained sand of the Cook Mountain Formation in borehole TN157_003511 (Figure 7). The lower part of the IRP, below −23 masl, in Figure 6a had low-quality data removed from the profile and not shown but likely includes high-resistivity Memphis Sand strata.

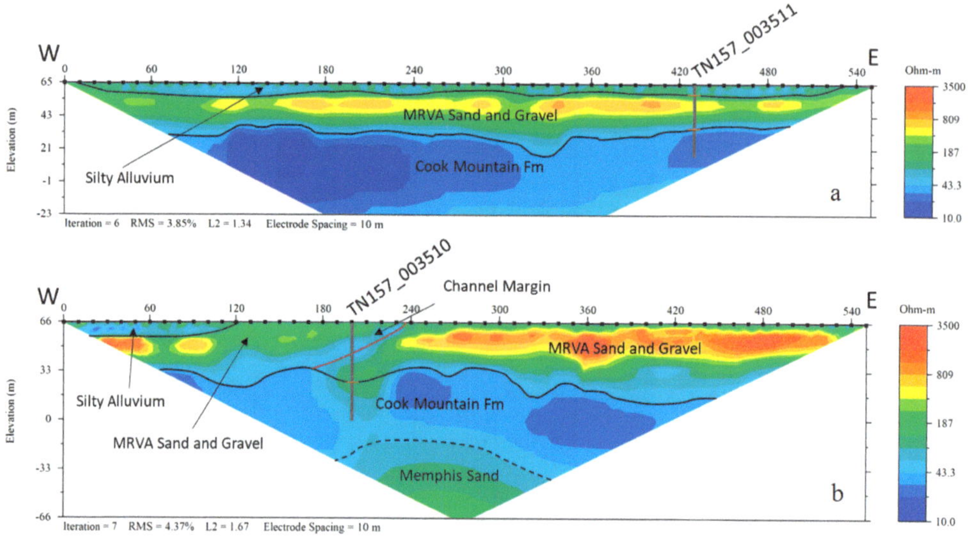

Figure 6. IRP of (**a**) Line A and (**b**) Line B at PI. The gray vertical lines represent boreholes. The solid black lines represent the stratigraphic boundaries, and the solid red line indicates the channel margin. The detailed descriptions of the two boreholes are shown in Figure 7.

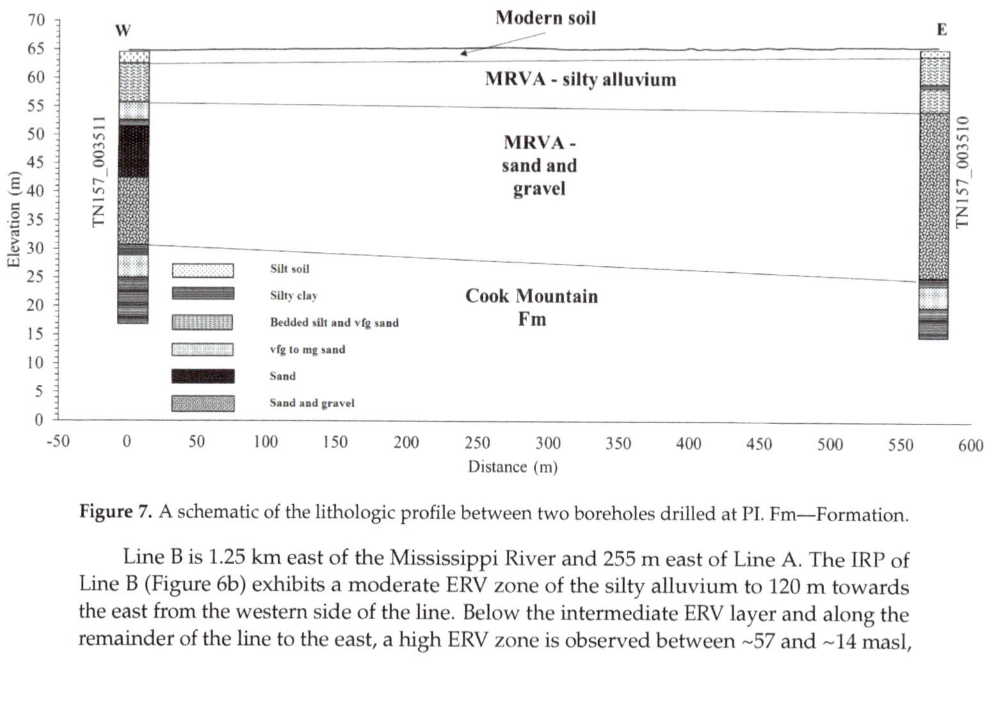

Figure 7. A schematic of the lithologic profile between two boreholes drilled at PI. Fm—Formation.

Line B is 1.25 km east of the Mississippi River and 255 m east of Line A. The IRP of Line B (Figure 6b) exhibits a moderate ERV zone of the silty alluvium to 120 m towards the east from the western side of the line. Below the intermediate ERV layer and along the remainder of the line to the east, a high ERV zone is observed between ~57 and ~14 masl, representing sand and gravel in the MRVA. A low-to-moderate ERV interval cuts across the high ERV zone from 240 to 170 m along the line. In borehole TN157_003510, this interval includes more silt and clay. The overall shape is consistent with a buried MR channel margin, similar to those inferred at the Tennessee Valley Authority (TVA) sites [14]. Like line A, the base of the high ERV zone has ~19 m of relief, with deepening of the contact to the east. Ref. [14] also noted the deepest level of incision of the MRVA in the underlying Eocene strata, which was deepest along the eastern side of the TVA sites. From ~34 masl to ~−40 masl, a low ERV layer underlies the MRVA that is represented by laminated-to-thinly-bedded silty clay and silty, very-fine-grained sand of the Cook Mountain Formation in borehole TN157_003510. The intermediate-to-high resistivity in the interval of ~26 to ~8 masl at the borehole site may represent finer sand in the upper part of the Cook Mountain Formation in that area. The intermediate-to-high ERV layer below the Cook Mountain Formation in the center of line B is the Memphis Sand, which is better resolved in line B than in line A.

The PI ER lines illustrate the stratigraphy beneath the Mississippi River floodplain and do not support what [11] previously interpreted as a broad breach. The silty alluvium is not represented as continuous based on the ER lines, which contrasts with borehole data collected at PI and data from the TVA sites. However, it is not uncommon to have poor resolution in the upper 5 m of an ER survey with 10 m electrode spacing, similar to that observed at the SF sites (Figure 4). The geometry of the sand and gravel of the lower MRVA at PI is consistent with that inferred at the TVA sites, with deeper incision and thickening of the MRVA to the east. Furthermore, line B illustrates the probable channel geometry of one of the channel complexes defined at the TVA sites. Like the TVA sites, Cook Mountain is below the MRVA, and the Cockfield Formation is absent. Although intermediate-to-high ER is observed locally in line B, the sediments representing these deposits are silty and clay-rich and do not represent well-sorted sand typical of incised paleo-channel deposits, such as that observed at SF (Figure 4). Although high-resistivity intervals are observed in AEM data beneath PI [29], they are in inaccessible areas west of lines A and B. The results of the ER study suggest that low-to-intermediate ERV intervals beneath the PI site do not represent breach sites. The extent of the breach at PI is likely much smaller than that [11] proposed initially.

3.4. Audubon Park

AP is situated in the heart of Memphis near the Sheahan well field, one of eleven Memphis' Light Gas and Water municipal-well fields. Numerous studies have been conducted using geochemical tracer, hydrologic, and borehole data on the leakage of modern water into the Memphis aquifer near the Sheahan well field [32,57,58]. The water table elevation is 60 masl [59] at the AP site and is depressed relative to the water table elsewhere in the Memphis area, suggesting water leakage from the shallow aquifer to the Memphis aquifer [32]. Two boreholes with geologic and geophysical logs are aligned with the ER line (TN157_000105 and TN157_000112), and a cored borehole is shown (TN157_003082) 0.5 km west of the line (Figure 8).

Within the uppermost 10 m, a layer with a variable ERV range of ~10 to 90 Ω-m is observed, with localized intermediate-to-high ERV zones of ~100 to 130 Ω-m (Figure 8). The low ERV zones likely represent loess with variable saturation; the lowest values represent more saturation. The spacing of low-resistivity maxima is periodic, with most separated by 10 to 30 m of intermediate-to-high resistivity, presumably representing drier zones. Below ~85 masl, a high ERV layer exceeding 300 Ω-m becomes prominent but is segmented by the intermediate-to-high ERV zones observed in the overlying loess. The upper part of the high-resistivity layer corresponds to sand and gravel of the Pliocene Upland Complex, a fluvial-terrace deposit that underlies most upland surfaces in Shelby County [37] or Pleistocene fluvial-terrace paleo-channel-fill deposits beneath the Sheahan well field area [60]. The source of the irregular variation in the high ERV layer is unclear. However, it may represent variable saturation or lithologic changes, especially clay accumulation due to paleosol development in the upper part of the sand and gravel. The base of the fluvial-terrace deposits is interpreted to be in the lower part of the highest ERV zones (>1000 Ω-m) based on geophysical and geologic logs. The lower part of the high ERV zone is the sand-rich Cockfield Formation based on the geologic borehole log from TN157_0003082 and geophysical logs from the other two boreholes. Between 160 and 250 m and 620 and 690 m, the fluvial terrace is underlain by a low-to-intermediate resistivity unit interpreted as the Cook Mountain Formation. The sand-rich Cockfield Formation is inset into the Cook Mountain Formation and from 290 to 570 m inset into the Memphis Sand. These characteristics are consistent with the Cockfield Formation being inset as paleochannel deposits into the underlying Eocene units. In all cases, the paleochannel has a concave-up geometry. The intermediate-to-high ERV interval beneath the Cockfield Formation represents the Memphis Sand based on the cored borehole log from TN157_0003082 and geophysical logs from the other two boreholes. The Memphis Sand in borehole TN157_0003082 has intervals of fine sand and clay and clay ball conglomerate, likely contributing to the intermediate ERV.

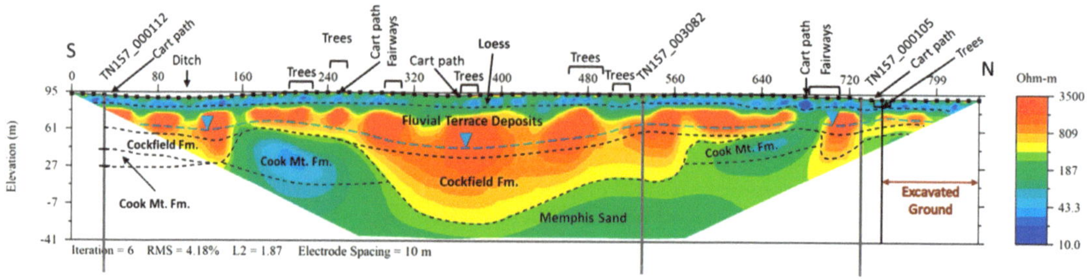

Figure 8. The IRP of the ER survey conducted at AP. The solid gray vertical lines represent wells near the survey line, and the dashed black lines are projected layer boundaries from the geologic logs. The water table is shown with a light blue dashed line with overlying inverted triangles.

The AP line generally follows the lithological variations expected from the stratigraphy determined from cored borehole TN157_0003082 and geophysical logs from other boreholes. However, it illustrates complicated patterns likely due to surface features and irregular saturation. Figure 8 shows numerous surface features along the line on the AP golf course. In addition, an underground sprinkler system and modification to the surface beneath putting greens, which are not shown, may also influence the signals. The distribution of the sprinklers, in particular, may create the periodic patterns of low resistivity observed in the upper 10 m of the line. The extension of the intermediate resistivity zones from the surface into the fluvial-terrace deposits may also have origins in the surface features, but as stated above, lithology and saturation could also play a role. The extensive modification of the surface and greater depth to the water table at the golf course, as opposed to the rather monotonous surface conditions with a shallow water table beneath the agricultural fields, especially at GC and PI, result in complications to the resulting IRPs, which require ample geologic control based on borehole data to constrain interpretation.

The IRP from the AP line supports the absence of a confining unit beneath much of the golf course, proximal to the ER line. It is consistent with borehole geophysical log data [58] and water level and hydrologic tracer studies [13,32,58]. Desaturation of the shallow aquifer and development of an anomalous depression in the water table [11] requires a direct hydrologic pathway from the fluvial-terrace deposits to the Memphis Sand, which is provided by the Cockfield paleo-channel system (Figure 8). Ref. [57] previously proposed that the confining unit was absent in a shallow borehole adjacent to borehole TN157_0003082, showing the influence of oxygenated shallow groundwater infiltrating the Memphis aquifer. Most recently, Ref. [58] used MODFLOW to model modern water through a breach beneath the AP golf course and found a favorable comparison to hydrologic tracer data. The AP ER survey imaged the physical manifestation of at least a part of the system of breaches beneath the Sheahan well field that have led to drainage of the fluvial-terrace aquifer and leakage of modern water into the Sheahan well field.

4. Discussion

4.1. Effect of Electrode Spacing

Analysis of the inverted resistivity data obtained from the ER surveys at SF reveals that electrode spacing can significantly impact the interpretation of hydrostratigraphic features. Notably, different electrode-spacing configurations lead to an apparent upward shift in the ERV layer boundaries representing various subsurface layers. The extent of this impact is more pronounced at SF (0.75–0.63 times) when the electrode spacing is reduced to half of its initial value. It is unclear why the ERV domain shifts boundaries for spacings at SF; however, the sensitivity appears to be greatest at depths < 10 m using the dipole–dipole array, similar to the results of [3]. Steel casing near the survey lines at SF may introduce errors in the depth calculation of the stratigraphic units [55], but this is likely only a local effect. The influence of the steel casing on the electrical resistivity values around the well needs to be further investigated, as well as the potential influence of other infrastructure at SF (e.g., [7]). The depth shift may also be possible with shorter electrode spacing due to a significant decrease in the apparent resistivity values, due to a shallow conductive layer at a low acquisition level [61].

Similar to the previous study [7], this study shows that an electrode spacing of 10 m yields ERV layer values and boundaries most consistent with local stratigraphy. This particular spacing configuration yields ERV layer boundary depths consistent with boundaries of subsurface stratigraphic units based on borehole data. Utilizing a larger electrode spacing yields resistivity measurements that predominantly reflect the deeper stratigraphic units, with commensurate data loss at depths < ~5 m. The results from the SF and AP surveys suggest, however, that surface and shallow subsurface features influence the ERV at depths > 5 m, suggesting that surveys using short (~2.5 m) and long (~10 m) electrode spacings with the dipole–dipole array will increase or decrease the overall layer resistivity.

4.2. Electrical Resistivity Data Analysis

At SF and AP (Figures 4a and 8), the IRPs indicate the absence of the Cook Mountain Formation or substantial clay in the Cockfield Formation between the shallow and Memphis aquifers, indicating confining-unit breaches in these areas. The SF profile reveals paleochannel features similar to those that [18] identified using shallow seismic methods. The similarity demonstrates the potential compatibility of the two methods to be used in tandem to resolve the subsurface structure and ERV domain that characterize the geometry of paleochannel features, either inset into low-resistivity clay or within predominantly high-resistivity sand. At SF, silty clay of the Cook Mountain Formation is present ~0.5 km west of the ER survey line at the same depth as the paleochannel features [12], suggesting incision and lateral migration of paleo-river systems in the Eocene as an origin of the hydrogeologic breaches. The pattern of paleo-channel overlap suggests channel migration towards the west or southwest through time, eroding the underlying Cook Mountain Formation and placing the sandy paleo-channel fill of the Cockfield Formation in direct contact with the Memphis Sand. However, the study of paleochannel migration is beyond this research's scope and requires future investigation. The AP profile incision, or partial incision of Cockfield paleochannels through the Cook Mountain Formation, is evident at three locations along the profile. Ref. [58] shows that the paleochannel beneath the AP golf course extends westward into a buried Eocene tributary drainage system. The ER profile indicates little evidence for lateral channel migration; however, perhaps shallow S-wave seismic analysis similar to that conducted at SF by [18] may be useful to resolve such depositional structure.

The ER surveys at GC and PI demonstrate consistency with regional Eocene stratigraphy and similarity of lithologic characteristics across the study area. The lithology of the

Cook Mountain Formation is similar across Shelby County [19], which is well supported by the similarity in the ERV range observed. The Cook Mountain Formation is thought to have been deposited across Shelby County as a shallow marine to brackish water prodelta clay [14,19]; thus, the continuity of lithology is expected. Thus, the absence of the Cook Mountain Formation implies removal rather than non-deposition. Paleochannel erosion of coastal-plain prodelta and marine clay units similar to the Cook Mountain Formation is also known from elsewhere in the Gulf and Atlantic coasts [62–64], suggesting further investigation may provide a broader mechanistic basis for their origin.

5. Conclusions

This study focuses on the investigation of electrical resistivity (ER) as a tool for investigating shallow stratigraphy (<100 m) in the Mississippi embayment, and defining the characteristics and geometry of breaches through a regional confining unit. ER surveys were conducted in the floodplain (GC, PI, and SF) and upland (AP) settings, some with excellent supporting geologic data and others with limited data. Initial investigations at SF, a well-studied site with numerous boreholes near the survey lines, focused on using electrode spacing (2.5, 5.0, and 10.0 m) to resolve stratigraphic units and compare them to borehole data. Electrode spacing was also investigated at GC, applying 7.0 and 10.0 m spacing. All surveys involved a dipole–dipole array, from which the apparent electrical resistivity values (ERV) were inverted to create inverted resistivity profiles (IRPs) using AGI EarthImager 2D—V. 2.1.5 software.

The electrode-spacing analysis suggests that varying the electrode spacing can result in an apparent vertical shift of ERV boundaries on IRPs. The 10 m electrode spacing yielded ERV boundaries that conform best to borehole-based stratigraphy. At SF, decreasing the electrode spacing by half caused the apparent depth of ERV boundaries to shift to shallower depth by a factor of ~0.63–0.75. A potential local source for the shift may be steel casing from a monitoring well < 5 m from the line. However, whether this is important at greater distances (50 or 100 m) from the steel casing is unclear. At GC, the shift in the depth for the two spacings was insignificant considering the m-scale resolution of the IRPs, suggesting that factors creating the shift may be specific to the SF site or greater deviation in electrode spacing. In both cases, the 10 m electrode spacing yields ERVs and ERV boundaries that best fit with known site stratigraphy. Furthermore, at SF, a channel-shaped ERV boundary noted in the 10 m survey correlated well with a similar structure identified by [18] using a shallow S-wave seismic survey, suggesting the potential to apply both survey tools to resolve structure and lithology of significant features better. Surface modifications, such as those at SF and the AP golf course, may create additional complications to interpreting IRPs and may be fruitful investigations for surveys with short (<5 m) electrode spacing.

The ER surveys at SF and AP identified paleochannel features within the sand-rich Cockfield Formation that have incised partially or entirely through the clay-rich Cook Mountain Formation and suggest that paleochannel incision may be a significant origin for breaches in the regional confining unit. The IRPs from SF show sand-filled paleochannel features within the Cockfield Formation directly overlying the Memphis Sand. The Cook Mountain Formation is present at the same depth in boreholes ~0.5 km away, suggesting that paleochannel incision removed the Cook Mountain Formation and the subsequent sand fill of the Cockfield Formation created the breach. More clearly at AP, the IRP shows sand-rich Cockfield Formation filling the paleo-channel structures incised into and through the Cook Mountain Formation and into the Memphis Sand. In both cases, the sand-rich Cockfield Formation in direct contact with the Memphis Sand creates a hydrogeologic connection between the alluvial- or fluvial-terrace aquifer and the regionally important Memphis aquifer. The paleo-channel origin implies that these hydrogeologic connections may have greater lateral extent, along the paleo-channel length, but such inference requires future investigation.

The GC and PI IRPs show a continuous Cook Mountain Formation with similar characteristics between the Memphis Sand and alluvial deposits, which supports the interpretation of the Cook Mountain as the primary confining unit for the Memphis aquifer. The regional extent and shallow marine origin of the Cook Mountain Formation imply that its absence is due to removal rather than non-deposition. Paleochannel incision through regional marine confining units in coastal-plain settings similar to those in the Eocene of southwestern Tennessee suggests that paleochannel breaches in marine confining units are more common than is currently recognized and may have a common origin.

Author Contributions: Conceptualization, M.R.H., D.L., S.S. and B.W.; methodology, M.R.H., D.L. and S.S.; software, M.R.H.; validation, M.R.H., D.L. and S.S.; formal analysis, M.R.H.; investigation, M.R.H.; resources, D.L. and S.S.; data curation, M.R.H.; writing—original draft preparation, M.R.H.; writing—review and editing, D.L., S.S. and B.W.; visualization, M.R.H. and D.L.; supervision, D.L.; project administration, S.S.; funding acquisition, B.W. All authors have read and agreed to the published version of the manuscript.

Funding: This study was part of the aquitard study project and was funded by Memphis Light, Gas, and Water (MLGW), contract number 12064.

Data Availability Statement: The data presented in this study are available on request from the corresponding author. The data are not publicly available due to compliance with the MLGW contract.

Acknowledgments: We thank MLGW for financing the research. We thank the Center for Applied Earth Science and Engineering Research (CAESER) and the Center for Earthquake Research and Information (CERI) for providing continuous support with the equipment and planning needed for this research. Special thanks to Kaushik Sarker, a graduate student at CERI, for his help and support during equipment setup and field data collection. We also thank the graduate students who helped us in the fields during data collection. We thank three anonymous reviewers for their helpful comments that improved the clarity and technical aspects of the manuscript.

Conflicts of Interest: The authors declare no conflict of interest.

Appendix A. Comparison of the Inverted Resistivity Profiles of the Data Collected Using Different Arrays

The inverted resistivity profile of the data collected using the dipole–dipole arrays resolves lateral variation better than the data collected using other arrays.

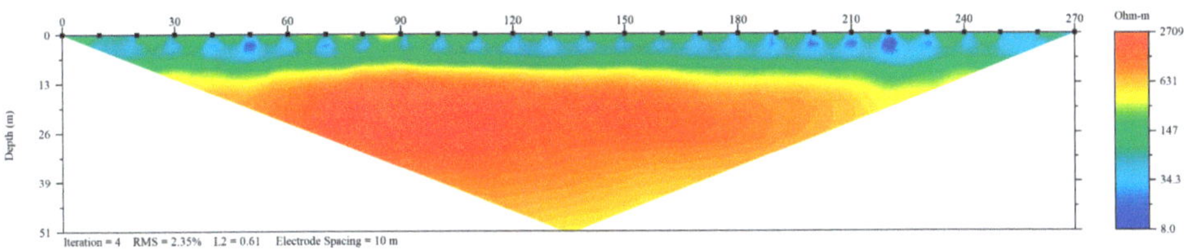

Figure A1. The inverted resistivity profile of the data was collected using the Wenner array with 10 m electrode spacing and 28 channels at SF.

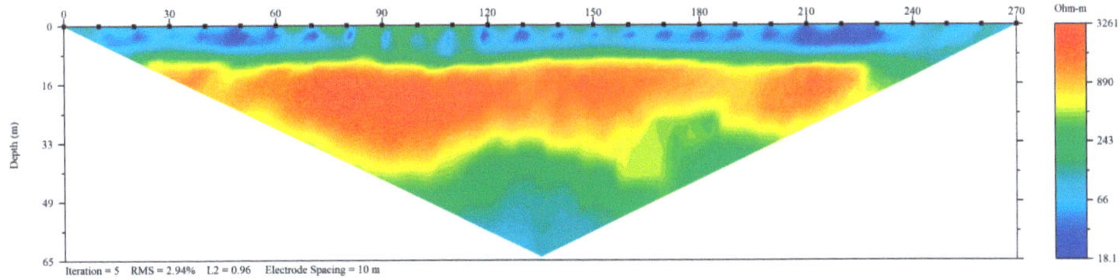

Figure A2. The inverted resistivity profile of the data was collected using the dipole–dipole array with 10 m electrode spacing and 28 channels at SF.

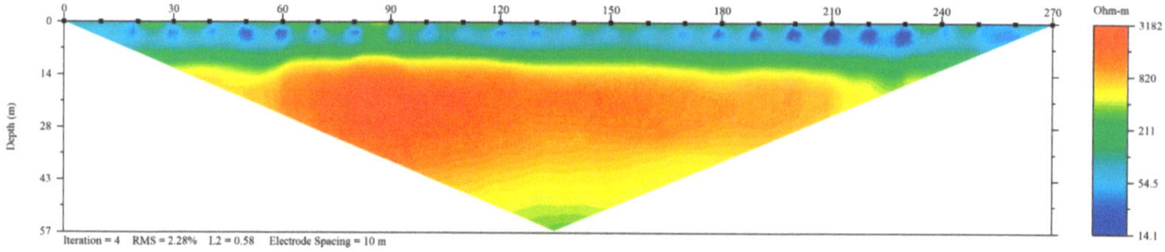

Figure A3. The inverted resistivity profile of the data was collected using the Schlumberger array with 10 m electrode spacing and 28 channels at SF.

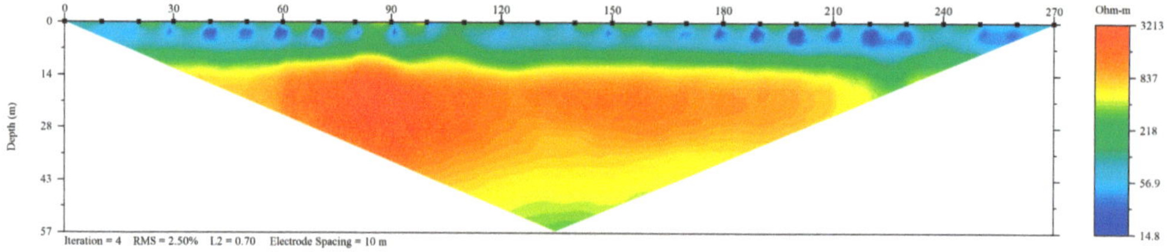

Figure A4. The inverted resistivity profile of the data was collected using the Inverse Schlumberger array with 10 m electrode spacing and 28 channels at SF.

Appendix B. Apparent Resistivity Data Was Collected Using a Roll-Along Survey at Audubon Park (AP), Showing a Triangulated Zone of No Data

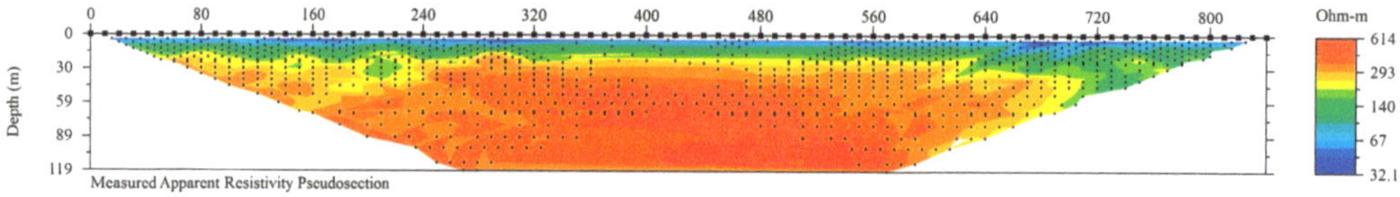

Figure A5. Apparent resistivity pseudosection of the data collected at AP using roll-along survey.

Appendix C. Verification of Maximum Depth of Investigation

Table A1. Analysis of the DOI and array length relationship at Shelby Farms.

Array Length, L (m)	Depth of the Deepest Data Point, D (m)	D/L
270	59	0.22
270	52	0.19
135	25.9	0.19

Table A2. Analysis of DOI and array length relationship at Grays Creek.

Array Length, L (m)	Depth of the Deepest Data Point, D (m)	D/L
550	119	0.22
385	78	0.20

Table A3. Analysis of DOI and array length relationship at President's Island.

Array Length, L (m)	Depth of the Deepest Data Point, D (m)	D/L
440	88	0.20
550	111	0.20
550	105	0.19

Appendix D. Gamma Logs of Boreholes Drilled at President's Island

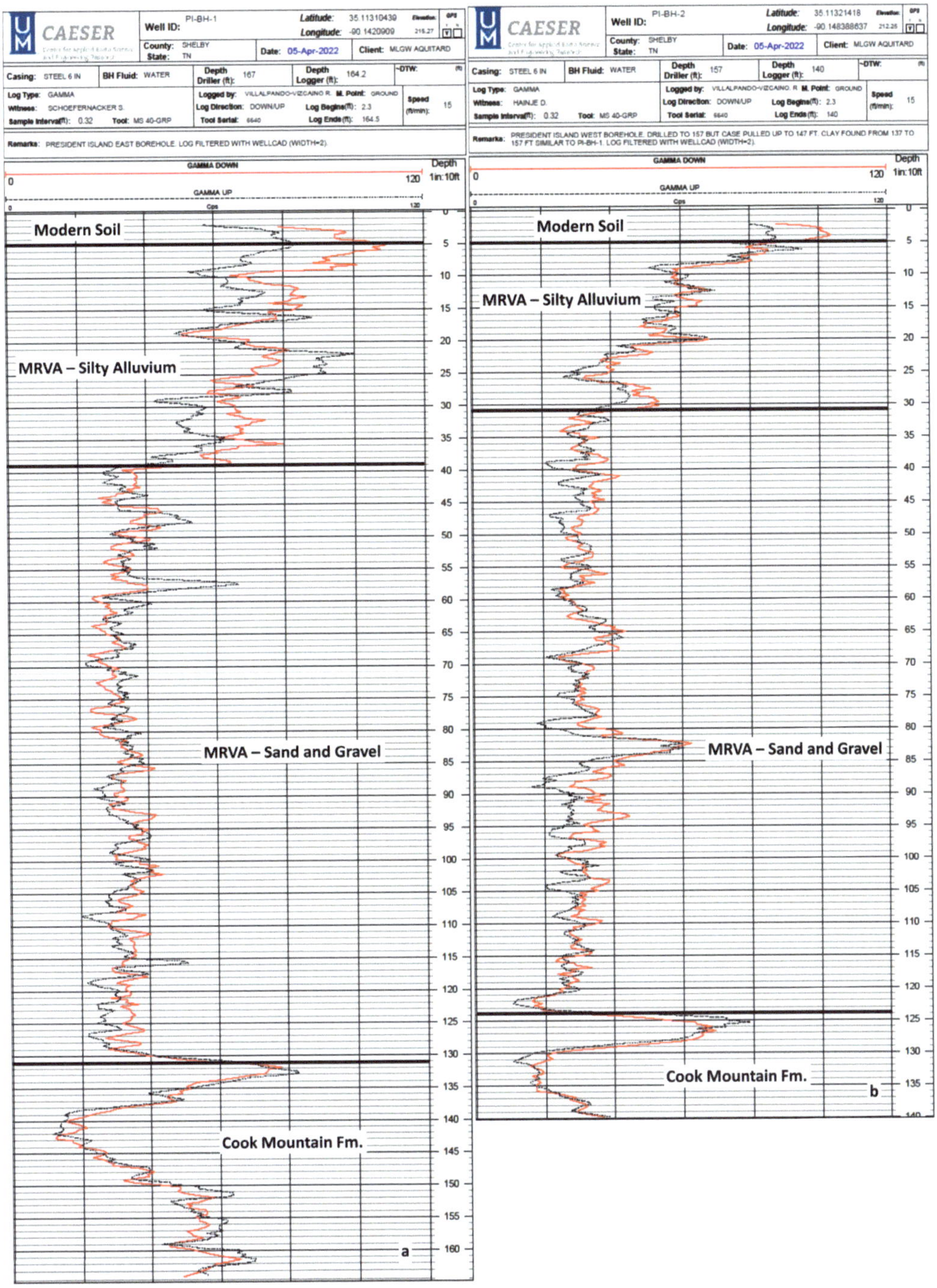

Figure A6. Gamma logs obtained at Pesident's Island. (**a**) Borehole TN157_003510, and (**b**) Borehole TN157_003511. Red line is gamma log down borehole and gray is up the borehole.

References

1. Kowalczyk, S.; Maślakowski, M.; Tucholka, P. Determination of the Correlation between the Electrical Resistivity of Non-Cohesive Soils and the Degree of Compaction. *J. Appl. Geophys.* **2014**, *110*, 43–50. [CrossRef]
2. Casagrande, M.F.S.; Furlan, L.M.; Moreira, C.A.; Rosa, F.T.G.; Rosolen, V. Non-Invasive Methods in the Identification of Hydrological Ecosystem Services of a Tropical Isolated Wetland (Brazilian Study Case). *Environ. Chall.* **2021**, *5*, 100233. [CrossRef]
3. Singh, U.; Sharma, P.K. *Non-Invasive Subsurface Groundwater Exploration Techniques BT—Environmental Processes and Management: Tools and Practices for Groundwater*; Shukla, P., Singh, P., Singh, R.M., Eds.; Springer International Publishing: Cham, Switzerland, 2023; pp. 1–16, ISBN 978-3-031-20208-7.
4. Fajana, A.O. Groundwater Aquifer Potential Using Electrical Resistivity Method and Porosity Calculation: A Case Study. *NRIAG J. Astron. Geophys.* **2020**, *9*, 168–175. [CrossRef]
5. Wahab, S.; Saibi, H.; Mizunaga, H. Groundwater Aquifer Detection Using the Electrical Resistivity Method at Ito Campus, Kyushu University (Fukuoka, Japan). *Geosci. Lett.* **2021**, *8*, 15. [CrossRef]
6. Wu, J.; Dai, F.; Liu, P.; Huang, Z.; Meng, L. Application of the Electrical Resistivity Tomography in Groundwater Detection on Loess Plateau. *Sci. Rep.* **2023**, *13*, 4821. [CrossRef] [PubMed]
7. Schoefernacker, S. Evaluation and Evolution of a Groundwater Contaminant Plume at the Former Shelby County Landfill, Memphis, Tennessee. Ph.D. Dissertation, The University of Memphis, Memphis, TN, USA, 2018.
8. Hussain, Y.; Uagoda, R.; Borges, W.; Nunes, J.; Hamza, O.; Condori, C.; Aslam, K.; Dou, J.; Cárdenas-Soto, M. The Potential Use of Geophysical Methods to Identify Cavities, Sinkholes and Pathways for Water Infiltration. *Water* **2020**, *12*, 2289. [CrossRef]
9. Thomas, J.C.; Spring, M.A.; Gruhn, L.R.; Bristow, E.L. *Application of Geophysical Methods to Enhance Aquifer Characterization and Groundwater-Flow. Model. Development, Des. Moines River Alluvial Aquifer, Des. Moines, Iowa, 2022*; US Geological Survey: Reston, VA, USA, 2023. [CrossRef]
10. Graham, D.D.; Parks, W.S. *Potential for Leakage among Principal Aquifers in the Memphis Area, Tennessee*; Water-Resources Investigations Report 85-4295; US Geological Survey: Reston, VA, USA, 1986. [CrossRef]
11. Parks, W.S. *Hydrogeology and Preliminary Assessment of the Potential for Contamination of the Memphis Aquifer in the Memphis Area, Tennessee*; Water-Resources Investigations Report 90-4092; US Geological Survey: Reston, VA, USA, 1990. [CrossRef]
12. Gentry, R.; McKay, L.; Thonnard, N.; Anderson, J.L.; Larsen, D.; Carmichael, J.K.; Solomon, K. *Novel Techniques for Investigating Recharge to the Memphis Aquifer*; American Water Works Association: Denver, CO, USA, 2006; p. 97.
13. Larsen, D.; Waldron, B.; Schoefernacker, S.; Gallo, H.; Koban, J.; Bradshaw, E. Application of Environmental Tracers in the Memphis Aquifer and Implication for Sustainability of Groundwater Resources in the Memphis Metropolitan Area, Tennessee. *J. Contemp. Water Res. Educ.* **2016**, *159*, 78–104. [CrossRef]
14. Carmichael, J.K.; Kingsbury, J.A.; Larsen, D.; Schoefernacker, S. *Preliminary Evaluation of the Hydrogeology and Groundwater Quality of the Mississippi River Valley Alluvial Aquifer and Memphis Aquifer at the Tennessee Valley Authority Allen Power Plants, Memphis, Shelby County, Tennessee*; Open-File Report 2018-1097; US Geological Survey: Reston, VA, USA, 2018.
15. Larsen, D.; Waldron, B.; Schoefernacker, S. Updated Map of Semi-Confined Conditions in the Memphis Aquifer, Shelby County, Tennessee: A Work in Progress. In Proceedings of the Virtual 2022 Tennessee Water Resources Symposium, Virtual, 20–21 April 2022; TN Section AWRA: Nashville, TN, USA, 2022; p. 2C-3.
16. Parks, W.S.; Carmichael, J.K. *Geology and Ground-Water Resources of the Memphis Sand in Western Tennessee*; Resources Investigations Report 88-4182; US Geological Survey: Reston, VA, USA, 1990. [CrossRef]
17. Waldron, B.; Larsen, D.; Hannigan, R.; Csontos, R.; Anderson, J.; Dowling, C.; Bouldin, J. *Mississippi Embayment Regional Ground Water Study*; US Environmental Protection Agency: Washington, DC, USA, 2011; p. 192.
18. Waldron, B.A.; Harris, J.B.; Larsen, D.; Pell, A. Mapping an Aquitard Breach Using Shear-Wave Seismic Reflection. *Hydrogeol. J.* **2009**, *17*, 505–517. [CrossRef]
19. Larsen, D.; Schoefernacker, S.R.; Waldron, B. Stratigraphy of Upper Claiborne Strata in Western Tennessee and Hydrogeologic Implications. In *Geological Society of America Abstracts with Programs 159-10*; The Geological Society of America: Boulder, CO, USA, 2022; Volume 54. [CrossRef]
20. Hong-Jing, J.; Shun-Qun, L.; Lin, L. The Relationship between the Electrical Resistivity and Saturation of Unsaturated Soil. *Electron. J. Geotech. Eng.* **2014**, *19*, 3739–3746. [CrossRef]
21. Lu, Y.; Abuel-Naga, H.; Al Rashid, Q.; Hasan, M.F. Effect of Pore-Water Salinity on the Electrical Resistivity of Partially Saturated Compacted Clay Liners. *Adv. Mater. Sci. Eng.* **2019**, *2019*, 7974152. [CrossRef]
22. Fatima Zohra, H.; Laredj, N.; Maliki, M.; Missoum, H.; Bendani, K. Laboratory Evaluation of Soil Geotechnical Properties via Electrical Conductivity Evaluación de Laboratorio de Las Propiedades Geotécnicas Del Suelo Mediante Conductividad Eléctrica. *Rev. Fac. De Ing.* **2019**, *90*, 101–112. [CrossRef]
23. Corwin, D.L.; Yemoto, K. Salinity: Electrical Conductivity and Total Dissolved Solids. *Soil. Sci. Soc. Am. J.* **2020**, *84*, 1442–1461. [CrossRef]
24. Ward, S.H. The Resistivity and Induced Polarization Methods. In Proceedings of the 1st EEGS Symposium on the Application of Geophysics to Engineering and Environmental Problems, Golden, CO, USA, 28–31 March 1988; Environment and Engineering Geophysical Society: Denver, CO, USA, 1998; pp. 109–250. [CrossRef]

25. Zonge, K.; Wynn, J.; Urquhart, S. Resistivity, Induced Polarization, and Complex Resistivity. In *Near-Surface Geophysics; Investigations in Geophysics*; Butler, D.K., Ed.; Society of Exploration Geophysicists: Tulsa, OK, USA, 2005; pp. 265–300, ISBN 978-1-56080-130-6.
26. Bedrosian, P.A.; Schamper, C.; Auken, E. A Comparison of Helicopter-Borne Electromagnetic Systems for Hydrogeologic Studies. *Geophys. Prospect.* **2016**, *64*, 192–215. [CrossRef]
27. Baldridge, W.S.; Cole, G.L.; Robinson, B.A.; Jiracek, G.R. Application of Time-Domain Airborne Electromagnetic Induction to Hydrogeologic Investigations on the Pajarito Plateau, New Mexico, USA. *Geophysics* **2007**, *72*, B31–B45. [CrossRef]
28. Wang, H.; Revil, A. Surface Conduction Model for Fractal Porous Media. *Geophys. Res. Lett.* **2020**, *47*, e2020GL087553. [CrossRef]
29. Minsley, B.J.; Rigby, J.R.; James, S.R.; Burton, B.L.; Knierim, K.J.; Pace, M.D.M.; Bedrosian, P.A.; Kress, W.H. Airborne Geophysical Surveys of the Lower Mississippi Valley Demonstrate System-Scale Mapping of Subsurface Architecture. *Commun. Earth Environ.* **2021**, *2*, 131. [CrossRef]
30. Nabi, A.; Liu, X.; Gong, Z.; Ali, A. Electrical Resistivity Imaging of Active Faults in Palaeoseismology: Case Studies from Karachi Arc, Southern Kirthar Fold Belt, Pakistan. *Nriag J. Astron. Geophys.* **2020**, *9*, 116–128. [CrossRef]
31. Porras, D.; Carrasco, J.; Carrasco, P.; González, P.J. Imaging Extensional Fault Systems Using Deep Electrical Resistivity Tomography: A Case Study of the Baza Fault, Betic Cordillera, Spain. *J. Appl. Geophys.* **2022**, *202*, 104673. [CrossRef]
32. Larsen, D.; Gentry, R.; Solomon, D. The Geochemistry and Mixing of Leakage in a Semi-Confined Aquifer at a Municipal Well Field, Memphis, Tennessee, USA. *Appl. Geochem.* **2003**, *18*, 1043–1063. [CrossRef]
33. Bradley, M.W. *Ground-Water Hydrology and the Effects of Vertical Leakage and Leachate Migration on Ground-Water Quality near the Shelby County Landfill, Memphis, Tennessee*; Water-Resources Investigations Report 90-4075; US Geological Survey: Reston, VA, USA, 1991; p. 47. [CrossRef]
34. Parks, W.S.; Mirecki, J.E. *Hydrogeology, Ground-Water Quality, and Potential for Water-Supply Contamination near the Shelby County Landfill in Memphis, Tennessee*; U.S. Geological Survey Water-Resources Investigations Report 91-4173; US Geological Survey: Reston, VA, USA, 1992; p. 79. [CrossRef]
35. Clark, B.R.; Hart, R.M. *The Mississippi Embayment Regional Aquifer Study (MERAS): Documentation of a Groundwater-Flow Model Constructed to Assess Water Availability in the Mississippi Embayment*; Scientific Investigations Report 2009–5172; US Geological Survey: Reston, VA, USA, 2009; p. 61. [CrossRef]
36. Dieter, C.A.; Maupin, M.A.; Caldwell, R.R.; Harris, M.A.; Ivahnenko, T.I.; Lovelace, J.K.; Barber, N.L.; Linsey, K.S. *Estimated Use of Water in the United States in 2015*; Circular 1441; US Geological Survey: Reston, VA, USA, 2018; p. 65. [CrossRef]
37. Van Arsdale, R.B.; Cox, R.T. The Mississippi's Curious Origins. *Sci. Am.* **2007**, *296*, 76B–82B. [CrossRef]
38. Brahana, J.V.; Broshears, R.E. *Hydrogeology and Ground-Water Flow in the Memphis and Fort Pillow Aquifers in the Memphis Area, Tennessee*; Water-Resources Investigations Report 89-4131; US Geological Survey: Reston, VA, USA, 2001; p. 56. [CrossRef]
39. Konduru Narsimha, V.K. Altitudes of Water Levels 2005, and Historic Water Level Change in Surficial and Memphis Aquifer, Memphis, Tennessee. Master's Thesis, The University of Memphis, Memphis, TN, USA, 2007.
40. Lloyd, O.B.; Lyke, W.L. *Ground Water Atlas of the United States: Segment 10, Illinois, Indiana, Kentucky, Ohio, Tennessee*; Hydrologic Atlas 730; US Geological Survey: Reston, VA, USA, 1995; p. 30. [CrossRef]
41. Vanderlip, C.A.; Cox, R.T.; Larsen, D.; Mitchell, J.; Harris, J.B.; Cearley, C.S. Newly Recognized Quaternary Surface Faulting and Folding Peripheral to the New Madrid Seismic Zone, Central United States, and Implications for Restraining Bend Models of Intraplate Seismic Zones. *J. Geol.* **2021**, *129*, 77–95. [CrossRef]
42. Martin, R.V.; Van Arsdale, R.B. Stratigraphy and Structure of the Eocene Memphis Sand above the Eastern Margin of the Reelfoot Rift in Tennessee, Mississippi, and Arkansas, USA. *GSA Bull.* **2017**, *129*, 970–996. [CrossRef]
43. Bursi, J.B. Recharge Pathways and Mechanisms to the Memphis Aquifer. Master's Thesis, University of Memphis, Memphis, TN, USA, 2015.
44. Brahana, J.V.; Parks, W.S.; Gaydos, M.W. *Quality of Water from Freshwater Aquifers and Principal Well Fields in the Memphis Area, Tennessee*; Water-Resources Investigations Report 87-4052; US Geological Survey: Reston, VA, USA, 1987; p. 26. [CrossRef]
45. Jamaluddin; Umar, E. Identification of Subsurface Layer with Wenner-Schlumberger Arrays Configuration Geoelectrical Method. *IOP Conf. Ser. Earth Environ. Sci.* **2018**, *118*, 12006. [CrossRef]
46. Clifford, J.; Binley, A. Geophysical Characterization of Riverbed Hydrostratigraphy Using Electrical Resistance Tomography. *Near Surf. Geophys.* **2010**, *8*, 493–501. [CrossRef]
47. Lowry, W.F. Geophysical Techniques. In *Field Sampling Procedures Manual*; The New Jersey Department of Environmental Protection (NJDEP): Trenton, NJ, USA, 2022; p. 46.
48. Wightman, W.E.; Jalinoos, F.; Sirles, P.; Hanna, K. *Application of Geophysical Methods to Highway Related Problems*; Tech Report FHWA-IF-04-021; US Federal Highway Administration: Lakewood, CO, USA, 2003; p. 744.
49. Palacky, G.J. Resistivity Characteristics of Geologic Targets. In *Electromagnetic Methods in Applied Geophysics–Theory Volume 1*; Investigations in Geophysics, Volume 3; Society of Exploration Geophysicists: Houston, TX, USA, 1988; pp. 52–129, ISBN 978-0-931830-51-8.
50. Samouëlian, A.; Cousin, I.; Tabbagh, A.; Bruand, A.; Richard, G. Electrical Resistivity Survey in Soil Science: A Review. *Soil. Tillage Res.* **2005**, *83*, 173–193. [CrossRef]
51. Friedman, S.P. Soil Properties Influencing Apparent Electrical Conductivity: A Review. *Comput. Electron. Agric.* **2005**, *46*, 45–70. [CrossRef]

52. Zhou, M.; Wang, J.; Cai, L.; Fan, Y.; Zheng, Z. Laboratory Investigations on Factors Affecting Soil Electrical Resistivity and the Measurement. *IEEE Trans. Ind. Appl.* **2015**, *51*, 5358–5365. [CrossRef]
53. Fondriest Environmental Inc. *Conductivity, Salinity and Total Dissolved Solids*; Fondriest Environmental Inc.: Fairborn, OH, USA, 2014.
54. Moore, K.M. Investigation of the Hydrogeology of the Memphis Light, Gas, and Water Shaw Wellfield, Shelby County, Tennessee. Master's Thesis, The University of Memphis, Memphis, TN, USA, 2021.
55. Heagy, L.J.; Oldenburg, D.W. Direct Current Resistivity with Steel-Cased Wells. *Geophys. J. Int.* **2019**, *219*, 1–26. [CrossRef]
56. Kingsbury, J.A.; Parks, W.S. *Hydrogeology of the Principal Aquifers and Relation of Faults to Interaquifer Leakage in the Memphis Area, Tennessee*; Water-Resources Investigations Report 93-4075; US Geological Survey: Reston, VA, USA, 1993; p. 18. [CrossRef]
57. Gentry, R.W.; Ku, T.-L.; Luo, S.; Todd, V.; Larsen, D.; McCarthy, J. Resolving Aquifer Behavior near a Focused Recharge Feature Based upon Synoptic Wellfield Hydrogeochemical Tracer Results. *J. Hydrol.* **2005**, *323*, 387–403. [CrossRef]
58. Hasan, K. Investigation of Modern Leakage Based on Numerical and Geochemical Modeling near a Municipal Well Field in Memphis, Tennessee. Ph.D. Dissertation, The University of Memphis, Memphis, TN, USA, 2023.
59. Lozano-Medina, D.; Waldron, B.; Schoefernacker, S.; Antipova, A.; Villalpando-Vizcaino, R. Stories of a Water-Table: Anomalous Depressions, Aquitard Breaches and Seasonal Implications, Shelby County, Tennessee, USA. *Environ. Monit. Assess.* **2023**, *195*, 953. [CrossRef]
60. Hossain, M.S.; Larsen, D.; Schoefernacker, S.R.; Vizcanio, R.; Hasan, M.R. Investigation of Pliocene and Pleistocene Fluvial-Terrace and Alluvial Deposits in Shelby County, Tennessee, and Their Relationship to the Shallow Aquifer System. In *Geological Society of America Abstracts with Programs 159-11*; The Geological Society of America: Boulder, CO, USA, 2022; Volume 54. [CrossRef]
61. Clément, R.; Descloitres, M.; Günther, T.; Ribolzi, O.; Legchenko, A. Influence of Shallow Infiltration on Time-Lapse ERT: Experience of Advanced Interpretation. *Comptes Rendus Geosci.* **2009**, *341*, 886–898. [CrossRef]
62. Morton, R.A.; Suter, J.R. Sequence Stratigraphy and Composition of Late Quaternary Shelf-Margin Deltas, Northern Gulf of Mexico1. *Am. Assoc. Pet. Geol. Bull.* **1996**, *80*, 505–530. [CrossRef]
63. Castle, J.W.; Miller, R.B. Recognition and Hydrologic Significance of Passive-Margin Updip Sequences: An Example from Eocene Coastal-Plain Deposits, USA. *J. Sediment. Res.* **2000**, *70*, 1290–1301. [CrossRef]
64. Galloway, W.E. Chapter 15 Depositional Evolution of the Gulf of Mexico Sedimentary Basin. In *The Sedimentary Basins of the United States and Canada*; Miall, A., Ed.; Elsevier: Amsterdam, The Netherlands, 2008; Volume 5, pp. 505–549, ISBN 1874-5997.

water

MDPI

Article

Coupled Geophysical and Hydrogeochemical Characterization of a Coastal Aquifer as Tool for a More Efficient Management (Torredembarra, Spain)

Alex Sendrós [1,2,*], Ingrid J. Cubides [3], Mahjoub Himi [1], Raúl Lovera [1,2], Aritz Urruela [1], Josefina C. Tapias [2,4], Lluís Rivero [1,2], Ruben Garcia-Artigas [5] and Albert Casas [1,2]

1 Department of Mineralogy, Petrology and Applied Geology, Universitat de Barcelona, 08028 Barcelona, Spain; himi@ub.edu (M.H.); rlovera@ub.edu (R.L.); aritz.urruela@ub.edu (A.U.); lrivero@ub.edu (L.R.); albert.casas@ub.edu (A.C.)
2 Water Research Institute, Universitat de Barcelona, 08001 Barcelona, Spain; jtapias@ub.edu
3 School of Geology, Industrial University of Santander, Bucaramanga 680003, Colombia; ingridcubides65@gmail.com
4 Department of Biology, Health and Environment, Universitat de Barcelona, 08028 Barcelona, Spain
5 Agència Catalana de l'Aigua, Generalitat de Catalunya, 08008 Barcelona, Spain; ruben.garciaa@gencat.cat
* Correspondence: alex.sendros@ub.edu

Abstract: The aquifers of the Spanish Mediterranean coast are generally subjected to intense exploitation to meet the growing water supply demands. The result of the exploitation is salinization due to the marine saltwater intrusion, causing a deterioration in the quality of the water pumped, limiting its use for community needs, and not always being well delimited. To prevent deterioration, a groundwater control network usually allows precise knowledge of the areas affected by saltwater intrusion but not the extent of the saline plumes. Moreover, the characterization of aquifer systems requires a model that defines the geometry of aquifer formations. For this objective, we integrated hydrogeological, hydrogeochemical, and electrical resistivity subsoil data to establish a hydrogeological model of the coastal aquifer of Torredembarra (Tarragona, NE Spain). In this research, we have carried out a regional and local-scale study of the aquifer system to define the areas prone to being affected by saline intrusion (electrical resistivity values below 10 Ω·m). The obtained results could be used as a support tool for the assessment of the most favorable areas for groundwater withdrawal, as well as enabling the control and protection of the most susceptible areas to be affected by saltwater intrusion.

Keywords: aquifer geometry; electrical resistivity tomography; saltwater intrusion; geoelectrical sounding; groundwater sustainability

1. Introduction

Surface water resources in coastal areas are often scarce, and the aquifers play a pivotal role in managing the complex issue of water supply [1–3]. Aquifers are an underground set of rock or sediment formations that are saturated and permeable enough to transmit economical amounts of water to springs and wells, generating for all living beings a greater possibility of access to a drinkable water resource [4]. Aquifers also have a primordial relationship with the wetlands and ecosystems present around them, greatly influencing the genesis and preservation of these waters, which is why they are of significant use for survival, especially for human beings [5].

However, the accelerated increase in economic and demographic development in worldwide coastal areas in general, and in Spain in particular, has generated a natural imbalance [6,7]. The increasing water demand, because of per capita demand and population increases, causes a strongly increasing exploitation of the available high-quality water resources, increasing the risk of salinization via the mixing of fresh and saltwater and generating marine intrusion [8]. Marine intrusion is a dynamic process in which saltwater from the sea moves inland during periods of lower aquifer recharge and recedes towards the aquifer when freshwater recharge from the continent increases, resulting in the mixing of these waters and the contamination of the aquifers that supply the need for water resources [9,10].

These trends can be further worsened in large areas by the enduring effects of climate change [11]. All of these conditions are causing stress on water resources, especially in terms of the risks of groundwater quality degradation in the case of coastal aquifers [12]. Over the decades, all of the described increasing difficulties, including environmental concerns, have highlighted the need to improve the management of groundwater resources [13–15].

Spain is a country that is mostly surrounded by continental and island coasts, and it is there where most of its population is located, as well as a series of aquifers of vital interest due to the volume and strategic nature of the water resources they store [16].

The aquifers of the Spanish and Catalan Mediterranean coast are generally subjected to intense exploitation to meet the growing water supply needs for domestic, tourist, industrial, and agricultural uses [17]. The result of the intense exploitation, mainly by sustained groundwater pumping, is salinization due to the marine saltwater intrusion, causing a deterioration in the quality of the water pumped and limiting its use for certain needs of the community [18].

To prevent the quality deterioration of groundwater resources, the Catalan Water Agency (ACA) has designed a piezometric network for monitoring salinity and the groundwater decline and depletion caused by intensive pumping. The network has 190 groundwater control points, and conductivity, pH, and temperature are analyzed in situ, as are the concentrations of the majority of anions and cations in the laboratory. The monitoring system allows precise knowledge of the zones affected by saltwater intrusion but not the extent of the saline plumes [19].

The use of wells and piezometers that have already been drilled and equipped, as well as the use of lithological logs, aquifer water sampling, and water table measurements, are all examples of direct hydrogeological procedures classically used. The logs and hydrochemical data only provide punctual information on the water quality, hydraulic data, and aquifer geometry. Additionally, a model that specifies the geometry and structural boundaries of aquifer formations is necessary for the hydrogeological and hydrogeochemical characterization of aquifer systems [20].

Indirect methods, such as near-surface geophysical techniques, have been extensively used in groundwater research in coastal locations to acquire this key information. Geophysical techniques are high-resolution methods that reveal details about the physical characteristics of the subsoil and its spatial distribution on a variety of working scales [21,22]. Moreover, the large electrical resistivity contrast between seawater (0.2 Ωm) and freshwater (>5 $\Omega \cdot$m) makes it possible to map the subsurface groundwater salinity distribution using geoelectrical techniques [23]. When lithological data are scarce or unable to provide the detailed subsurface knowledge needed for groundwater modeling, these non-invasive techniques become beneficial. Vertical electrical soundings (VES) [24–27] and Electrical resistivity tomography (ERT) are geoelectrical methods widely used to characterize coastal aquifer characteristics and properties such as on the Morocco coastal rift [24], Southwest Portugal [28], Pucket (Thailand) coastal aquifer [25], and delineating seawater intrusion on Monterrey (Mexico) [29], Bela Plain (Pakistan) [30], and also in Mediterranean coastal aquifers such as Cap-Bon (Tunisia) [31], South-Western Sicily (Italy) [32], River Nil Delta (Egypt) [33,34]; Rhodope (Greece) [35], Port de la Selva (Spain) [36] and Vélez Málaga (Spain) [37] among others.

The procedure used in this research combines the acquisition of new geophysical data with the use of publicly available geophysical data from the Spanish Geological and Mining Survey (IGME) and hydrochemical data from the Water Catalan Agency (ACA), and we consider it to have three fundamental advantages for developing new assessment tools: speed of application, resolution, and an efficient cost-benefit ratio.

We have conducted a regional-scale qualitative and quantitative study of the aquifer system to define a hydrogeological conceptualization of the coastal aquifer, to characterize the hydrogeochemistry of groundwater, and to identify the areas prone to being affected by the saline intrusion. The results could be utilized as a decision-supporting tool to evaluate the most favorable locations for groundwater extraction as well as control the most susceptible zones to be affected by saltwater intrusion. In addition, the use of indirect geophysical methods will limit the number of new piezometers and the subsequent risk of coming into contact with different aquifer units during drilling operations.

2. Study Area

The studied zone is situated northeast of the Iberian Peninsula and southeast of the Camp de Tarragona basin. The area is limited to the north by the Bonastre Massif and to the south by the Mediterranean Sea (Figure 1).

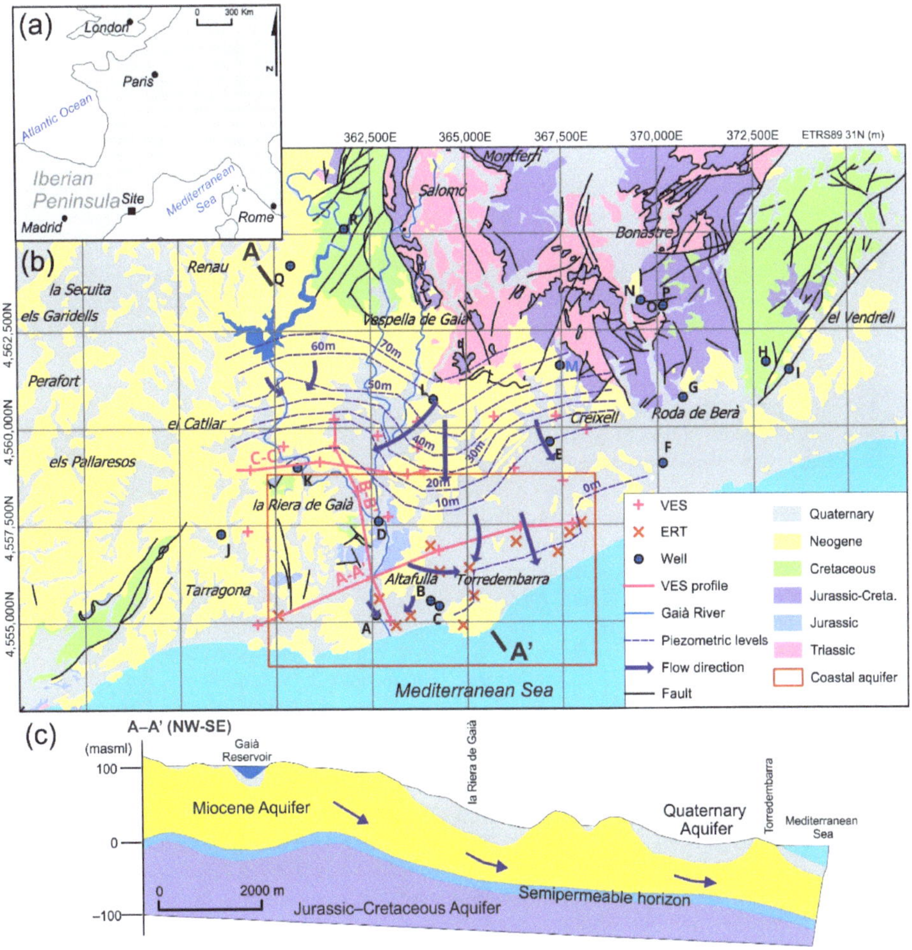

Figure 1. (**a**) Location sketch of the site; (**b**) Hydrogeological setting of the Catllar-Torredembarra site and geophysical (Vertical Electrical Sounding-VES and Electrical Resistivity-ERT) and hydrogeochemical (well) data used for the present research. Piezometric surfaces are in meters above mean sea level. Modified from [38,39]; (**c**) hydrogeological cross-section. Modified from [40].

The region has a Mediterranean climate with a warm thermal regime in the summer (average of 26 °C in July) and a moderately cold thermal regime in the winter (average of 8 °C in February). The yearly rainfall ranges, on average, from 550 to 650 mm/year. The wettest months are reported in the autumn, whereas July is the driest month and has the highest potential evapotranspiration values [41].

Surface water resources are consequently scarce. The Gaià River is the main resource, having an average discharge rate of 0.3 m^3/s (2015–2020) [42], and its 60.5 hm^3 reservoir is used for petrochemical industrial and agriculture supply [43].

The municipalities of the studied area use groundwater for water supply, similar to other Mediterranean coastal regions, and have undergone extensive urbanization along with a concomitant shift away from an agricultural and fishing economy to one governed by the expansion of tourism from the 1960s to the present. Due to the surge in water use for recreational purposes and tourists, particularly during the summer, the area's water demands have therefore significantly increased, when the Torredembarra and Altafulla populations reach 475,000 inhabitants compared to 21,000 inhabitants during the winter or non-tourist season [44].

The Camp de Tarragona zone is the result of the sedimentary filling of an Oligocene tectonic trench between the Pre-Coastal and Coastal Catalan mountain ranges [45]. It is filled by sediments from the Miocene to the Quaternary of marine or continental origin due to denudation of the southern edge of the Pre-Coastal Mountain Range [46]. This arrangement forms a multilayer system in which, basically, there are two hydrogeological units with different potentials but which coexist with other deep aquifers with independent hydrogeological functioning subject to a hydraulic connection in favor of fractures in deep calcareous levels.

The Quaternary and Miocene aquifers would form the same hydrogeological unit in this study area since they are hydraulically connected; this would therefore be the most important aquiferous unit in the area in terms of exploitation and outcrop extension. The Quaternary aquifer is formed by Plio-Quaternary gravels of fluvial-torrential and piedmont origin in hydraulic contact with conglomerates, continental sandstones, and Miocene calcarenites in the sector of the Francolí River (located west of the studied zone). The shallow aquifer has a free aquifer behavior with an average thickness of 10–15 m and high transmissivity values of up to 2.3–3.5 × 10^{-2} m^2/s, where clastic sediments dominate, favoring its intense exploitation [47]. The underground flow is perpendicular to the Mediterranean, except in the Francolí alluvial, whose bed drains laterally to the aquifer.

The Miocene Aquifer, on the other hand, has multilayer behavior. The average thickness of the Miocene infills is 50 to 300 m, and transmissivities are variable from medium to high, with values of 5 to 20 × 10^{-4} m^2/s [47]. The piezometric surface ranges from 110 m above mean sea level to the north of the Gaià Reservoir and less than 1 m above sea level on the coast, which in the Torredembarra areas even reaches below sea level [39].

The lower aquifer is called the Jurassic-Cretaceous Aquifer, which is made up of two hydraulically interconnected formations: basal Miocene conglomeratic breccias and Mesozoic calcareous-dolomitic materials, located at an average depth of 100–140 m, with minimum depths of 250–300 m depending on the tectonic effect. Its permeability is very high, of a secondary-fractured type, with which transmissivities are also high, up to 20 × 10^{-4} m^2/s.

The elevation of the piezometric surface ranges between 2 and 200 m above sea level; the greatest gradients are around Bonastre Massif, becoming smoother towards the western sector and the coastline of the unit, maintaining an N-S flow except in the recharge zone, where there is a small divergence that distributes the flow towards the southwest and towards El Vendrell.

In general terms and according to the authors of [40], recharge totals 41.6 hm^3/year, basically from the infiltration of rainfall and contributions from other lateral units, infiltration from rivers and reservoirs, or the return of irrigation. The most important discharge is towards the sea, with a flow of 17.8 hm^3/year, although a flow of 20 hm^3/year between the Gaià Cretaceous block and the neighboring sub-units and pumping of 6.3 hm^3/year must be taken into account.

3. Materials and Methods

3.1. Groundwater Quality Assessment

Hydrochemical characterization of this study area was carried out using groundwater physicochemical parameters from the database of the Water Catalan Agency control network [42]. We have selected hydrochemical data from the 18 control groundwater points that are available in this study area. The control points are water wells and piezometers from 6 to 450 m deep with an average of 100 m depth and are used for monitoring three different water bodies defined in 2002 to fulfill the Water Framework Directive (Directive 2000/60/EC) (Table 1).

Table 1. The Catalan Water Agency controlled the control network of the studied site during the period 2105–2020 [42]. Water type classification is based on major ion content.

Id	Water Body	Well Depth (m)	Number of Samples	Water Type
A	Lower Gaià	12	6	chloride-calcium (2015–2017)/calcium-bicarbonate (2020)
B	Garraf	13	2	chloride-sodic
C	Lower Gaià	6	1	chloride-sodic
D	Garraf	120	2	calcium-bicarbonate
E	Lower Gaià	100	4	magnesium-bicarbonate
F	Lower Gaià	31	3	magnesium-bicarbonate
G	Lower Gaià	118	6	calcium-bicarbonate
H	Not defined	10	4	calcium-bicarbonate
I	Garraf	100	2	calcium-bicarbonate
J	Lower Gaià	80	6	chloride-sodic
K	Lower Gaià	158	3	magnesium-bicarbonate
L	Lower Gaià	100	6	calcium-bicarbonate
M	Not defined	180	4	calcium-bicarbonate
N	Not defined	118	6	calcium-bicarbonate
O	Gaià-Bonastre Massif	120	2	calcium-bicarbonate
P	Not defined	140	4	calcium-bicarbonate
Q	Gaià-Bonastre Massif	450	5	calcium-bicarbonate
R	Not defined	64	6	calcium-bicarbonate

The water samples were collected and transported by the own staff of ACA to their laboratory following a quality assurance management system [48]. For the analysis of the inorganic elements, the ACA laboratory uses an inductively coupled plasma mass spectrometry instrument) and follows the criteria and specifications of the international standards [49].

The parameters that are commonly analyzed in all of them (pH, temperature, electrical conductivity (EC), Ca^{2+}, Na^{2+}, Mg^{2+}, K^{+}, HCO_3^{-}, Cl^{-}, SO_4^{2-}, and NO_3^{+}). and periodic monitoring were selected for the present research. The network is monitored twice a year: in the dry seasons between June and August (Summer) when the water demand is considerably high, and in September and November (Autumn) when the demand has decreased and they are preparing for the winter. Both summer and autumn data were evaluated over the last five years with complete and available data (from 2015 to 2020) (Table 1).

The quality and representativeness of the analytical results have been assessed by calculating the percentage error for all chemical analyses, considering only the data provided for major ions.

$$error(\% = 100\frac{\sum cations - \sum anions}{\sum cations + \sum anions}$$

We have obtained, as a result, a relative error of ionic balance between 0.75% and 3.87% with a mean of 2.3%, which we consider an admitted value and in accordance with the geological and environmental conditions present at this study site.

The hydrogeochemical data were sorted and plotted out using the software EASYQUIM v5.0 [50], which generates the main chemical diagrams such as Piper [51], Schöeller-Berkaloff [52], and Stiff [53] for further analysis and interpretation.

3.2. Geoelectrical Surveys

In this research, we have used Vertical Electrical Soundings (VES) data acquired in the 1980s for the Spanish Mining and Geological Survey (IGME) and Electrical Resistivity Tomography (ERT) data acquired in 2021. Electrical surveys generally have the scope of identifying the subsurface resistivity distribution through surface measurements. The true resistivity of the subsoil can be calculated from these observations. In unconsolidated sediments, the porosity—presuming all pores are water-saturated—and clay concentration both affect electrical resistivity (typically, sandy soil has a higher resistivity than clayey soil). Nevertheless, there is an overlap in the values for different types of soils and rocks due to the fact that the resistivity of a particular soil or rock sample depends on several properties, including porosity, the degree of water saturation, and the concentration of dissolved salts [54].

The ERT is a geophysical method that is regarded as the contemporary development of traditional geoelectrical techniques such as VES [26] and electrical trenching. The basic basis is the same; however, in this technique, computer-controlled multi-electrodes that change automatically are employed in place of the conventional four electrodes fixed in the soil surface with a common basic spacing (two for energizing and two for monitoring the voltage generated) [55].

The larger number of electrodes arranged on a line and the larger volumes of soil in which properties and boundaries can be identified in space and time have made ERT surveys less labor-intensive and more cost-effective than previous VES campaigns [56]. The current is injected into the ground through electrodes 1 and 4 in the simplest geometry (that of a Wenner array), and the potential difference between electrodes 2 and 4 is measured. The apparent resistivity value gathered is attributed to being below the midpoint of the four electrodes. A trapezium of measurements is created by choosing various electrode combinations and employing multiples of the base electrode spacing. Typically, this is represented as a pseudosection in which the vertical axis is related to the survey length.

Measurements were made in this research using the Syscal Pro multi-electrode system (IRIS instruments, Orléans, France). Multielectrode systems generate a significant amount of data, necessitating automated data management and processing. There are three steps involved in working with the resistivity data. The first is creating a pseudosection, which is an initial approximation of a picture created by plotting each acquired apparent resistivity value. The second is the removal of geometrical effects from the mathematical inversion processing, which transforms the observed data into a picture of genuine depths and true formation resistivities. The geological interpretation of the resulting physical parameters is the last phase.

The VES survey apparent resistivity data from IGME were inverted in IPI2WIN software v3.0.1 [57], allowing for the obtaining of the 1D geoelectric response of the subsoil.

The apparent resistivity values of the ERT data were inverted using the RES2DINV program v3.54.44 [58]. This software uses a non-linear optimization technique via least-squares fitting to divide the subsurface into cells with specified dimensions, for which the resistivities are changed iteratively until a satisfactory agreement between the input data and the model responses is reached [59]. The root mean square (RMS) value of the difference between experimental data and the revised model response is used as a criterion to gauge convergence at each iteration step throughout the inversion process. The inversion is considered to converge, and the procedure is finished if the data error RMS value, or the relative decrease of the data error RMS value, falls below a pre-defined level.

4. Results and Discussion

4.1. Groundwater Quality Assessment

Seventy-three water samples with physicochemical parameters and major ionic concentration results were available in the Water Catalan Agency database (18 control points and the 2015–2020 period). The ionic content of the samples ranged from medium to high (EC between 559 and 3043 μS/cm) (Table 2). Based on the concentrations provided in the European Water Directive (Directive 98/83/EC), which coincides with the limits laid down by Spanish legislation (R.D. 140/2003), 25% of the water samples could not be directly used for drinking water purposes. The main water issues, according to the Directive and major ion concentration, are related to chloride concentration (14% of the samples > 250 mg/L), nitrate concentration (9% of the samples are above 50 mg/L), and sulfate concentration (8% of the samples > 250 mg/L).

Table 2. Descriptive statistics for the physical-chemical parameters of the control groundwater network (2015–2020 period). The number of samples considered is 73.

	T	pH	EC	Na^+	K^+	Cl^-	HCO_3^-	SO_4^{2-}	NO_3^-	Ca^+	Mg^{2+}
	°C		μS/cm	mg/L	mg/L	mg/L	mg/L	mg/L	mg/L	mg/L	mg/L
Min	13.7	6.7	599	12.3	1.0	21.6	162.0	54.5	4.9	36.7	31.8
Max	23.5	8.3	3043	431.0	10.0	610.1	385.8	292.5	170.8	168.0	99.0
Mean	19.5	7.5	1298	83.5	4.1	147.5	326.6	145.4	36.54	97.3	54.3
SD	2.5	0.3	658	109.3	2.6	172.1	50.4	74.0	40.5	29.0	18.1

Nitrate concentrations had the most widespread impact and were probably the most difficult to solve in the region of Catalonia (northeast of Spain). This is essentially a result of fertilization practices, particularly the application of livestock manure [60].

Chloride concentrations are usually used as salinity indicators and could increase as a result of over-pumping [61], especially in coastal areas due to the effect of marine intrusion [62].

For the analysis of the marine intrusion, a series of ionic ratio plots have been carried out to evidence the presence of ionic relations and to characterize the presence of marine intrusion in the sector:

- As can be identified in Figure 2a, a clear linear relationship (Coefficient of Determination $R^2 = 0.9015$) is identified where the increase in EC is directly related to the increase in chloride concentration. Since the chloride ion is a conservative ion and there are no geological units in the sector that could contribute this ion to the water, its origin, and concentration are related to anthropogenic origin and/or marine intrusion events.
- The samples located upstream of the Riera de Gaià show lower concentrations of Na and Cl with respect to the more coastal waters (Figure 2b). However, it is important to mention that most of the samples are aligned, with a value of the ratio between these ions close to 0.85, a value of 0.85 corresponding to seawater composition.
- The majority of the water samples have a $rCl/rHCO_3$ ratio between 0.1 and 5; therefore, their characteristics are more like those of inland waters (Figure 2c). Nevertheless, the ratio is increasing at points close to the shoreline, indicating a slight marine influence [54].

- From Figure 2d, it can be seen that the samples tend to have a ratio of 0.1, which indicates the value of the seawater ratio; however, there is a degree of dispersion in the samples, probably associated with the wide origin of the sulfate [63].

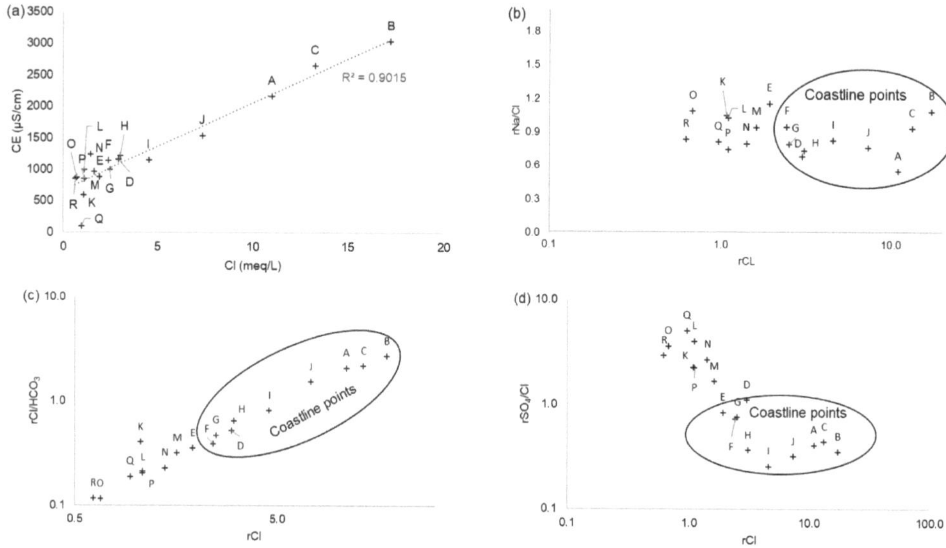

Figure 2. (**a**) Electrical conductivity vs chloride content graph; (**b**) rNaCl/rCl vs rCL chloride content graph; (**c**) $rCl/rHCO_3$ vs. rCL chloride content graph; (**d**) rSO_4/rCl vs. rCl graph.

The Piper, Schöeller-Berkaloff, and Stiff diagrams were used to classify and identify the main characteristics of the four types of water existing in this study zone (Figure 3).

The groundwater at control point A corresponds to a chloride-calcium (Ca-Cl) water type. Point A is monitoring the Lower Gaià waterbody and is one of the closest positions towards the SW to the coastline (0.6 km). However, it is important to highlight that hydrogeochemical point A is the only point in which a significant radical change is seen in 2020, where concentrations show a large increase of 58% in the ion HCO_3^- and a decrease in the major ions Cl^- (89%), NO_3^- (94%), Mg^{2+} (44%), Na^+ (83%), and EC (50%), compared to the years 2015 to 2019, and it is classified as calcium-chloride water in 2020. This fact is probably due to the decrease in industrial and economic activities' water demands in the area during the 2020 lockdown (5,103,657 m^3 in 2020 and 6,347,000 m^3 yearly average in the 2015–2019 period [64]) and the possible reduction of saltwater intrusion in this coastal area.

The second type of water is chloride-sodic (Na-Cl), corresponding to points B (Garraf Water Body), C, and J (Lower Gaià). Points B and C are placed close to the coast (approximately 0.8 km) and J further to the SW, approximately 2.75 km from the coastal zone.

The third type of water is magnesium-bicarbonate ($Mg\text{-}HCO_3$), corresponding to Lower Gaià water body control points E, F, and K. Points E and F are located approximately to the SE, between 3 and 1 km close to the coastal zone, and K further to the SW, approximately 4.8 km away.

The fourth type of water is the calcium bicarbonate type ($Ca\text{-}HCO_3$), which comprises most of the control points (D, G, H, I, L, M, N, O, P, Q, and R) and the three water bodies defined in the area. Most of these positions (G, H, I, M, N, O, P, and R) are located in the topographically highest part of the site, starting from the NW towards the NE.

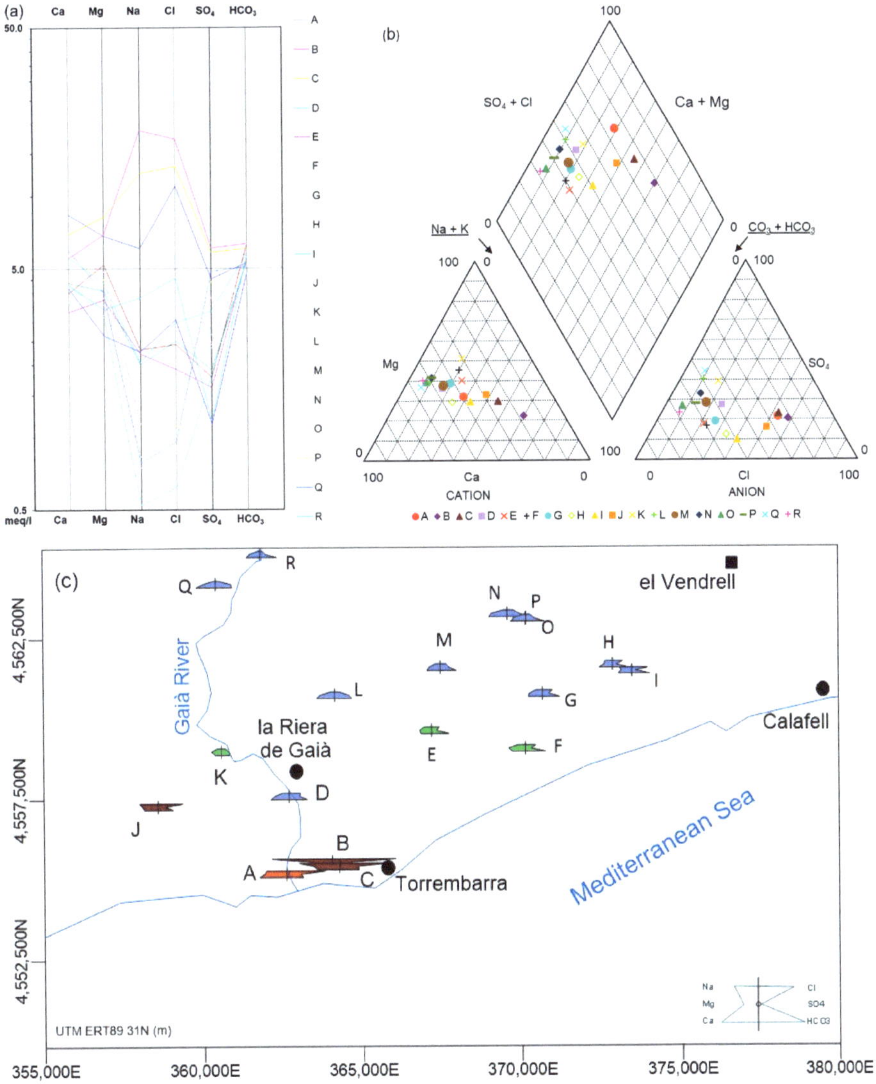

Figure 3. (**a**) Schöeller-Berkalof diagram, (**b**) Piper diagram, (**c**) Stiff Diagrams of groundwater control points at the studied site: (Ca-Cl) water types are represented in light red, (Na-Cl) types in dark red, (Mg-HCO_3) types in green, and (Ca-HCO_3) water types in blue.

4.2. Geophysical Surveys

To assess the lateral extent and thickness of the surface aquifer, geoelectrical data from 25 VES with 685 to 1000 m of survey lines was gathered within the studied zone. The VES data were acquired in 1985 by IGME using a Wenner-Schlumberger array, which is well suited to unveiling vertical geological changes [65]. Thanks to the topographic and less urbanized conditions of the terrain present in 1985, the VES acquired at that time were able to reach great extensions on the surface of the terrain on the horizontal (between 800 m and 1000 m), allowing them to also reach greater depths in the vertical (between 150 m and 200 m).

The VES results, once inverted (2.14–10.8% root mean square error), were also interpolated using IP2WIN software v3.0.1, obtaining a 2D distribution of subsurface resistivity values. We have used the lithological information available from boreholes and carried out three profile lines interpolating VES data perpendicular to the main geological structure direction to unveil the geometry of the main aquifer's units and zones with brackish/salty water (Figure 1).

The IPI2Win software v3.0.1 [57] has also been used for their representation as cross-sections.

The VES cross-sections show substantial variation in resistivity values (mainly in the 5–2000 Ω·m range). Large areas with higher resistivity values are interpreted as coarse-grained sediments and rock responses, and areas with lower resistivity values indicate the prevalence of Miocene fine-grained sediments. A general trend of decreasing electrical resistivities of the subsurface from the north to the south (coastline) has also been identified in this study area (Figure 4). The lowest resistivity values (lower than 10 Ω·m) are mainly observed in cross-section A-A′ from −20 mamsl. In cross-section B, the values are found 8 m below sea level in the southern part and 120 m below sea level, 1500 m away from the coastline. The areas with electrical resistivity values below 10 Ω·m are interpreted as saturated sediments, probably with brackish/salty water, according to the coastal setting. Several authors also reported saltwater-saturated sediment responses below 10 Ω·m in different coastal settings, such as Cap-Bon, [31] Tunisia, Málaga, Spain [37], the Delta Nil Aquifer [34], and Monterrey [29].

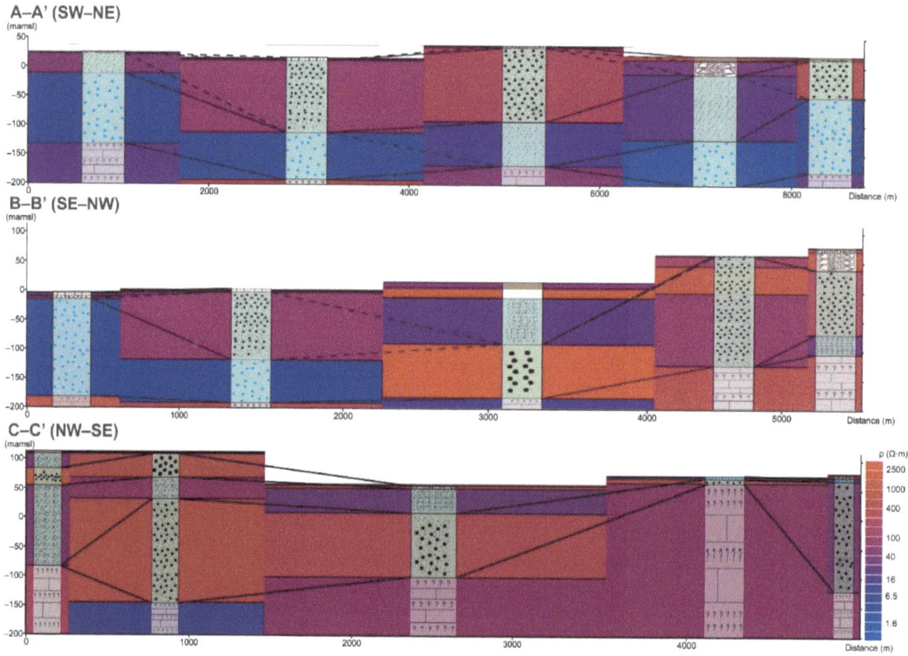

Figure 4. Electrical resistivity cross-sections and their geological interpretation were obtained from 15 VES models at the studied site. The reddish tones indicate high electrical resistivities, and the blue tones indicate lower electrical resistivities. A large dot size indicates the prevalence of Miocene coarse-grained sediments and rocks, and a low dot size indicates the prevalence of Miocene fine-grained sediments.

Moreover, we carried out a 2D ERT campaign acquiring 13 profiles—235 m in length—focused on the coastal area and the zones in which VES surveys indicated saltwater sediments (Figure 1).

We have used a mixed Wenner-Schlumberger array with 48 electrodes spaced 5 m apart for measurements. This array was chosen because it provides a moderately strong signal, is moderately sensitive to both horizontal and vertical structures [66], and provides about 40 m of maximum depth in the current research.

The ERT resistivity results (3.8–26.4% root mean square error) were also interpreted using borehole lithological information, water table measurements in piezometers, and concentrations of salinity indicators (water electrical conductivity and chloride concentration) from groundwater control points located close to the Torrembarra and Altafulla sites (Figures 5 and 6).

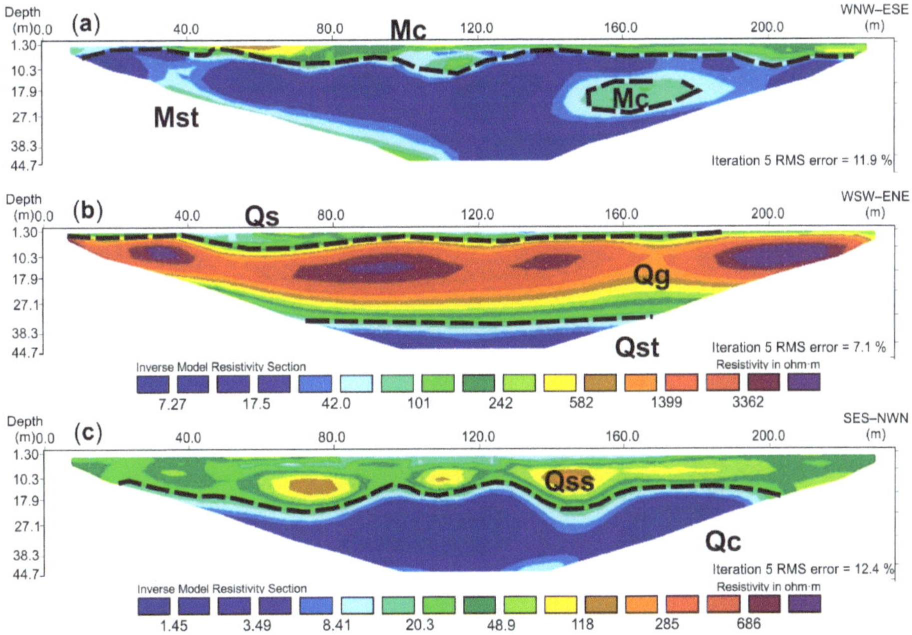

Figure 5. (**a**) Interpreted ERT cross-section 2 over consolidated sediments and rocks. Mc: Miocene sand and calcarenite, Ms: Miocene silt, (**b**) Interpreted ERT cross-section 5 over unconsolidated sediments. Qs: Quaternary sand, Qg: Quaternary gravel and sand, Qst: Quaternary silt, (**c**) Interpreted ERT cross-section 7 close to Torrembarra coastline. Qss: Quaternary sand and silt, Qc: Quaternary clay.

We have acquired the ERT cross-sections over mainly consolidated sediments (1, 2, 3, 9, and 10), within quaternary sediments (4, 5, 6, 11, 12, and 13), and close to the coastline (7 and 8) (see ERT locations in Figure 6).

In profiles acquired over consolidated sediments, two geoelectric units can be identified (Figure 5). The shallowest layer is characterized by 200–1500 Ω·m values and could be identified at the upper part of the geoelectrical cross-sections. The layer has an average thickness of 10 m and is interpreted as unsaturated Miocene-age sandstones and calcarenites.

Underneath, the ERT geoelectrical cross-sections show a single layer with variable and fluctuating thickness that is defined by values less than 100 Ω·m. The resistivity values are interpreted as Micoene siltstone and saturated Miocene sedimentary rock responses.

The water level is interpreted in these sections from 10 m depth (ERT cross-section 10) to 35 m depth (ERT cross-section 3).

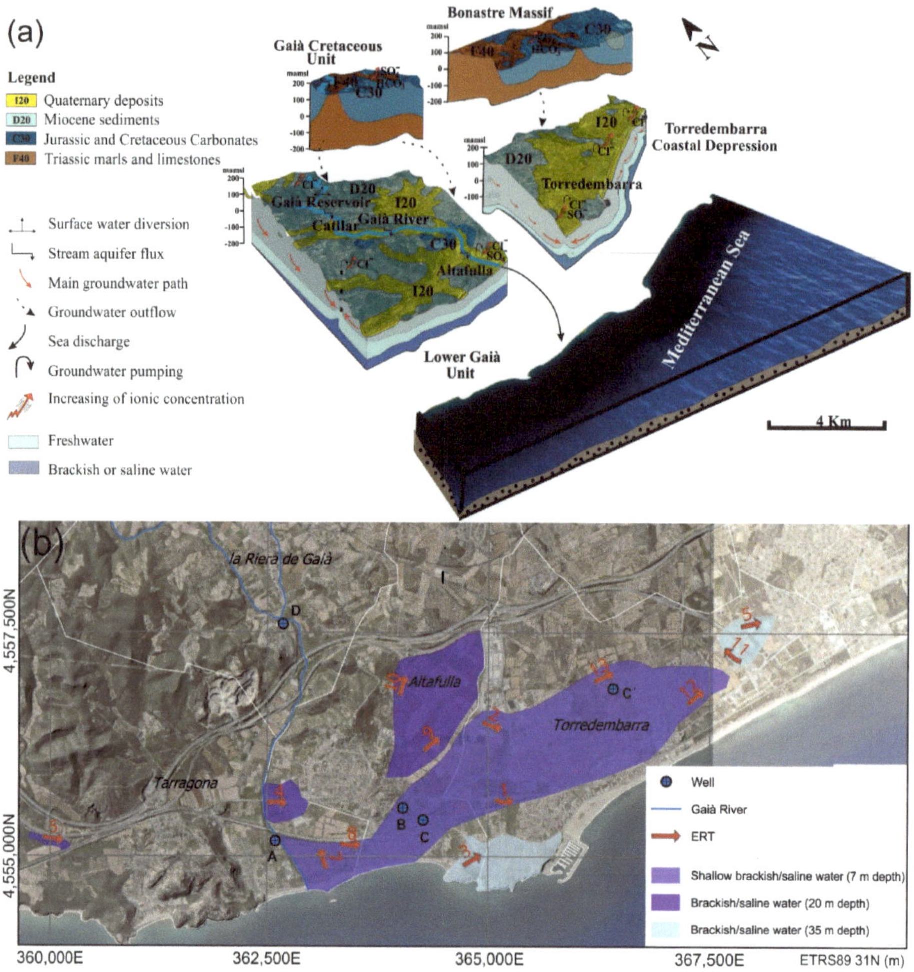

Figure 6. (**a**) Bonastre-Torredembarra hydrogeological model; (**b**) Altafulla-Torredembarra groundwater salinity map inferred from ERT results, topography, and hydrogeochemical parameters from nearby groundwater control points.

In profiles acquired over unconsolidated/slightly consolidated sediments, three geoelectric units can be identified:

- 500–1000 Ω·m layer interpreted as gravels and/or carbonate rocks (Upper Pleistocene).
- 5–50 Ω·m levels corresponding to clay and silt responses (Upper Pleistocene).
- 100–200 Ω·m interpreted as sand and gravel (Holocene–Upper Pleistocene).

In profiles acquired close to sea level, the electrical values are generally lower than 300 Ω·m, and two geoelectric units can be identified. The upper layer is characterized by values from 20–500 Ω·m, has a thickness of 7–25 m, and is interpreted as sandy and silty sediments.

Below, the ERT geoelectrical cross-sections show a single layer with variable thickness that is distinguished by values lower than 20 Ω·m. These figures are interpreted as clay and silt responses and/or brackish/salty saturated sediments (Holocene).

4.3. Hydrogeological Conceptual Model and Groundwater Salinization Map

From the joint interpretation of all the data and results, a hydrogeological conceptual model and a groundwater salinization map have been made (Figure 6). The underground flows that recharge the system come from the upper part, upstream of the Gaià Reservoir and part of the Gaia Cretaceous unit, while in the eastern sector, the waters coming from the southern pre-litoral unit and south pre-coastal unit feed the coastal aquifer system of the Torredembarra area.

A large part of the aquifer system drains towards the sea, except for the el Catllar town sector, where the flows drain towards the east, which corresponds to the Tarragona city coastal unit.

Concerning anthropic interventions, a significant volume is extracted from the pumping wells, mainly for agricultural and industrial use, and according to the background hydrochemical data and the data presented in Section 4.1, there is evidence of a decrease in the marine intrusion effect on the water quality of the Torredembarra-Altafulla coastal aquifer system. The effect of the marine intrusion as described by [39,67] on the sector has receded, with a general decrease in the parameters of electrical conductivity, chlorides, and sulfates. This, to the extent that the origin of the sulfate comes not only from marine intrusion but also from the groundwater recharge water upstream, has a significant and important mark.

Nevertheless, it is important to highlight that some marine intrusion effects remain, and we have performed a groundwater salinization map based on ERT, topography, and hydrochemical data of the Torredembara-Altafulla zone. The map divides the studied area into four groups:

- High-depth groundwater salinity. ERT cross sections show lower base resistivity values, which could indicate water with a high salt concentration.
- Medium-depth groundwater salinity. ERT cross sections show values related to brackish and/or saline water responses not exceeding 20 m depth.
- Low-depth groundwater salinity. Saltwater intrusion is present in most of the cross-section and is located at a depth of approximately 7 m.

5. Conclusions

We have carried out a regional-to-local-scale quantitative and qualitative study of the aquifer system to define a hydrogeological conceptual model of the coastal aquifer, characterize the hydrogeochemistry of groundwater, and identify the areas prone to being affected by the saline intrusion in the Torredembarra-Altafulla zone. The use of VES surveys enables a regional characterization of the aquifer system and the zones with the potential to be affected by saltwater intrusion, while ERT surveys are well suited to provide a highly detailed characterization of the coastal aquifer and delimit the zone both in extension and depth.

The obtained results could be used as a support tool for the evaluation of the most favorable areas for groundwater withdrawal, as well as enabling the protection and control of the most susceptible areas to be polluted by saltwater intrusion. Anyway, the use of indirect geophysical methods will limit the number of new piezometers for improving the groundwater model and the subsequent risk of coming into contact with different water bodies during drilling operations.

Author Contributions: Conceptualization, A.C.; methodology, A.S., A.C., M.H. and I.J.C.; formal analysis, A.S. and A.C.; data curation, I.J.C., A.U., A.S. and R.L.; writing—original draft preparation, A.S.; writing—review and editing, I.J.C., A.U., R.L., L.R., M.H., J.C.T., A.C., R.G.-A. and A.S. All authors have read and agreed to the published version of the manuscript.

Funding: This research received no external funding.

Data Availability Statement: The data presented in this study are available on request from the corresponding author. We sincerely appreciate the comments and criticisms made to earlier versions of the paper by two reviewers, which greatly helped to improve it.

Acknowledgments: We thank Ariadna Garcia Bertran for their considerable contribution to the collection of field data and data interpretation.

Conflicts of Interest: The authors declare no conflict of interest.

References

1. Van der Gun, J. *Groundwater and Global Change: Trends, Opportunities and Challenges | International Groundwater Resources Assessment Centre*; United Nations Educational, Scientific and Cultural Organization: Paris, France, 2012; ISBN 9789230010492.
2. Aeschbach-Hertig, W.; Gleeson, T. Regional strategies for the accelerating global problem of groundwater depletion. *Nat. Geosci.* **2012**, *5*, 853–861. [CrossRef]
3. Akbarpour, S.; Niksokhan, M.H. Investigating effects of climate change, urbanization, and sea level changes on groundwater resources in a coastal aquifer: An integrated assessment. *Environ. Monit. Assess.* **2018**, *190*, 579. [CrossRef] [PubMed]
4. Fetter, C.W. *Applied Hydrogeology*, 4th ed.; Prentice Hal: Upper Saddle River, NJ, USA, 2001; ISBN 0-13-088239-9.
5. López-Geta, J.A.; de la Orden, J.A.; Gómez, J.D.; Ramos, G.; Mejías, M.; Rodriguez, L. *Coastal Aquifers Intrusion Technology: Mediterranean Countries*; Instituto Geológico y Minero de España: Madrid, Spain, 2003; ISBN 84-7840-469-4.
6. Mastrocicco, M.; Colombani, N. The Issue of Groundwater Salinization in Coastal Areas of the Mediterranean Region: A Review. *Water* **2021**, *13*, 90. [CrossRef]
7. Custodio Gimena, E. *Evaluación de la Gestión y Gobernanza del Agua Subterránea: Aplicación a Áreas Seleccionadas Españolas: (EGASE)*; Universidad Politécnica de Cataluña: Barcelona, Spain, 2022; ISBN 9788498804324.
8. Polemio, M.; Zuffiano, M. Review of Utilization Management of Groundwater at Risk of Salinization. *J. Water Resour. Plan. Manag.* **2020**, *146*, 3120002. [CrossRef]
9. Losada, I.J.; Izaguirre, C.; Díaz, P. *Cambio Climático en la Costa Española*; Ministerio de Agricultura, Alimentación y Medio Ambiente, Gobierno de España: Madrid, Spain, 2014; ISBN 978-84-491-1403-8.
10. Alfarrah, N.; Walraevens, K. Groundwater overexploitation and seawater intrusion in coastal areas of arid and semi-arid regions. *Water* **2018**, *10*, 143. [CrossRef]
11. Alexander, L.; Allen, S.; Bindoff, N.; Breon, F.-M.; Church, J.; Cubasch, U.; Emori, S.; Forster, P.; Friedlingstein, P.; Gillett, N.; et al. Climate change 2013: The physical science basis, in contribution of Working Group I (WGI) to the Fifth Assessment Report (AR5) of the Intergovernmental Panel on Climate Change (IPCC). In *Climate Change 2013: The Physical Science Basis*; Cambridge University Press: Cambridge, UK, 2013.
12. Post, V.E.A. Fresh and saline groundwater interaction in coastal aquifers: Is our technology ready for the problems ahead? *Hydrogeol. J.* **2005**, *13*, 120–123. [CrossRef]
13. UNESCO. Framework for groundwater monitoring in (semi-)arid regions. In *Studies in Hydrology 92-3-103679-7*; Van Lanen, H., Carrillo-Rivera, J., Eds.; Paris, France, 1998; Volume 57, pp. 7–20, ISBN 92-3-103679-7.
14. UNESCO. *Submarine Groundwater Discharge: Management Implications, Measurements and Effects. Groundwater Series n. 5.*; IHP: Paris, France, 2004; ISBN 92-9220-006-2.
15. Margat, J.; van der Gun, J. *Groundwater around the World: A Geographic Synopsis*; CRC Press Taylor & Francis Group: Boca Raton, FL, USA, 2013; ISBN 978-1-138-00034-6.
16. Sabater, S.; Barceló, D. *Water Scarcity in the Mediterranean: Perspectives under Global Change*; Springer: Berlin, Germany, 2010; ISBN 978-3-642-03970-6(H).
17. López-Geta, J.A. La gestión de los acuíferos costeros como fuente de un recurso importante y estratégico. progreso y futuro; In *Las Aguas Subterráneas en la Ley de Aguas Española: Un Decenio de Experiencia*; AIH: Madrid, Spain, 1995; pp. 221–237.
18. Custodio, E. Coastal aquifers of Europe: An overview. *Hydrogeol. J.* **2010**, *18*, 269–280. [CrossRef]
19. ACA. *Memòria Tècnica del Programa de Seguiment i Control (PSiC) del Districte de Conca Fluvial de Catalunya. Període 2019–2024*; Agència Catalana de l'Aigua: Barcelona, Spain, 2019.
20. Lekula, M.; Lubczynski, M.W.; Shemang, E.M. Hydrogeological conceptual model of large and complex sedimentary aquifer systems—Central Kalahari Basin. *Phys. Chem. Earth, Parts A/B/C* **2018**, *106*, 47–62. [CrossRef]
21. Binley, A.; Hubbard, S.S.; Huisman, J.A.; Revil, A.; Robinson, D.A.; Singha, K.; Slater, L.D. The emergence of hydrogeophysics for improved understanding of subsurface processes over multiple scales. *Water Resour. Res.* **2015**, *51*, 3837–3866. [CrossRef]
22. Goldman, M.; Kafri, U. Hydrogeophysical applications in coastal aquifers. In *Applied Hydrogeophysics*; Vereecken, H., Binley, A., Cassiani, G., Revil, A., Titov, K., Eds.; Springer: Dordrecht, The Netherlands, 2006; pp. 233–254.
23. Werner, A.D.; Bakker, M.; Post, V.E.A.; Vandenbohede, A.; Lu, C.; Ataie-Ashtiani, B.; Simmons, C.T.; Barry, D.A. Seawater intrusion processes, investigation and management: Recent advances and future challenges. *Adv. Water Resour.* **2013**, *51*, 3–26. [CrossRef]

24. Benabdelouahab, S.; Salhi, A.; Himi, M.; Stitou El Messari, J.E.; Casas Ponsati, A. Geoelectrical investigations for aquifer characterization and geoenvironmental assessment in northern Morocco. *Environ. Earth Sci.* **2019**, *78*, 209. [CrossRef]
25. Vann, S.; Puttiwongrak, A.; Suteerasak, T.; Koedsin, W. Delineation of seawater intrusion using geo-electrical survey in a coastal aquifer of Kamala beach, Phuket, Thailand. *Water* **2020**, *12*, 506. [CrossRef]
26. Koefoed, O. *Geosounding Principles 1: Resistivity Sounding Measurements*; Elsevier Science Publishing Company: Amsterdam, The Netherlands, 1979.
27. Cimino, A.; Cosentino, C.; Oieni, A.; Tranchina, L. A geophysical and geochemical approach for seawater intrusion assessment in the Acquedolci coastal aquifer (Northern Sicily). *Environ. Geol.* **2008**, *55*, 1473–1482. [CrossRef]
28. Paz, M.C.; Alcalá, F.J.; Medeiros, A.; Martínez-Pagán, P.; Pérez-Cuevas, J.; Ribeiro, L. Integrated MASW and ERT imaging for geological definition of an unconfined alluvial aquifer sustaining a coastal groundwater-dependent ecosystem in southwest Portugal. *Appl. Sci.* **2020**, *10*, 5905. [CrossRef]
29. Goebel, M.; Pidlisecky, A.; Knight, R. Resistivity imaging reveals complex pattern of saltwater intrusion along Monterey coast. *J. Hydrol.* **2017**, *551*, 746–755. [CrossRef]
30. Hasan, M.; Shang, Y.; Jin, W.; Shao, P.; Yi, X.; Akhter, G. Geophysical Assessment of Seawater Intrusion into Coastal Aquifers of Bela Plain, Pakistan. *Water* **2020**, *12*, 3408. [CrossRef]
31. Kouzana, L.; Benassi, R.; Ben mammou, A.; Sfar felfoul, M. Geophysical and hydrochemical study of the seawater intrusion in Mediterranean semi arid zones. Case of the Korba coastal aquifer (Cap-Bon, Tunisia). *J. Afr. Earth Sci.* **2010**, *58*, 242–254. [CrossRef]
32. Cosentino, P.; Capizzi, P.; Fiandaca, G.; Martorana, R.; Messina, P.; Pellerito, S. Study And Monitoring Of Salt Water Intrusion In The Coastal Area Between Mazara Del Vallo And Marsala (South-Western Sicily). In *Methods and Tools for Drought Analysis and Management*; Rossi, G., Vega, T., Bonaccorso, B., Eds.; Springer: Dordrecht, The Netherlands, 2007; pp. 303–321, ISBN 978-1-4020-5924-7.
33. El Osta, M.; Masoud, M.; Badran, O. Aquifer hydraulic parameters estimation based on hydrogeophysical methods in West Nile Delta, Egypt. *Environ. Earth Sci.* **2021**, *80*, 344. [CrossRef]
34. Abu Salem, H.S.; Gemail, K.S.; Junakova, N.; Ibrahim, A.; Nosair, A.M. An Integrated Approach for Deciphering Hydrogeochemical Processes during Seawater Intrusion in Coastal Aquifers. *Water* **2022**, *14*, 1165. [CrossRef]
35. Galazoulas, E.C.; Mertzanides, Y.C.; Petalas, C.P.; Kargiotis, E.K. Large Scale Electrical Resistivity Tomography Survey Correlated to Hydrogeological Data for Mapping Groundwater Salinization: A Case Study from a Multilayered Coastal Aquifer in Rhodope, Northeastern Greece. *Environ. Process.* **2015**, *2*, 19–35. [CrossRef]
36. Sendrós, A.; Urruela, A.; Himi, M.; Alonso, C.; Tapias, J.C.; Rivero, L.; Garcia-Artigas, R.; Casas, A. Characterization of a Shallow Coastal Aquifer in the Framework of a Subsurface Storage and Soil Aquifer Treatment Project Using Electrical Resistivity Tomography (Port de la). *Appl. Sci.* **2021**, *11*, 2448. [CrossRef]
37. Martínez, J.; Benavente, J.; García-Aróstegui, J.L.; Hidalgo, M.C.; Rey, J. Contribution of electrical resistivity tomography to the study of detrital aquifers affected by seawater intrusion-extrusion effects: The river Vélez delta (Vélez-Málaga, southern Spain). *Eng. Geol.* **2009**, *108*, 161–168. [CrossRef]
38. ICGC. *Tarragonès - Mapa Geològic Comarcal de Catalunya - 1:50000*; Institut Cartogràfic i Geològic de Catalunya: Barcelona, Spain, 2005.
39. FCIHS. *Estudio Hidrogeológico del Bajo Gaià Desde el Embalse del Catllar Hasta el Mar (Tarragonès, Tarragona)*; 46 Curso Internacional de Hidrología Subterránea: Barcelona, Spain, 2012.
40. IGME. *Estudio de Los Recursos Hídricos Subterráneos del Sistema Hidrogeológico 74. Camp de Tarragona*, 1st ed.; Instituto Geológico y Minero de España: Madrid, Spain, 1986; ISBN 84-7474-347-8.
41. AEMet. *Guía Resumida del Clima en España (1981–2010)*; Ministerio de Agricultura, Alimentación y Medio Ambiente; Gobierno de España: Madrid, Spain, 2012.
42. ACA. Catalonia Water Quality Control Data. Agència Catalana de l'Aigua. Departament de Territori i Sosteniblitat. Generalitat de Catalunya. Available online: http://aca-web.gencat.cat/sdim21/ (accessed on 2 November 2021).
43. Sanz, J.; Suescun, J.; Molist, J.; Rubio, F.; Mujeriego, R.; Salgado, B. Reclaimed water for the Tarragona petrochemical park. *Water Supply* **2014**, *15*, 308–316. [CrossRef]
44. IDESCAT. *Estimacions Població Estacional Torredembarra-Altafulla 2016*; Institut d'Estadística de Catalunya; Generalitat de Catalunya: Barcelona, Spain, 2022.
45. Cabrera, L.; Calvet, F. Onshore Neogene record in NE Spain: Vallès–Penedès and El Camp half-grabens (NW Mediterranean). In *Tertiary Basins of Spain: The Stratigraphic Record of Crustal Kinematics*; Friend, P.F., Dabrio, C.J.E., Eds.; World and Regional Geology; Cambridge University Press: Cambridge, UK, 1996; pp. 97–105.
46. SGE. *Geología de España*; Vera, J.A., Ed.; Instituto Geológico y Minero de España: Madrid, Spain, 2004; ISBN 84-7840-546-1.
47. IGME. *Mapa Hidrogeológico de España. E 1:50.000. Tarragona*; Instituto Geológico y Minero de España: Madrid, Spain, 1989.
48. *ISO/IEC 17020:2012*; Conformity Assessment. Requirements for the Operation of Various Types of Bodies Performing Inspection. Asociación Española de Normalización y Certificación: Madrid, Spain, 2012.
49. APHA. *Standard Methods for the Examination of Water and Wastewater*; American Public Health Association: Washington, DC, USA, 2012; Volume 10.
50. Vazquéz-Suñe, E.; Serrano-Juan, A. *EasyQuim*; Institute of Environmental Assessment and Water Research: Barcelona, Spain, 2012.

51. Piper, A.M. A graphic procedure in the geochemical interpretation of water-analyses. *Eos Trans. Am. Geophys. Union* **1944**, *25*, 914–928. [CrossRef]
52. Schoeller, H. Qualitative Evaluation of Ground Water Resources. In *Water Resource Series No. 33*; UNESCO: Paris, France, 1967; pp. 44–52.
53. Custodio, E.; Llamas, M.R. *Hidrología Subterránea*, 2nd ed.; Omega: Barcelona, Spain, 2001; Volume I, ISBN 978-84-282-0447-7.
54. Sendrós, A.; Díaz, Y.; Himi, M.; Tapias, J.C.; Rivero, L.; Font, X.; Casas, A. An evaluation of aquifer vulnerability in two nitrate sensitive areas of Catalonia (NE Spain) based on electrical resistivity methods. *Environ. Earth Sci.* **2014**, *71*, 77–84. [CrossRef]
55. Casas, A.; Himi, M.; Díaz, Y.; Pinto, V.; Font, X.; Tapias, J.C. Assessing aquifer vulnerability to pollutants by electrical resistivity tomography (ERT) at a nitrate vulnerable zone in NE Spain. *Environ. Geol.* **2008**, *54*, 515–520. [CrossRef]
56. Dahlin, T. 2D resistivity surveying for environmental and engineering applications. *First Break* **1996**, *14*, 275–284. [CrossRef]
57. Bovachev, A.A.; Modin, I.N.; Shevnin, V.A. 1D Interpretation of VES profile (IPI2Win). 2011, Volume 33. Available online: http://geophys01.geol.msu.ru/ipi2win.htm (accessed on 2 November 2021).
58. Loke, M.H.; Dahlin, T. A comparison of the Gauss–Newton and quasi-Newton methods in resistivity imaging inversion. *J. Appl. Geophys.* **2002**, *49*, 149–162. [CrossRef]
59. Loke, M.H.; Barker, R.D. Rapid least-squares inversion of apparent resistivity pseudosections by a quasi-Newton method. *Geophys. Prospect.* **1996**, *44*, 131–152. [CrossRef]
60. ACA. *Avaluació de la Problemàtica Originada Per L'excés de Nitrats D'origen Agrari en Les Masses D'aigua Subterrània a Catalunya: Informe Tècnic*; Agència Catalana de l'Aigua: Barcelona, Spain, 2016.
61. El Osta, M.; Masoud, M.; Alqarawy, A.; Elsayed, S.; Gad, M. Groundwater Suitability for Drinking and Irrigation Using Water Quality Indices and Multivariate Modeling in Makkah Al-Mukarramah Province, Saudi Arabia. *Water* **2022**, *14*, 483. [CrossRef]
62. Setiawan, I.; Morgan, L.; Doscher, C.; Ng, K.; Bosserelle, A. Mapping shallow groundwater salinity in a coastal urban setting to assess exposure of municipal assets. *J. Hydrol. Reg. Stud.* **2022**, *40*, 100999. [CrossRef]
63. Fontalva, J.M.G.; Calvache, M.L.; Duque, C. Origen de la salinidad de las aguas subterráneas del sistema acuífero costero de Torrevieja: Aspectos Hidroquímicos. *Geogaceta* **2010**, *48*, 127–130.
64. ACA. *Agència Catalana de l'aigua, Consumption and Volume Studies*; Agència Catalana de l'Aigua: Barcelona, Spain, 2021.
65. Dahlin, T.; Zhou, B. A numerical comparison of 2D resistivity imaging with 10 electrode arrays. *Geophys. Prospect.* **2004**, *52*, 379–398. [CrossRef]
66. Martorana, R.; Fiandaca, G.; Casas Ponsati, A.; Cosentino, P.L. Comparative tests on different multi-electrode arrays using models in near-surface geophysics. *J. Geophys. Eng.* **2009**, *6*, 1–20. [CrossRef]
67. Puigserver, D.; Carulla, N.; Torras, B.; Iglesias, M.; Pérez, A.; Borrás, X.; Carmona, J. Hydrogeological Recovery of the Over-exploited Baix Francoli-Gaia Block Aquifer System (Spain). In *Groundwater-Present Status and Future Task, Proceedings of the 34th Congress of International Association of Hydrogeologists, Beijing, China, 9 September 2006*; The International Association of Hydrogeologists: Beijing, China, 2006.

water

MDPI

Article

Geophysical Constraints to Reconstructing the Geometry of a Shallow Groundwater Body in Caronia (Sicily)

Alessandro Canzoneri [1], Patrizia Capizzi [1,*], Raffaele Martorana [1], Ludovico Albano [2], Alessandro Bonfardeci [1], Nunzio Costa [2] and Rocco Favara [2]

1 Department of Earth and Marine Sciences (DISTEM), Università degli Studi di Palermo, 90123 Palermo, Italy; alessandro.canzoneri@unipa.it (A.C.); raffaele.martorana@unipa.it (R.M.); alessandro.bonfardeci@unipa.it (A.B.)

2 Istituto Nazionale di Geofisica e Vulcanologia, Sezione di Palermo, 90146 Palermo, Italy; ludovicoalbano80@gmail.com (L.A.); n.costa@hotmail.it (N.C.); rocco.favara@ingv.it (R.F.)

* Correspondence: patrizia.capizzi@unipa.it

Abstract: The characterization of a groundwater body involves the construction of a conceptual model that constitutes the base knowledge for monitoring programs, hydrogeological risk assessment, and correct management of water resources. In particular, a detailed geological and geophysical approach was applied to define the alluvial Caronia Groundwater Body (CGWB) and to reconstruct a hydrogeological flow model. The analysis of the CGWB, located in north-eastern Sicily, was initially approached through a reanalysis of previous stratigraphic (boreholes) and geophysical (vertical electrical soundings and seismic refraction profiles) data, subsequently integrated by new seismic acquisitions, such as Multichannel Analysis of Surface Waves (MASW) and horizontal-to-vertical seismic ratio (HVSR). The analysis and reinterpretation of geoelectrical data allowed the construction of a preliminary 3D resistivity model. This initial modeling was subsequently integrated by a geophysical data campaign in order to define the depth of the bottom of the shallow CGWB and the thickness of alluvial deposits. Finally, a preliminary mathematical model flow was generated in order to reconstruct the dynamics of underground water. The results show that integration of multidisciplinary data represent an indispensable tool for the characterization of complex physical systems.

Keywords: groundwater body; VES; HVSR; MASW; 3D model; hydrogeological flow model

1. Introduction

The definition of the hydrogeological features of a specific territory represents an important topic because the protection and management of water resources are matters of public interest, and there is growing concern about underground water resources quality and quantity [1–3]. Until a few decades ago, the study of hydrogeological structures was carried out mainly using stratigraphic data derived from field geology, drilling, and logging [4,5] as well as from aquifer test methods [6,7]. Many authors instead used geochemical information, obtained from sampling of wells or springs, indicating the origin and the path followed by the groundwaters, to study the hydrogeological features of a natural system [8–10].

Moreover, the integration of stratigraphic logs and geophysical surveys allows characterization and description of heterogeneity of surface and subsurface sediments, over large areas [11,12], especially in alluvial environments where deposits can be characterized by strong lateral and vertical variations, linked to fluvial dynamics [13,14]. The non-invasive geophysical techniques, initially limited to vertical electric sounding [15–20] and seismic refraction methods [21–23], have always represented an important tool in hydrogeological research. In the past few decades, geophysical investigation applied to hydrology and hydrogeology added quantitative analysis to a simple qualitative approach, aimed at defining the general architecture and geometry of underground aquifers [24,25].

Furthermore, with the development of new methods of investigation and with the improvement in the pre-existing techniques and instrumentation, geophysical investigations have acquired a crucial role for the knowledge and characterization of underground coastal aquifers [26–33], especially regarding water pollution detection [34,35] and seawater intrusion [15,28,36–40]. Different seismic methodologies have been used to improve the definition of geometric features of the underground aquifers, also resolving some ambiguities about the subsurface geological structures deduced from geoelectrical investigation in particular environments, such as coastal areas [41,42]. In particular, the Multichannel Analysis of Surface Waves (MASW) method [43] helps to recognize some properties of subsurface materials and the depth of important buried interfaces, analyzing seismic shear-wave velocity (Vs) variations [40,44].

An alternative approach to estimate the depth of the seismic bedrock and thickness of surface deposits involves the use of the horizontal-to-vertical spectral ratio (HVSR) methodology [45–47], which is also applied to the groundwater exploration and aquifer geometry characterization [12,41,48]. Many authors demonstrated that reliable S-wave-velocity models can be obtained by inverting HVSR data with other stratigraphic and geophysical constraints [47,49–52]. In particular, the uncertainty about the HVSR stratigraphic interpretation [53] can be addressed or at least limited using S-wave velocity models obtained by MASW as constraints [46,47,54].

The correct definition of the advanced and refined flow model, which helps in the management of water resources, also represents a powerful tool for governments and public administrations. Different guidelines have been developed in order to control, protect, and manage water resources following a quantitative and qualitative approach [8,55]. In particular, for the study of groundwater bodies and monitoring of their quantitative status, the definition of a conceptual model is required [56]. This approach also requires the definition of a surface map of the aquifer, together with an evaluation of hydraulic connectivity between adjacent groundwater bodies and the underground waters' flow rate and direction [57]. Moreover, all these elements are extremely useful for realizing a synthetic scheme of hydrogeological basin balance.

In this view, in 2018/2019 the INGV (National Institute of Geophysics and Volcanology—Palermo Section) in collaboration with the Sicilian Regional Water and Waste Department (DAR) carried out a study (hereafter INGV-DAR 2019), with the aim of analyzing the groundwater bodies of Eastern Sicily [58]. Part of this study was focused on the characterization of coastal alluvial aquifers and relative groundwater bodies of this Sicilian sector, such as that related to the Caronia area [59].

The aim of this work is to analyze the Caronia Groundwater Body (CGWB), integrating stratigraphic, geophysical, and hydrogeological data, in order to quantitatively reconstruct the pattern of this underground structure and obtain a numerical flow model. Furthermore, the multidisciplinary approach adopted can allow the correct definition and future management of the water resources in the study area.

2. Geology and Hydrogeology

2.1. Geomorphological and Geological Setting

The Caronia Groundwater Body (CGWB) is located on the northern Sicilian coast, near the Caronia village. The CGWB falls in the north-eastern sector of the Nebrodi Mounts, representing a morphologically and geologically important area, adjacent to the orographic boundary between the Nebrodi and Peloritani Mounts.

Sicily is located in the central-western Mediterranean region, at the boundary between the African and European plates, and represents the connecting element between the African Maghrebian and the Southern Apennine Chains [60–63]. As part of the Appennine-Tyrrhenian System, the geological setting of Sicily (Figure 1a) is characterized by very articulate "collisional" complex (Figure 1a), which presents the typical configuration of a foreland–foredeep–chain system [63–66]. The growth of this complex, led by the convergence between African and European plates and the coeval roll-back of the subduction hinge of the Ionian Slab (Figure 1a), started in the late Oligocene-early Miocene and persisted through different tectonic phases (compressional, extensional, and transcurrent) until the late Quaternary [60,62,67,68].

The upstream sector of the CGWB area is characterized by smoothed outline hills, formed in the flyschoid sedimentary sequences and dissected by a narrow streambed, which gradually widens towards the mouth area. The river presents a characteristic delta shape, on which a wide coastal and alluvial plain rises.

The Nebrodi Mountains, together with the Peloritani Mountains, occupy the central-eastern sector of the Sicilian Maghrebian Chain (SMC), which represents the orogenic domain of the so-called "collisional" complex of Sicily [60,62,67–74]. In particular, the Sicilian orogen presents the maximum axial depression in the north-eastern sector [68], through a first-order thrust surface. This important structure, known as "Taormina Line" [75,76], outcrops along the San Fratello-Alcantara alignment, where the overlap of the Peloritani Tectonic Units on the Nebrodi ones occurs [61].

Furthermore, the SMC is also displaced and fragmented by a complex strike- and dip-slip "Neotectonic" fault grid (Plio-Quaternary), expressed by right-lateral, synthetic, NW–SE and W–E oriented, and left-lateral, antithetic, NE–SW and N–S oriented structures [68,77–80]. This complicated structural setting is well observable in the Caronia area (Figure 1b) [72], where recent high-angle normal and transcurrent faults are present, often reactivating and re-orienting older fault surfaces [61,68,72,76].

The tectono-stratigraphic sequence outcropping in the area adjacent to the CGWB (Nebrodi Mts.) consists of units derived from the deformation of the so-called Sicilide [81] and Numidian domains [82–84], generally detached from their original substrate and dismembered in first-order tectonic units, unconformably covered by early-middle Miocene thrust-top basin deposits and by Plio-Quaternary marine, transitional, and continental successions, as shown in Figure 1b [61,72,76,85,86].

2.2. Hydrogeological Setting

Starting from the analysis of the Potential Infiltration Coefficients [8,41,87], it is observed that the lithotypes with higher permeability are concentrated along the coastal plain and the Caronia stream. These sectors are entirely characterized by recent alluvial deposits and middle-late Quaternary deposits, mainly consisting of gravels, sands, and silts (Figure 1b) [61,76]. These lithotypes are characterized by a high permeability due to primary porosity and generally host unconfined aquifers, with transmissivity increasing in the direction of the outlet of the rivers and on the coastal plains, proportionally with the increase in the thickness of the aquifer [15,88]. These types of aquifers are generically in hydraulic connection with the recharging areas to the coastal plains [3,15,89,90]. Despite having a low capacitive role, due to its limited extension, the CGWB has been considered to be able to satisfy the local water needs (INGV-DAR 2019).

The CGWB is hydrogeologically superimposed on flyschoid lithotypes with lower permeability [91]. Laterally, with the exception of the northern margin represented by the sea, the CGWB is delimited by boundaries with no or very low flow, represented by the flyschoid lithotypes of the Reitano-Monte Castellaci and Pizzo Michele-Monte Castelli [41,59]. Regarding the hydrogeological characteristics of the lithotypes outcropping in the CGWB [92–94], the following hydrogeological complexes have been distinguished [41,91,95,96]:

- Present and recent alluvial hydrogeological complex (Holocene): deposits with high permeability (10^{-2} m/s > k > 10^{-4} m/s) which form the main part of the CGWB, extending in length for about 6 km and just under a kilometer wide.
- Hydrogeological complex constituted by river and marine terrace deposits (middle-late Pleistocene) and "Ghiaie e sabbie di Messina" (middle Pleistocene): it does not

outcrop as it is covered by recent alluvial deposits. These deposits have high permeability (10^{-2} m/s > k > 10^{-4} m/s) due to primary porosity, even if very heterogeneous.

- Hydrogeological complex of the Reitano Flysch (early-middle Miocene): it presents localized and variable permeability, mainly due to fracturing. It is composed of micaceous sandstone and lithic arkoses with medium-large grain, lightly cemented, with decimetric intercalations of silty-clays. In the upper part of this complex, decametric conglomeratic bodies relative to the so-called "Conglomerati di Caronia", intercalated with the arenaceous deposits of Reitano Flysch, are often present. These lithotypes vary from moderately permeable (10^{-4} m/s > k > 10^{-5} m/s) to semi-permeable (10^{-6} m/s > k > 10^{-7} m/s) and constitute the main substrate of the CGWB.
- Hydrogeological complex of the Numidian Flysch belonging to the Monte Maragone Unit (Late Oligocene-early Miocene): it consists of argillites alternated with silty-clays, followed upwards by quartzarenites and quartzosiltites in large banks. These lithotypes have low and very low permeability (10^{-7} m/s > k> 10^{-9} m/s). For these characteristics, the Numidian Flysch deposits constitute the impermeable substrate of the southernmost part of CGWB.

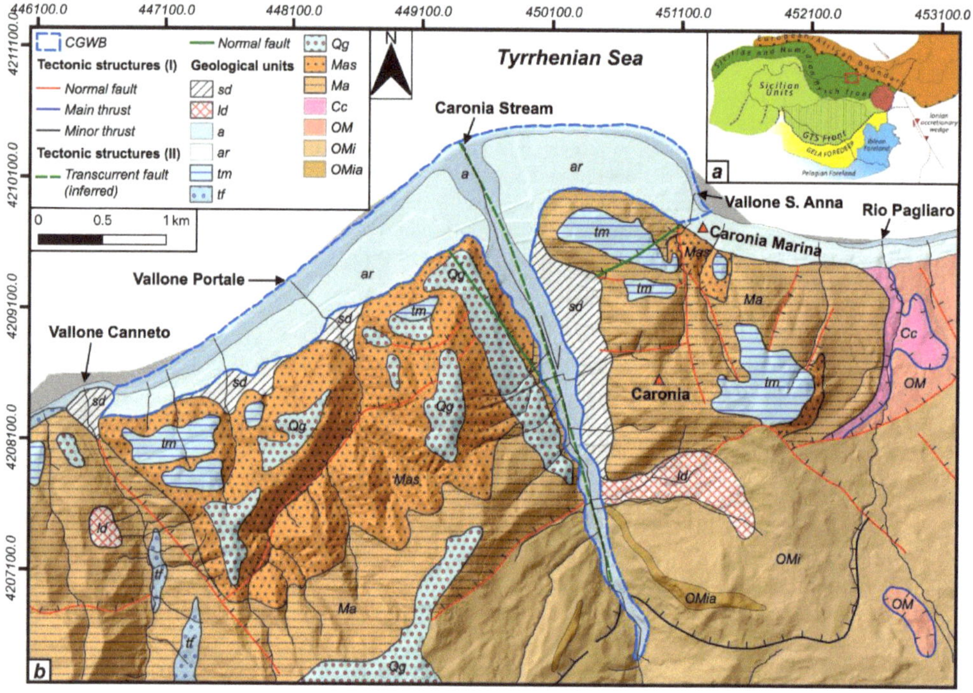

Figure 1. Geological sketch map of the area adjacent to the CGWB (blue dashed line). Inside (**a**), the tectonic map illustrating the main elements of the "collisional" complex of Sicily. The geological sketch map (**b**) was elaborated with GIS software (QGIS 3.16), using a DEM image as topographic support. The stratigraphic and tectonic information for this map are derived from previous work [85] and from the official geological map [86], as well as from the accompanying explanatory notes [61,76]. Moreover, the tectonic structures described by previous authors [76,85] are reported as I in this map, while those derived from the official geological chart [86] are indicated as II. OMi-Omia = Numidian Flysch-Monte Maragone Tectonic Unit; OM = Numidian Flysch-Monte Salici-Monte Castelli Tectonic Unit; Cc = "Argille Scagliose Superiori" Tectonic Unit; Ma = Reitano Flysch; Mas = Reitano Flysch "Conglomerati di Caronia" Member; Qg = "Ghiaie e sabbie di Messina"; tf = stream terrace deposits;

tm = marine terrace deposits; ar = recent alluvial deposits; a = present alluvial deposits; ld = landslide deposits; sd = slope deposits.

3. Materials and Methods

3.1. Previous Geognostic and Geophysical Investigation

Six boreholes, falling within the studied area, were analyzed in order to better characterize the alluvial deposits and geological substrate features and thickness (Figure 2). Four of these boreholes (BH1, BH2, BH3, and BH6) were drilled using the core destruction method, in relation to the geological and hydrogeological surveys for the construction of an aqueduct between the villages of Santo Stefano di Camastra and Caronia. The remaining two boreholes (BH4 and BH5) were instead drilled using the continuous coring method, during the geological surveys for the construction of the A20 highway (Palermo-Messina). Similar stratigraphic succession was found in the analyzed boreholes, except for the BH2 entirely drilled in the alluvial deposits of the Caronia River (up to 25 m depth). In the other boreholes, the cumulative thickness of the alluvial deposits ranges from a minimum of 9 m up to a maximum of 15 m (Figure 2). Below the alluvial deposits, arkoses interbedded with silty clays are present. These latter deposits, ascribed to the Reitano Flysch, often present a surface-deteriorated and -fractured cap, the thickness of which varies between 4 and 5 m. In general, the maximum thickness of the Reitano Flysch deposits, observed in the analyzed boreholes, ranges between 8 and 29 m (Figure 2).

Geoelectric surveys and seismic refraction surveys (Figure 3) were realized within the Caronia water body in the 1970s in order to describe the aquifer present in the Caronia area. These acquisitions were made as part of a project sponsored by [97] CASMEZ (1978). This project produced maps concerning the geophysical (geoelectric and seismic) characterization of the area. In particular, 17 vertical electrical sounding (VES) and three refraction seismic lines were carried out inside the CGWB. Each line consists of several segments of measurements arranged in continuity.

The seismic lines were positioned, distancing them from each other by about 200 m, along a direction almost perpendicular to the axis of the Caronia stream so as to intercept increasing volumes of alluvial deposits proceeding from upstream to downstream. The seismic refraction data show a two-layer model, with the thickness of the first layer varying from a few meters up to a maximum of 30 m in the northernmost area. However, original data were not available and only interpretive models could be evaluated.

3.2. New Processing and Interpretation of Vertical Electrical Soundings

Although the vertical electrical sounding (VES) technique allows, in theory, the solution of the inverse problem for electrical resistivity one-dimensional models, in many cases this method is applied to reconstruct 3D resistivity models in the subsoil, through the interpolation of 1D models which are mutually constrained, each with the adjacent models, compatibly with the tectonic structures present in the studied area. Practically, during the interpolation phase, optimization criteria must be followed to guarantee the acceptability of the model from a geological point of view. In fact, especially for large-scale geoelectric investigations, the application of 2D and 3D techniques is often impractical, for economic or logistical reasons, as the data acquisition would be impracticable or at least considerably difficult to carry out.

The use of this method of reconstruction is advisable only when the resistivity gradient is small along the directions parallel to the ground surface. In these cases, therefore, 2D or 3D resistivity models can be made by performing VES surveys on more or less regular meshes or along profiles, possibly reserving the application of 2D and 3D acquisition techniques to sectors of limited extension within the investigation area, when the resistivity varies with high horizontal gradients.

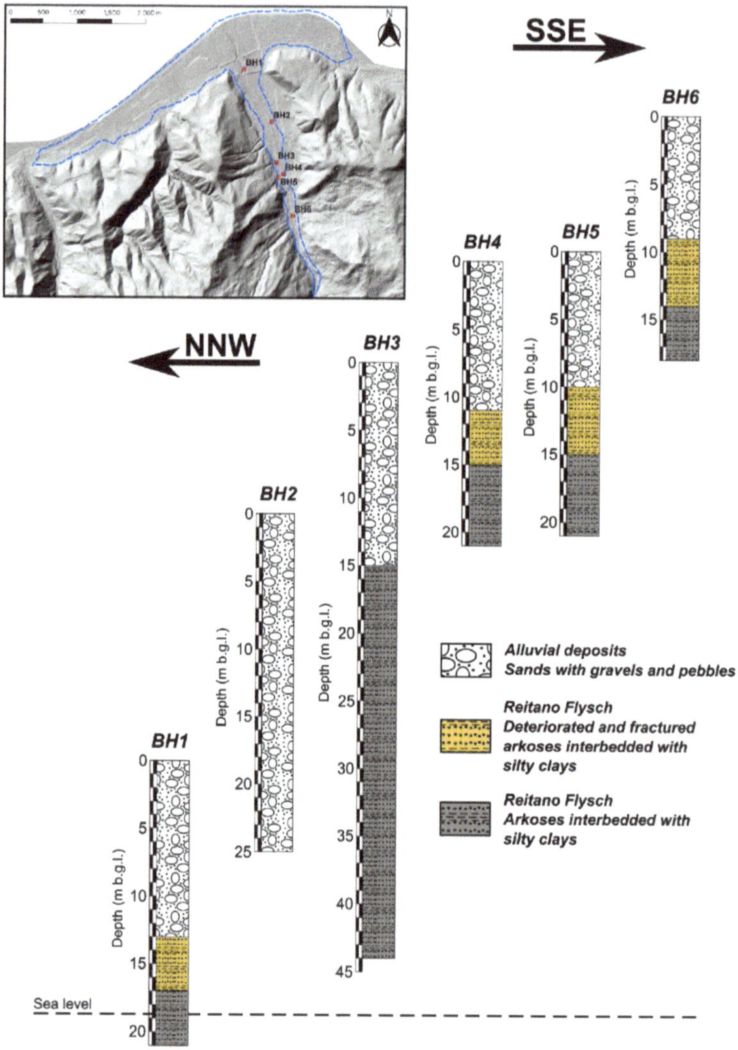

Figure 2. Stratigraphic successions of the six boreholes analyzed for this study. In the map on the top of the figure, the location of each borehole within the CGWB area is shown. Stratigraphic logs in this scheme are positioned with respect to their altitude above sea level, even if the horizontal distance among them is not to scale.

All the VESs collected were available only in paper format, given only by the graph of the apparent resistivity curves as a function of the half-distance AB/2 between the current electrodes. For each VES, an ASCII file was created containing the geographic coordinates and the altitude of the center of the survey, the array type, the number of measurements considered for each curve (corresponding to the number of points picked on the graph), and, for each measurement, the value of AB/2 and apparent resistivity.

Subsequently, using a MATLAB script, data were imported into the ZondIP1D software (v. 5.2) and converted into the proprietary format. By grouping them into a single database, all the VESs were carried out within the water body of the Caronia torrent.

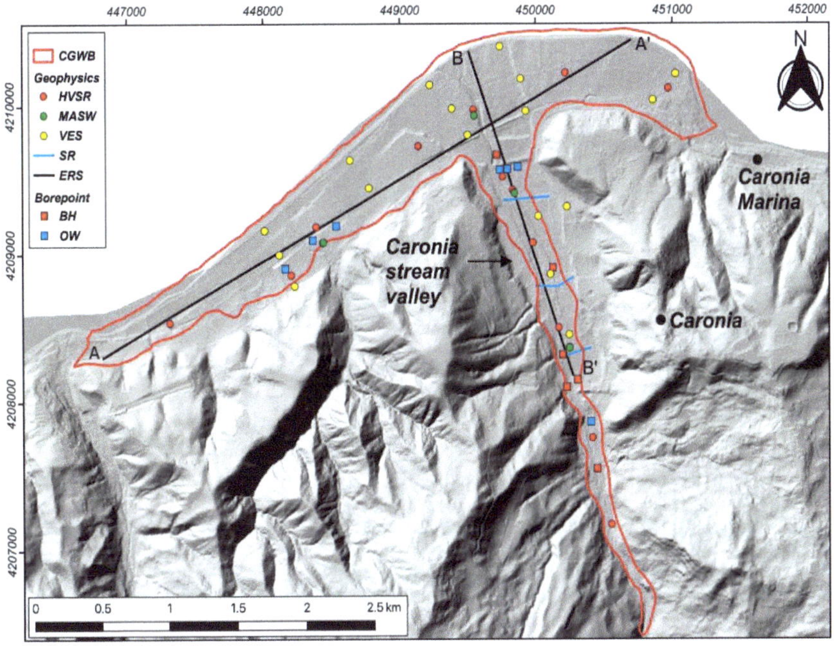

Figure 3. DEM image of the area adjacent to the CGWB (solid red line), illustrating the location of the geophysical surveys, boreholes, and wells analyzed for this study. In particular, the positions of horizontal-to-vertical spectral ratio (HVSR), Multichannel Analysis of Surface Waves (MASW), vertical electrical sounding (VES), seismic refraction (SR), electrical resistivity sections (ERS), boreholes (BH), and observation wells (OW) are shown.

ZondIP1D software allowed inversions with the least squares method. The inversions were carried out taking care to limit the lateral heterogeneities and constraining the inversions, where possible, with coring stratigraphic data. In the areas characterized by quite regular geological structures, a 1.5D inversion algorithm was used, in which the deeper layer of the resistivity section is considered almost horizontal, while the most superficial layers of the section can have sharper lateral variations. In consideration of this, a window comprising three or more adjacent VESs is inverted simultaneously, giving greater weight to the points of the central curve for the poorly fitting calculation.

The 1D inverse models obtained were therefore interpolated with each other to derive the electrical resistivity sections (Figure 4). The choice of the section traces was made considering the VES areal distribution and, at the same time, the shape of the geological structures.

The preliminary resistivity sections were useful for the making of the geological sections of the river basin. In fact, from the analysis of these resistivity sections, a first description of the characteristics of the subsoil was obtained within the Caronia water body. A preliminary geological interpretation was given to each "resistivity pattern" recognized in the sections, based on the mapped lithology and the geometry of the sedimentary bodies expected, given the structural and geomorphological asset of the area.

3.3. New Geophysical Surveys

Previous geophysical studies were improved with new geophysical measurements to define with greater precision the depth of the substrate of the groundwater body and the thicknesses of the lithologies present in the subsoil. The new geophysical data were acquired with active seismic techniques using the Multichannel Analysis of the Surface

Waves (MASW) technique and microtremor recordings elaborated with the horizontal-to-vertical spectral ratio (HVSR) technique. The location of seismic investigations was based on the position of previous investigations, with the aim of integrating the available data mesh (Figure 3). The HVSR (horizontal-to-vertical spectral ratio) methodology [45,98] is a useful technique for the analysis of seismic noise. This methodology consists of calculating the ratio between the horizontal components with respect to the vertical component of the spectrum of seismic ambient noise. A 3D velocimeter records microtremor signals along the three directions and the H/V spectral ratio is calculated. The inversion of this curve returns an S-wave velocity profile and can be used to recover stratigraphical information if the inversion process is constrained by some other variable and soil features.

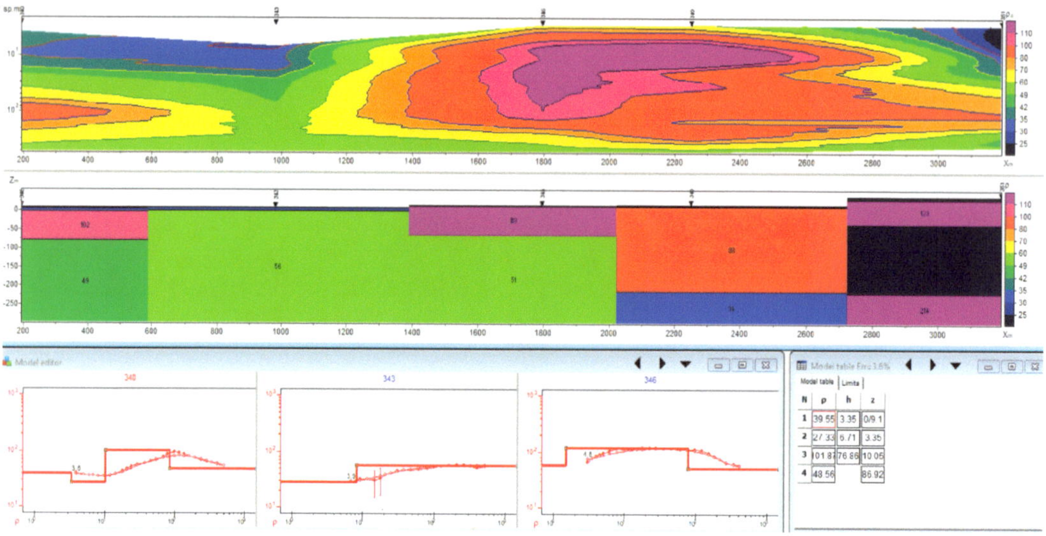

Figure 4. Example of the inversion of a series of aligned VESs and construction of the electrical resistivity section, using the ZondIP1D software: (**top**) pseudo-section of apparent resistivity, (**middle**) preliminary section obtained by aligning the inverse models relating to each VES, (**bottom**) apparent resistivity curves and related inverse models.

An total of 15 seismic microtremor acquisitions were made using a 3D velocimeter specially designed for ambient noise studies. This instrument provides good accuracy in the frequency range from 0.1 to about 30 Hz. The location of the seismic microtremor measures concerned not only the main arm of the water body but also the area of the flood plain, where other acquisitions were realized both on the left hydrographic and on the right hydrographic of the Caronia torrent. The duration of each ambient noise recording was 18 min.

MASW methodology [43] is based on the analysis of the surface waves dispersion curve. This is derived by picking the maximum oscillation on the phase velocity vs. frequency graph, derived by applying a Fourier transform to seismic signals registered at the geophones. The seismic source is generated by a heavy sledgehammer, which hits the ground at a determined offset from the geophone spread. The dispersion curve obtained is inverted by starting from an initial conceivable S-wave velocity model and solving the inverse problem via the Gauss–Newton least squares inversion algorithm.

Four MASW surveys were carried out to obtain more detailed information on the trend of Vs near surface and to constrain the inversion of the HVSR curves [53]. Indeed, constraining the HVSR inversion with independent parameters, such as shear wave velocity or the thickness of upper layers derived from MASW inversion, is useful for reducing the number of equivalent inverse HVSR models. In this way, four active seismic measurements were selected in correspondence with the same number of HVSR measurements chosen in different positions of the Caronia water body, and their inverse models served to constrain the most superficial part of the corresponding inverse HVSR models.

MASW surveys were carried out along the road of the Caronia stream and in the coastal plain, at four points distant from each other in an attempt to better describe the changes in S-wave velocity and in flood thickness along the river path (Figure 5).

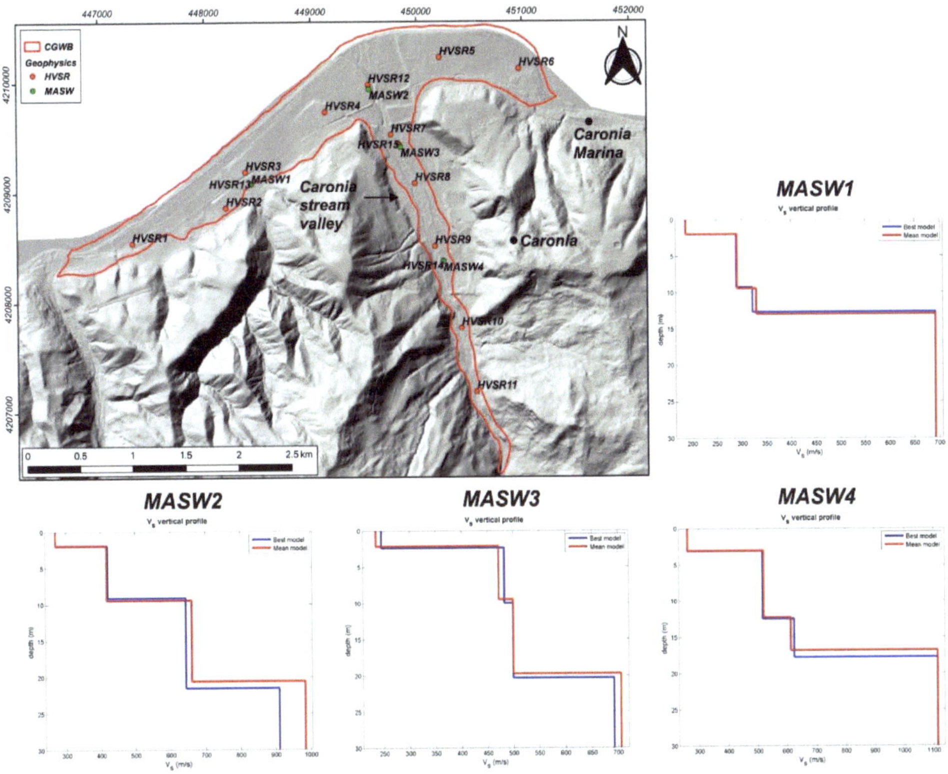

Figure 5. Inverse models resulting from the four MASW surveys performed to study the CWGB. Top left, location of the MASW surveys (green dot) and of the HVSR recordings (red dot).

All acquisitions were realized using a high-resolution multichannel seismograph. The MASW array generally consisted of twelve 4.5 Hz vertical geophones, 2 m spaced and at an offset of 5 m. A single measurement (MASW 3) was carried out using 24 vertical geophones spaced 2 m and at an offset of 8 m. HVSR surveys carried out in the water body were divided into clusters on the basis of their geographical proximity and lithological similarity to the nearest MASW survey, whose inverse model was used as a constraint.

4. Results and Discussion

4.1. MASW and HVSR Results

MASW results show the velocity of the S-wave compatible with the outcropping succession present in the investigated sites (Figure 5). MASW 2, 3, and 4, carried out along the main axis of the Caronia torrent from the southern sector towards the north, show an average velocity value equal to 259 m/s, attributed to the current alluvial deposits.

Results of 15 seismic passive measurements (HVSR), constrained by four MASW acquisitions, allowed reconstruction of seismostratigraphic patterns. Figure 6 shows the HVSR curves related to those measures carried out near the four MASW surveys, whereas their inverse 1D models, constrained by the corresponding MASW models, are presented in Figure 7.

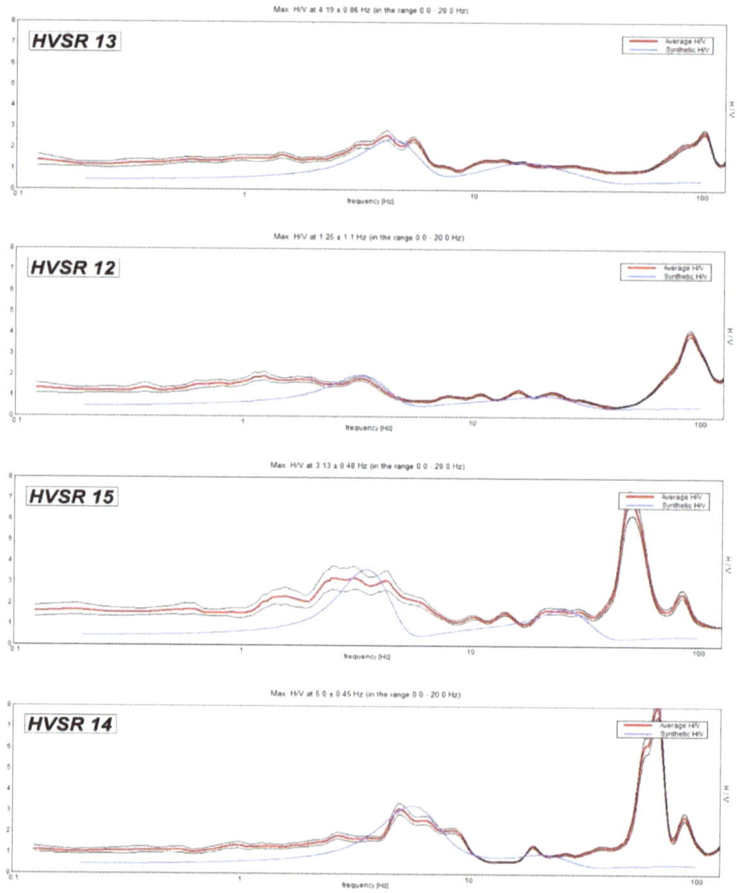

Figure 6. Comparison between the observed (red) and predicted (blue) HVSR curves, relating to the four microtremor recordings performed in proximity to the MASW surveys. The gray lines indicate the standard deviation of the observed curve.

Referring to these models, a seismostratigraphic level is shown with an average value of 272 m/s for the top of the sedimentary deposits. The thickness of this upper layer varies from 1.5 m to 4 m. Based on the geological survey, these Vs characteristics can be associated with the alluvial deposits located near the current river for HVSR 12, HVSR 14, and HVSR 15. Very similar Vs features have been described in previous works: similar values, slightly higher and confirmed to 300 m/s, have been reported for conglomerates, gravels, and sands from alluvial deposits [99]. The lowest shear wave velocity calculated for this upper layer corresponds to values close to 180 m/s, as shown for HVSR 13. This latter value is attributable to the recent alluvial plain deposits present in the wide plain next to the sea. A second deeper seismostratigraphic level is recognized for each of the four HVSR models shown in Figure 7. For this layer, Vs varies from 300 to 500 m/s. Its thickness varies from 10 m in HVSR 15 to 25 m in HVSR 12. This layer could be associated with permeable non-surfacing alluvial deposits. These would constitute the largest volume of the CWB.

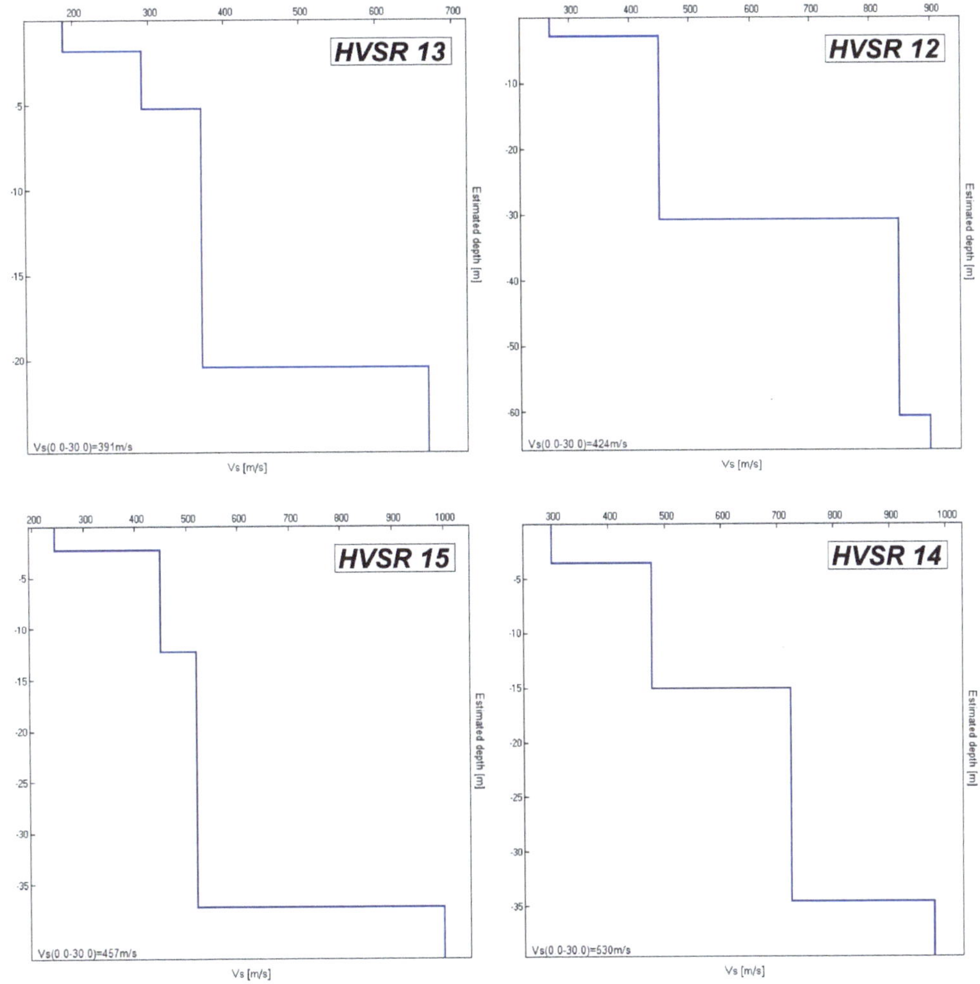

Figure 7. Inverse models related to the HVSR curves shown in Figure 6.

A third layer is characterized by a higher S-wave velocity, ranging from 400 m/s to 800 m/s. The thickness of this layer is greater in HVSR 13, where it is equal to 30 m. The Vs values are attributable to the Reitano Flysch. The wide range of velocities of this layer is due to the heterogeneous nature of these deposits, which are more altered in the upper part and more compact in depth.

A last seismic stratigraphic layer is present in all velocity models except HVSR 13. Resulting velocities between 900 and 1000 m/s could be associated with clayey deposits of varicolored clays.

4.2. A Tridimensional Model of the Electrical Resistivity

The inverse models of electrical resistivity, obtained by the above-discussed re-inversion of the available VES in the area, were interpolated using Voxler software (Golden Software, v. 4.0), to obtain a 3D graphic representation of the trend of electrical resistivity. This is limited at the top by the Digital Elevation Model of the area, at the sides by the boundaries of the water body, and at the base by the depths of the investigation reached by each VES (Figure 8). To take into account the high contrasts of the geophysical parameter, the logarithm of electrical resistivity was preferred.

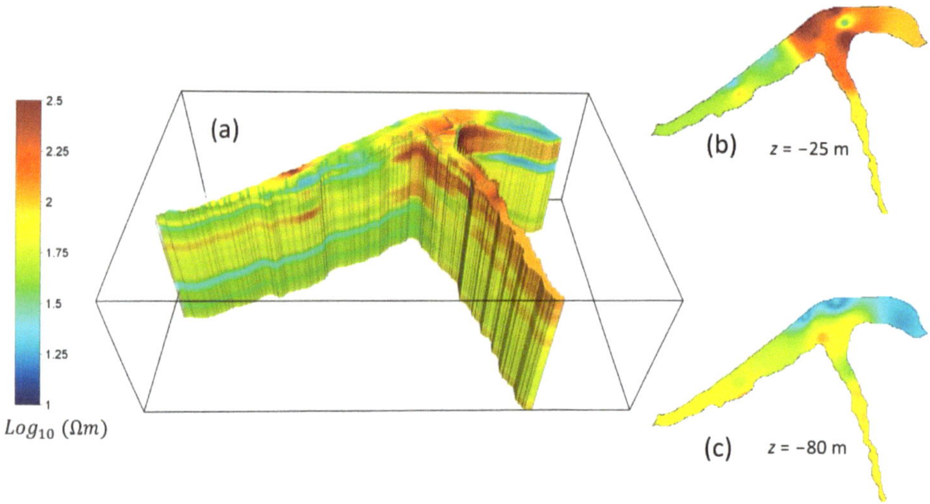

Figure 8. (**a**) Three-dimensional resistivity model obtained by the interpolation of the 1D inverse models related to the VES surveys in the Caronia water body and horizontal slides of the model at altitude of (**b**) 25 m b.s.l. and (**c**) 80 m b.s.l.

The tridimensional model describes the presence of a resistive overburden (60–80 Ωm), interpreted as alluvial materials, above a conductive substrate (7–20 Ωm), corresponding to the clayey-marly matrix soils of the Reitano Flysch surfacing near the river. In order to have a clearer view of the electrical resistivity trend within the model, horizontal slices were extracted at predefined levels below sea level. Two of them are presented and discussed here as examples. The first horizontal slice at 25 m b.s.l. (Figure 8b) shows high values of resistivity at the possible marine intrusion (therefore located in the most superficial portion) and a strong contrast of resistivity located at a transcurrent fault in evidence in the mountain zone. The second slice at 80 m b.s.l. (Figure 8c) instead shows very low resistivity values in the north-eastern part. This could be indicative of a marine intrusion phenomenon, probably located in the most porous portion of the alluvial material of the plain.

Starting from the 3D model, two vertical sections are also presented, useful for geological interpretation. The first section (A-A′) extends for 4400 m along the coastal strip, in a direction subparallel to the coastline. The second section (B-B′) extends along the axis of the stream for a length of 2300 m (Figure 9). The location of the two sections is shown in Figure 3.

The electrical resistivity features of the sedimentary deposits observable in sections A-A′ and B-B′ show horizontal and vertical variations, ranging from 20 to 500 Ωm (Figure 9). In particular, these electrical resistivity variations allowed different horizons and layers to be distinguished (Figure 9). These latter features, characterized by precise resistivity values, were attributed to the geological deposits outcropping around the Caronia area (Figure 1) and used to reconstruct the stratigraphic and structural setting.

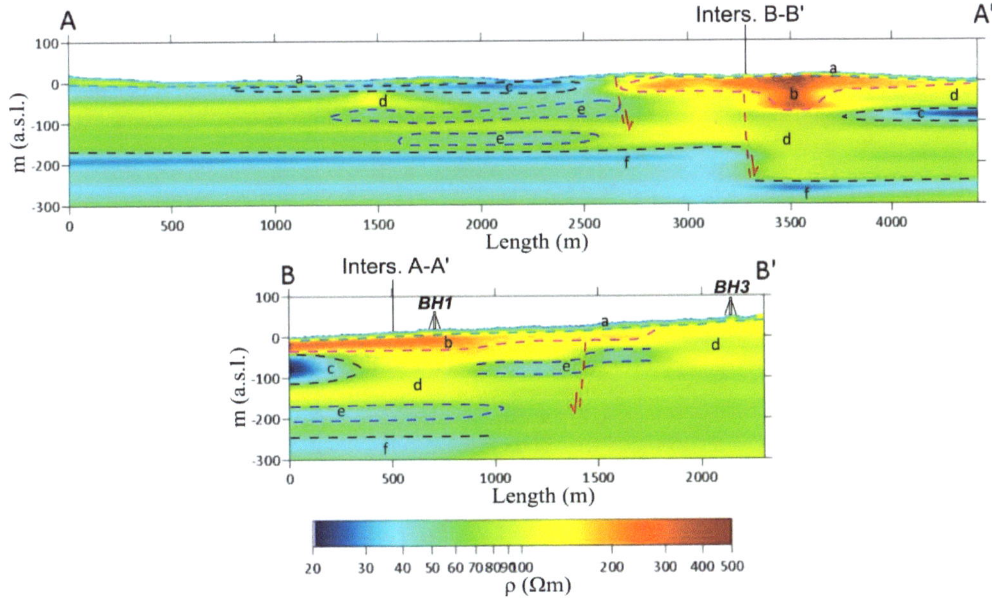

Figure 9. Vertical sections of electrical resistivity obtained from the 3D model: (**top**) section A-A′ subparallel to the coastline; (**bottom**) section B-B′ along the axis of the Caronia stream. Letters a to f indicate the different layers, bounded above and below by relative dashed lines, identified on the basis of the electrical resistivity variations. Red dashed lines indicate the presence of normal and transtensional faults. BH1 and BH3 are relative to the boreholes located along the B-B′ section (for further details see Figures 2 and 3).

From the bottom to the top of the two sections, layer f, with resistivity values of about 40 Ωm, could refer to the clays and marls of the "Argille Scagliose Superiori" unit, outcropping to the east of the Caronia area. This layer has been continuously recognized in section A-A′ and only in the northernmost part of section B-B′. Above these deposits, the increasing resistivity values, here identified by the layer d (60–100 Ωm), can be associated with the arenaceous facies with intercalations of silty-clays belonging to the Reitano Flysch formation. In this view, layers labeled e (50–60 Ω m) could represent thicker clayey portions of the Miocene lithological unit. The highest resistivity value portions (200–500 Ω m) observed in both sections, named layer b, can be related to the "Conglomerati di Caronia" member deposits and/or to conglomeratic channeled structures in correspondence with the central sector of the streambed. From another point of view, these high resistivity values, coinciding with the mouth sector of the Caronia stream, could also indicate the presence of freshwaters flowing below the streambed. The low resistivity lenticular and cuneiform bodies (20–40 Ω m) observed in both sections, named layer c, probably describe the seawater intrusion in the Caronia sedimentary succession, in correspondence with sectors with high porosity and/or minor freshwater hydraulic load. Finally, observing the BH1 and BH3 stratigraphic logs and the resistivity values (40–70 Ωm), the uppermost layer a was ascribed to the alluvial deposits, filling the Caronia streambed and the relative coastal plain.

The three stepped forms recognized in the sections (Figure 6), indicated by red dashed lines, could represent normal tectonic structures. These faults are compatible in terms of location, shape, and type of dislocation with those recognized and already mapped [86] (APAT, 2013) in the hills immediately west and east to the terminal part of the Caronia stream valley, and with the inferred transtensional structure where the streambed develops

(Figure 1). Moreover, this latter structure could also be responsible for the thickening of the succession, especially as regard to layers a and b, in correspondence with the central sector of the Caronia stream mouth (Figure 9). In this view, space created by the net-slip movement linked to this fault may have been gradually filled by Reitano Flysch conglomeratic bodies (i.e., "Conglomerati di Caronia" member and/or channeled structures) and by Quaternary deposits.

4.3. Estimate of the Bottom of the CGWB

All the geophysical surveys performed (VES, SR, HVSR, and MASW) were geologically interpreted using the stratigraphic logs (from BH1 to BH6) as calibration. This made it possible to transform the information of the geophysical parameters in terms of the thickness of the water body in the investigated points. These values were interpolated by applying a kriging-type algorithm using the Surfer software (v. 18). In correspondence with the water body superficial limit, the thickness of the groundwater body was set equal to zero. The resulting thickness map (Figure 10a) shows two areas with significant thicknesses: the smallest is located in the extreme eastern portion, where 30 m thicknesses are reached; the largest is located west of the current delta, where thicknesses reach about 40 m. This part coincides with the area already identified as a transcurrent fault in the 3D resistivity model. The thicknesses of the CGWB are larger in the eastern part of this probable fault area.

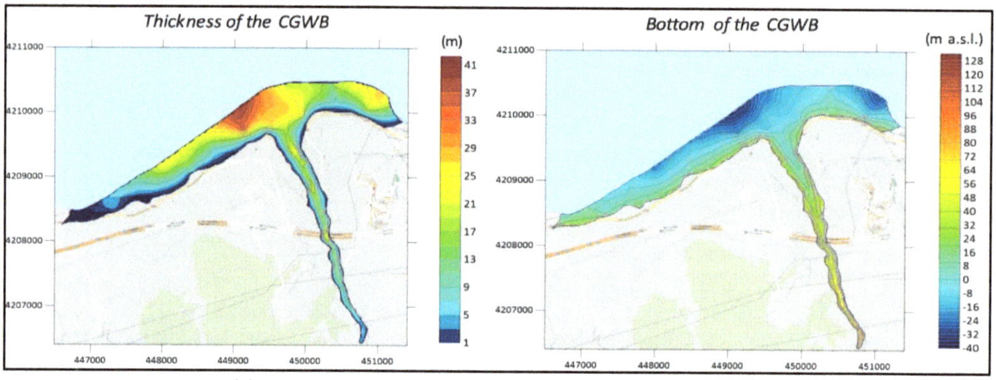

(a) (b)

Figure 10. (**a**) Thickness of the Caronia Groundwater Body; (**b**) altitude of the bottom of the Caronia Groundwater Body (m a.s.l.).

Furthermore, the bottom of the water body was rebuilt (Figure 10b) because it is needed as an input for the construction of a mathematical flow model.

4.4. Numerical Flow Model for the CGWB

The reconstruction of the lower boundary of the water body plays a key role in the construction of the mathematical model of static flow. This was carried out to implement the hydrogeological model of the CGWB. The finite difference calculation code MODFLOW [100] in steady state was used to create the flow model. This kind of analysis solves the groundwater flow differential equation using the finite difference approach within the groundwater systems through a gridded spatial discretization [101]. In order to create a numerical model, which describes the hypothetical trend of flow within a groundwater body, it is necessary to define the model in the spatial domain. This must be divided into cells, to each of which is assigned both the initial and boundary conditions and the physical and hydrogeological properties. These conditions and properties are considered homogeneous within the cell itself, and attributed to the center of the cell [102].

The sectors considered by the numerical flow model correspond to the coastal plain and the Caronia stream mouth. In the first phase, the top and bottom levels were defined. For the top, corresponding to the topographic surface, altitudes derived from the DEM of Sicily were assumed. The hydrogeological substrate, defined in the digital model of the CGWB, was adopted as the bottom of the flow model. For the horizontal discretization, the modeled area (coastal plain) was subdivided using a mesh of 50 × 50 m cells, while 12 layers were identified for the vertical subdivision (Figure 11). Subsequently, the hydrogeological properties were defined to be assigned to the various sectors of the model. Hydraulic conductivity values along the three spatial directions (Kx = 3.5×10^{-5} m/s, Ky = 3.5×10^{-5} m/s, and Kz = 3.5×10^{-6} m/s) were attributed to layers 1 to 11, describing the part of the succession in which the groundwater flows. Finally, layer 12, corresponding to the bottom, was defined as null flow. A further step for the realization of a flow model is related to the definition of the boundary conditions [103].

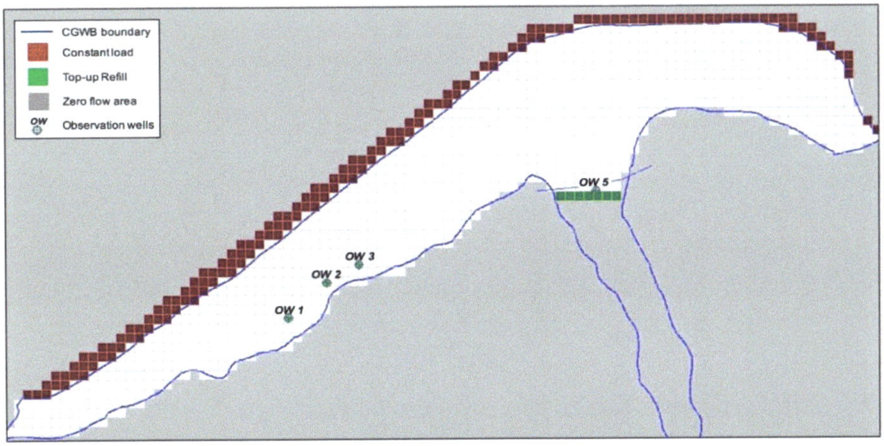

Figure 11. Horizontal discretization of the modeled area subdivided using a mesh of 50 × 50 m cells and boundary conditions used for the numerical flow model.

These features define the physical constraints in the perimeter cells that are imposed on the flow model (Figure 11). As boundary conditions, the following parameters were adopted:

(A) Refill value: 161 mm/year for 365 days. This refill was applied only to layer 1, which defines the portion of the model identifiable with the topographic surface.
(B) Constant load: An area representing the portion of the model (red cells in Figure 11) where the flow stops being considered as having a constant hydraulic load. It was defined by assigning the value 0 to the coastline downstream of the domain.
(C) Top-up refill: This parameter represents the sub-alveal contributions. This feature describes the direct contributions provided by the upstream hydrological basin. These values were assigned, as saturated cells (green cells in Figure 11), corresponding to the groundwater depth measured in the observation well n.5.
(D) The area outside the modeled area was considered to be zero flow (gray sectors in Figure 11).

The mathematical model obtained was finally calibrated and validated thanks to the piezometric values detected during the field activities performed in the area under study. Four wells were considered, one of which (OW 5) was inside the Caronia streambed. These wells have an undisturbed piezometric level and are sufficiently far from other wells in use. The simulation results are shown in Figure 12. The trend of the simulated piezometry shows how the aquifer is fed by the inputs coming from the mountain. Furthermore, the stream portion considered in this model provides an important contribution to feed the coastal plain aquifer.

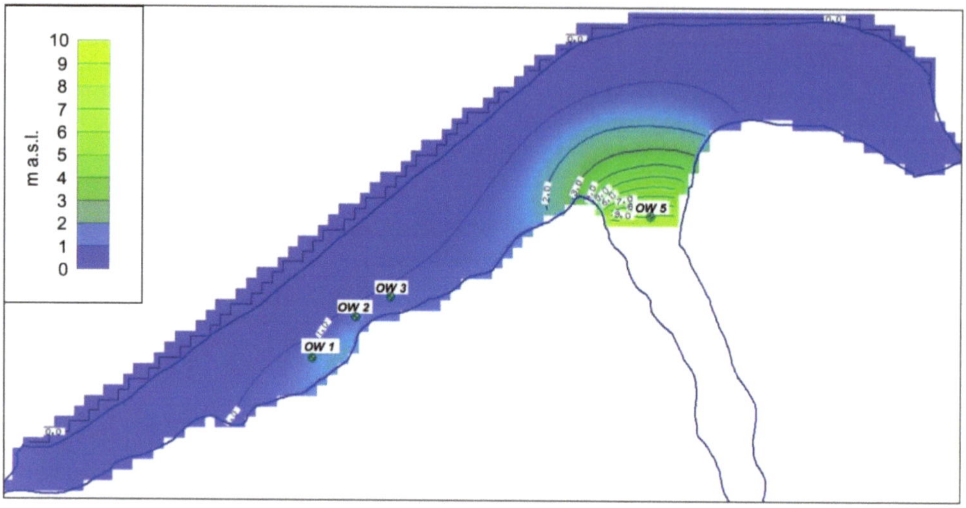

Figure 12. Results of the mathematical flow model with relative values of simulated piezometry.

Finally, the calculated vs. observed piezometric values are shown in Figure 13. A normalized RMS of 7.49% is obtained, with an average square deviation of 0.63 m. This is considered to be an acceptable value considering the width of the simulated area.

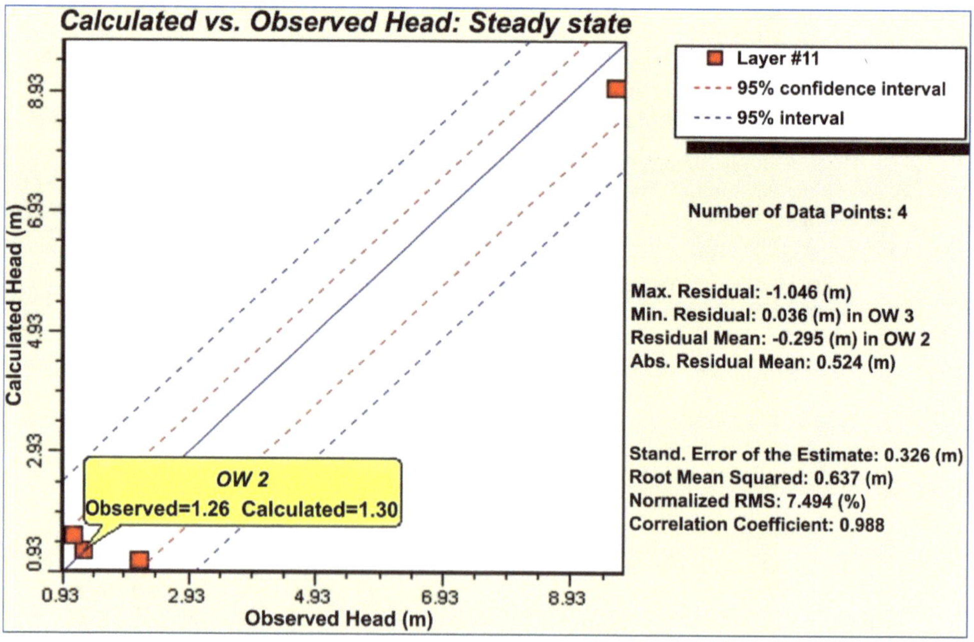

Figure 13. Calculated vs. observed piezometric values.

5. Conclusions

The characterization of the Caronia Groundwater Body (CGWB), in north-eastern Sicily, was realized thanks to the construction of a conceptual physical model and the related hydrogeological flow model, which is fundamental for the definition of monitoring programs, hydrogeological risk assessment, and correct water resources management.

A series of previous geophysical data (vertical electrical sounding surveys and refraction seismic profiles) were re-inverted using constraints obtained from stratigraphic data. The new geophysical models obtained, together with the information obtained from new surveys with surface wave and microtremor techniques (MASW and HVSR), made it possible to obtain three-dimensional models of physical parameters of the subsoil (electrical resistivity, seismic velocity of pressure, and shear waves). These models were used to define the depth of the bottom of the CGWB and the thickness of the alluvial deposits that characterize the streambed. This allowed shape, volume, and physical characteristics of the CGWB to be defined in sufficient detail. This numerical information served as a basis to generate a mathematical model of groundwater flow in order to simulate the spatial and temporal variability of the flow and, consequently, to make forecast estimates of the water supplies to the coastal plain. The underground flow model was preliminarily tested by comparing the theoretical results with some piezometric measurements performed. The main result of the calibration of the model was the refinement and validation of the assignment of the average hydrodynamic parameters of the modeled aquifer. In this way, it was possible to carry out realistic and reliable simulations of the underground flow in the CGWB. The results showed a good correspondence between predicted and observed data, confirming the reliability of the model obtained for the description of the aquifer dynamics.

Author Contributions: Conceptualization, P.C., R.M. and R.F.; Methodology, A.C., P.C., R.M. and A.B.; Software, R.M.; Validation, P.C. and R.M.; Formal analysis, A.C., P.C., R.M., A.B. and N.C.; Investigation, A.C., L.A. and A.B.; Data curation, P.C., R.M. and N.C.; Writing—original draft, A.C., P.C., R.M., L.A., A.B. and N.C.; Writing—review & editing, A.C., P.C., R.M. and A.B.; Visualization, A.C. and R.M.; Supervision, P.C. and R.F.; Project administration, R.F.; Funding acquisition, R.F. All authors have read and agreed to the published version of the manuscript.

Funding: This research was funded by Regione Siciliana, Commissario Delegato per l'Emergenza Bonifiche e la tutela delle Acque in Sicilia.

Data Availability Statement: Not applicable.

Conflicts of Interest: The authors declare no conflict of interest.

References

1. Chenini, I.; Zghibi, A.; Kouzana, L. Hydrogeological investigations and groundwater vulnerability assessment and mapping for groundwater resource protection and management: State of the art and a case study. *J. Afr. Earth Sci.* **2015**, *109*, 11–26. [CrossRef]
2. Erdoğan, M.; Karagüzel, R. A new hydrogeologically based approach to determining protected areas in drinking water supply reservoirs: A case study in the Ağlasun sub-basin (Burdur, Turkey). *Environ. Earth Sci.* **2016**, *75*, 126. [CrossRef]
3. Käser, D.; Hunkeler, D. Contribution of alluvial groundwater to the outflow of mountainous catchments. *Water Resour. Res.* **2016**, *52*, 680–697. [CrossRef]
4. Paillet, F.L.; Reese, R.S. Integrating Borehole Logs and Aquifer Tests in Aquifer Characterization. *Groundwater* **2000**, *38*, 713–725. [CrossRef]
5. Zuffetti, C.; Comunian, A.; Bersezio, R.; Renard, P. A new perspective to model subsurface stratigraphy in alluvial hydrogeological basins, introducing geological hierarchy and relative chronology. *Comput. Geosci.* **2020**, *140*, 104506. [CrossRef]
6. Straface, S.; Yeh, T.-C.J.; Zhu, J.; Troisi, S.; Lee, C.H. Sequential aquifer tests at a well field, Montalto Uffugo Scalo, Italy. *Water Resour. Res.* **2007**, *43*. [CrossRef]
7. Alexander, M.; Berg, S.J.; Illman, W.A. Field Study of Hydrogeologic Characterization Methods in a Heterogeneous Aquifer. *Groundwater* **2011**, *49*, 365–382. [CrossRef]
8. Fontana, M.; Grassa, F.; Cusimano, G.; Favara, R.; Hauser, S.; Scaletta, C. Geochemistry and potential use of groundwater in the Rocca Busambra area (Sicily, Italy). *Environ. Geol.* **2008**, *57*, 885–898. [CrossRef]
9. Cangemi, M.; Madonia, P.; Albano, L.; Bonfardeci, A.; Di Figlia, M.G.; Di Martino, R.M.R.; Nicolosi, M.; Favara, R. Heavy Metal Concentrations in the Groundwater of the Barcellona-Milazzo Plain (Italy): Contributions from Geogenic and Anthropogenic Sources. *Int. J. Environ. Res. Public Health* **2019**, *16*, 285. [CrossRef]

10. Chen, Q.; Jia, C.; Wei, J.; Dong, F.; Yang, W.; Hao, D.; Jia, Z.; Ji, Y. Geochemical process of groundwater fluoride evolution along global coastal plains: Evidence from the comparison in seawater intrusion area and soil salinization area. *Chem. Geol.* **2020**, *552*, 119779. [CrossRef]
11. Massey, A.C.; Taylor, G.K. Coastal evolution in south-west England, United Kingdom: An enhanced reconstruction using geophysical surveys. *Mar. Geol.* **2007**, *245*, 123–140. [CrossRef]
12. Sauret, E.S.G.; Beaujean, J.; Nguyen, F.; Wildemeersch, S.; Brouyere, S. Characterization of superficial deposits using electrical resistivity tomography (ERT) and horizontal-to-vertical spectral ratio (HVSR) geophysical methods: A case study. *J. Appl. Geophys.* **2015**, *121*, 140–148. [CrossRef]
13. Larkin, R.G.; Sharp, J.M., Jr. On the relationship between river-basin geomorphology, aquifer hydraulics, and ground-water flow direction in alluvial aquifers. *Geol. Soc. Am. Bull.* **1992**, *104*, 1608–1620. [CrossRef]
14. Gregory, K.; Benito, G.; Downs, P. Applying fluvial geomorphology to river channel management: Background for progress towards a palaeohydrology protocol. *Geomorphology* **2008**, *98*, 153–172. [CrossRef]
15. Cimino, A.; Cosentino, C.; Oieni, A.; Tranchina, L. A geophysical and geochemical approach for seawater intrusion assessment in the Acquedolci coastal aquifer (Northern Sicily). *Environ. Geol.* **2007**, *55*, 1473–1482. [CrossRef]
16. Hamzah, U.; Samsudin, A.R.; Malim, E.P. Groundwater investigation in Kuala Selangor using vertical electrical sounding (VES) surveys. *Environ. Geol.* **2006**, *51*, 1349–1359. [CrossRef]
17. Alile, O.M.; Ujuanbi, O.; Evbuomwan, I.A. Geoelectric investigation of groundwater in Obaretin Iyanomon locality, Edo state, Nigeria. *J. Geol. Min. Res.* **2011**, *3*, 13–20. [CrossRef]
18. Coker, J.O. Vertical electrical sounding (VES) methods to delineate potential groundwater aquifers in Akobo area, Ibadan, South-western, Nigeria. *J. Geol. Min. Res.* **2012**, *4*, 35–42. [CrossRef]
19. Fadele, S.I.; Sule, P.O.; Dewu, B.B.M. The use of vertical electrical sounding (VES) for groundwater exploration around Nigerian College of Aviation Technology (NCAT), Zaria, Kaduna State, Nigeria. *Pac. J. Sci. Technol.* **2013**, *14*, 549–555.
20. Abdullahi, M.G.; Toriman, M.E.; Gasim, M.B. The Application of Vertical Electrical Sounding (VES) for Groundwater Exploration in Tudun Wada Kano State, Nigeria. *J. Geol. Geosci.* **2015**, *4*, 1–3. [CrossRef]
21. van Overmeeren, R.A. A combination of electrical resistivity, seismic refraction, and gravity measurements for groundwater exploration in Sudan. *Geophysics* **1981**, *46*, 1304–1313. [CrossRef]
22. Haeni, F.P. Application of seismic refraction methods in groundwater modeling studies in New England. *Geophysics* **1986**, *51*, 236–249. [CrossRef]
23. Grelle, G.; Guadagno, F.M. Seismic refraction methodology for groundwater level determination: "Water seismic index". *J. Appl. Geophys.* **2009**, *68*, 301–320. [CrossRef]
24. Dafflon, B.; Irving, J.; Holliger, K. Use of high-resolution geophysical data to characterize heterogeneous aquifers: Influence of data integration method on hydrological predictions. *Water Resour. Res.* **2009**, *45*. [CrossRef]
25. Khan, U.; Faheem, H.; Jiang, Z.; Wajid, M.; Younas, M.; Zhang, B. Integrating a GIS-Based Multi-Influence Factors Model with Hydro-Geophysical Exploration for Groundwater Potential and Hydrogeological Assessment: A Case Study in the Karak Watershed, Northern Pakistan. *Water* **2021**, *13*, 1255. [CrossRef]
26. Bowling, J.C.; Rodriguez, A.B.; Harry, D.L.; Zheng, C. Delineating Alluvial Aquifer Heterogeneity Using Resistivity and GPR Data. *Groundwater* **2005**, *43*, 890–903. [CrossRef]
27. Kim, J.-W.; Choi, H.; Lee, J.-Y. Characterization of hydrogeologic properties for a multi-layered alluvial aquifer using hydraulic and tracer tests and electrical resistivity survey. *Environ. Geol.* **2005**, *48*, 991–1001. [CrossRef]
28. Goldman, M.; Kafri, U. Hydrogeophysical Applications in Coastal Aquifers. In *Applied Hydrogeophysics*; Springer: Dordrecht, The Netherlands, 2007; pp. 233–254. [CrossRef]
29. Wattanasen, K.; Elming, S. Direct and indirect methods for groundwater investigations: A case-study of MRS and VES in the southern part of Sweden. *J. Appl. Geophys.* **2008**, *66*, 104–117. [CrossRef]
30. Doro, K.O.; Leven, C.; Cirpka, O.A. Delineating subsurface heterogeneity at a loop of River Steinlach using geophysical and hydrogeological methods. *Environ. Earth Sci.* **2013**, *69*, 335–348. [CrossRef]
31. Dickson, N.E.M.; Comte, J.-C.; McKinley, J.; Ofterdinger, U. Coupling ground and airborne geophysical data with upscaling techniques for regional groundwater modeling of heterogeneous aquifers: Case study of a sedimentary aquifer intruded by volcanic dykes in Northern Ireland. *Water Resour. Res.* **2014**, *50*, 7984–8001. [CrossRef]
32. Vilhelmsen, T.N.; Behroozmand, A.A.; Christensen, S.; Nielsen, T.H. Joint inversion of aquifer test, MRS, and TEM data. *Water Resour. Res.* **2014**, *50*, 3956–3975. [CrossRef]
33. Michael, H.A.; Post, V.E.A.; Wilson, A.M.; Werner, A.D. Science, society, and the coastal groundwater squeeze. *Water Resour. Res.* **2017**, *53*, 2610–2617. [CrossRef]
34. Daniels, J.J.; Roberts, R.; Vendl, M. Ground penetrating radar for the detection of liquid contaminants. *J. Appl. Geophys.* **1995**, *33*, 195–207. [CrossRef]
35. Benson, A.K.; Payne, K.L.; Stubben, M.A. Mapping groundwater contamination using dc resistivity and VLF geophysical methods–A case study. *Geophysics* **1997**, *62*, 80–86. [CrossRef]
36. Frohlich, R.K.; Urish, D.W.; Fuller, J.; O'Reilly, M. Use of geoelectrical methods in groundwater pollution surveys in a coastal environment. *J. Appl. Geophys.* **1994**, *32*, 139–154. [CrossRef]

37. Cosentino, P.; Capizzi, P.; Fiandaca, G.; Martorana, R.; Messina, P.; Pellerito, S. Study and Monitoring of Salt Water Intrusion in The Coastal Area Between Mazara Del Vallo And Marsala (South-Western Sicily). In *Methods and Tools for Drought Analysis and Management*; Springer: Berlin/Heidelberg, Germany, 2007; pp. 303–321. [CrossRef]
38. Capizzi, P.; Cellura, D.; Cosentino, P.; Fiandaca, G.; Martorana, R.; Messina, P.; Schiavone, S.; Valenza, M. Integrated hydrogeochemical and geophysical surveys for a study of sea-water intrusion. *Boll. Geofis. Teor. Appl.* **2010**, *51*, 285–300.
39. Di Napoli, R.; Martorana, R.; Orsi, G.; Aiuppa, A.; Camarda, M.; De Gregorio, S.; Candela, E.G.; Luzio, D.; Messina, N.; Pecoraino, G.; et al. The structure of a hydrothermal system from an integrated geochemical, geophysical, and geological approach: The Ischia Island case study. *Geochem. Geophys. Geosyst.* **2011**, *12*. [CrossRef]
40. Martorana, R.; Lombardo, L.; Messina, N.; Luzio, D. Integrated geophysical survey for 3D modelling of a coastal aquifer polluted by seawater. *Near Surf. Geophys.* **2013**, *12*, 45–59. [CrossRef]
41. Capizzi, P.; Martorana, R.; Favara, R.; Albano, L.; Bonfardeci, A.; Catania, M.; Costa, N.; Gagliano, A. Geophysical Contribution to the Reconstruction of the Hydrological Model of "Barcellona-Milazzo Plain" Groundwater Body, Northen Sicily. In *25th European Meeting of Environmental and Engineering Geophysics*; European Association of Geoscientists & Engineers: Moscow, Russia, 2019; pp. 1–5. [CrossRef]
42. Paz, M.C.; Alcalá, F.J.; Medeiros, A.; Martínez-Pagán, P.; Pérez-Cuevas, J.; Ribeiro, L. Integrated MASW and ERT Imaging for Geological Definition of an Unconfined Alluvial Aquifer Sustaining a Coastal Groundwater-Dependent Ecosystem in Southwest Portugal. *Appl. Sci.* **2020**, *10*, 5905. [CrossRef]
43. Park, C.B.; Miller, R.D.; Xia, J. Multimodal Analysis of High Frequency Surface Waves. In Proceedings of the Symposium on the Application of Geophysics to Engineering and Environmental Problems, Oakland, CA, USA, 14–18 June 1999.
44. Alcalá, F.J.; Martínez-Pagán, P.; Paz, M.C.; Navarro, M.; Pérez-Cuevas, J.; Domingo, F. Combining of MASW and GPR Imaging and Hydrogeological Surveys for the Groundwater Resource Evaluation in a Coastal Urban Area in Southern Spain. *Appl. Sci.* **2021**, *11*, 3154. [CrossRef]
45. Nakamura, Y. A method for dynamic characteristics estimation of subsurface using microtremor on the ground surface. *Quat. Rep. Railw.* **1989**, *30*, 25–33.
46. Martorana, R.; Capizzi, P.; Avellone, G.; D'alessandro, A.; Siragusa, R.; Luzio, D. Assessment of a geological model by surface wave analyses. *J. Geophys. Eng.* **2016**, *14*, 159–172. [CrossRef]
47. Martorana, R.; Agate, M.; Capizzi, P.; Cavera, F.; D'Alessandr, A. Seismo-stratigraphic model of "La Bandita" area in the Palermo Plain (Sicily, Italy) through HVSR inversion constrained by stratigraphic data. *Ital. J. Geosci.* **2018**, *137*, 73–86. [CrossRef]
48. Sorensen, C.; Asten, M. Microtremor methods applied to groundwater studies. *Explor. Geophys.* **2007**, *38*, 125–131. [CrossRef]
49. Delgado, J.; Casado, C.L.; Estévez, A.; Giner, J.; Cuenca, A.; Molina, S. Mapping soft soils in the Segura river valley (SE Spain): A case study of microtremors as an exploration tool. *J. Appl. Geophys.* **2000**, *45*, 19–32. [CrossRef]
50. Martorana, R.; Capizzi, P.; D'Alessandro, A.; Luzio, D.; Di Stefano, P.; Renda, P.; Zarcone, G. Contribution of HVSR measures for seismic microzonation studies. *Ann. Geophys.* **2018**, *61*, 225. [CrossRef]
51. Pino, P.; D'Amico, S.; Orecchio, B.; Presti, D.; Scolaro, S.; Torre, A.; Totaro, C.; Farrugia, D.; Neri, G. Integration of geological and geophysical data for revaluation of local seismic hazard and geological structure: The case study of Rometta, Sicily (Italy). *Ann. Geophy.* **2018**, *61*, SE227. [CrossRef]
52. Scolaro, S.; Pino, P.; D'Amico, S.; Orecchio, B.; Presti, D.; Torre, A.; Totaro, C.; Farrugia, D.; Neri, G. Ambient noise measurements for preliminary microzoning studies in the city of Messina, Sicily. *Ann. Geophys.* **2018**, *61*, 228. [CrossRef]
53. Dal Moro, G. *Surface Wave Analysis for Near Surface Applications*; Elsevier: Amsterdam, The Netherlands, 2014.
54. Cafiso, F.; Canzoneri, A.; Capizzi, P.; Carollo, A.; Martorana, R.; Romano, F. Joint interpretation of electrical and seismic data aimed at modelling the foundation soils of the Maredolce monumental complex in Palermo (Italy). *Archaeol. Prospect.* **2020**, *30*, 69–85. [CrossRef]
55. Gisser, M.; Sánchez, D.A. Competition versus optimal control in groundwater pumping. *Water Resour. Res.* **1980**, *16*, 638–642. [CrossRef]
56. Percopo, C.; Brandolin, D.; Canepa, M.; Capodaglio, P.; Cipriano, G.; Gafà, R.; Iervolino, D.; Marcaccio, M.; Mazzola, M.; Mottola, A.; et al. Criteri tecnici per l'analisi dello stato quantitativo e il monitoraggio dei corpi idrici sotterranei. *ISPRA Man. E Linee Guid.* **2017**, *157*, 1–108.
57. Heilweil, V.M.; Brooks, L.E. *Conceptual Model of the Great Basin Carbonate and Alluvial Aquifer System: US Geological Survey Scientific Investigations Report*; U.S. Geological Survey: Reston, VA, USA, 2011.
58. Granata, A.; Castrianni, G.; Pasotti, L.; Gagliano Candela, E.; Scaletta, C.; Madonia, P.; Morici, S.; Bellomo, S.; La Pica, L.; Gagliano, A.L.; et al. Studio per la definizione dei modelli concettuali dei corpi idrici sotterranei di Peloritani, Nebrodi e ragusano e indagini geofisiche correlate. Atti del 37° Convegno del Gruppo Nazionale di Geofisica della Terra Solida-GNGTS. *OGS* **2018**, *37*, 104–108.
59. Canzoneri, A.; Capizzi, P.; Martorana, R.; Albano, L.; Bonfardeci, A.; Costa, N.; Gagliano, A.L.; Favara, R. Modello idrogeoligoco del corpo idrico "Fiumara di Caronia" (Sicilia Settentrionale). Atti del 38°Convegno del Gruppo Nazionale di Geofisica della Terra Solida-GNGTS. *OGS* **2019**, *38*, 636–640. Available online: https://gngts.ogs.it/atti-del-38-convegno-nazionale-2/ (accessed on 1 July 2023).
60. Catalano, R.; Di Stefano, P.; Sulli, A.; Vitale, F. Paleogeography and structure of the central Mediterranean: Sicily and its offshore area. *Tectonophysics* **1996**, *260*, 291–323. [CrossRef]

61. Giunta, G.; Giorgianni, A. *Note Illustrative del F.598 "S. Agata di Militello" Della Carta Geologica d'Italia Alla scala 1:50.000. SystemCart*; ISPRA—Università degli Studi di Palermo: Rome, Italy, 2013.
62. Catalano, R.; Valenti, V.; Albanese, C.; Accaino, F.; Sulli, A.; Tinivella, U.; Gasparo Morticelli, M.; Zanolla, C.; Giustiniani, M. Sicily's fold/thrust belt and slab rollback: The SI.RI.PRO. seismic crustal transect. *J. Geol. Soc.* **2013**, *170*, 451–464. [CrossRef]
63. Basilone, L. *Lithostratigraphy of Sicily: UNIPA Springer Series*; Springer International Publishing: Berlin/Heidelberg, Germany, 2018; pp. 1–349.
64. Amodio-Morelli, L.; Bonardi, G.; Colonna, V.; Dietrich, D.; Giunta, G.; Ippolito, F.; Liguori, V.; Lorenzoni, F.; Paglionico, A.; Perrone, V.; et al. L'Arco Calabro-Peloritano nell'orogene Appenninico-Maghrebide. *Mem. Soc. Geol. Ital.* **1976**, *17*, 1–60.
65. Bigi, G.; Cosentino, D.; Parotto, M.; Sartori, R.; Scandone, P. Structural Model of Italy (scale 1:500.000): Geodinamic Project: Consiglio Nazionale delle Ricerche, S.EL.CA: Rome, Italy. 1990. Available online: https://www.socgeol.it/438/structural-model-of-italy-scale-1-500-000.html/ (accessed on 1 July 2023).
66. Giunta, G.; Nigro, F. Tectono-sedimentary constraints to the Oligocene-to-Miocene evolution of the Peloritani thrust belt (NE Sicily). *Tectonophysics* **1999**, *315*, 287–299. [CrossRef]
67. Nigro, F.; Renda, P. Evoluzione geologica ed assetto strutturale della Sicilia centro- settentrionale. *Boll. Soc. Geol. Ital.* **1999**, *118*, 375–388.
68. Giunta, G.; Luzio, D.; Agosta, F.; Calò, M.; Di Trapani, F.; Giorgianni, A.; Oliveri, E.; Orioli, S.; Perniciaro, M.; Vitale, M.; et al. An integrated approach to investigate the seismotectonics of northern Sicily and southern Tyrrhenian. *Tectonophysics* **2009**, *476*, 13–21. [CrossRef]
69. Bianchi, F.; Carbone, S.; Grasso, M.; Invernizzi, G.; Lentini, F.; Longaretti, G.; Merlini, S.; Mostardini, F. Sicilia orientale: Profilo geologico Nebrodi-Iblei. *Mem. Soc. Geol. Ital.* **1987**, *38*, 429–458.
70. Catalano, R.; Franchino, A.; Merlini, S.; Sulli, A. Central western Sicily structural setting interpreted from seismic reflection profiles. *Mem. Soc. Geol. Ital.* **2000**, *55*, 5–16.
71. Catalano, R.; Gatti, V.; Avellone, G.; Basilone, L.; Frixa, A.; Ruspi, R.; Sulli, A. Subsurface Geometries in Central Sicily FTB as a Premise for Hydrocarbon Exploration. In Proceedings of the 70th EAGE Conference and Exhibition Incorporating SPE EUROPEC, Nuova Fiera Di Roma, Italy, 9–12 June 2008. [CrossRef]
72. Lentini, F.; Carbone, S. Geologia della Sicilia (Geology of Sicily). *Mem. Descr. Carta Geol. d'It.* **2014**, *XCV*, 7–414.
73. Morticelli, M.G.; Valenti, V.; Catalano, R.; Sulli, A.; Agate, M.; Avellone, G.; Albanese, C.; Basilone, L.; Gugliotta, C. Deep controls on foreland basin system evolution along the Sicilian fold and thrust belt. *BSG-Earth Sci. Bull.* **2015**, *186*, 273–290. [CrossRef]
74. Basilone, L.; Bonfardeci, A.; Romano, P.; Sulli, A. Natural Laboratories for Field Observation About Genesis and Landscape Effects of Palaeo-Earthquakes: A Proposal for the Rocca Busambra and Monte Barracù Geosites (West Sicily). *Geoheritage* **2018**, *11*, 821–837. [CrossRef]
75. Giunta, G.; Bellomo, D.; Carnemolla, S.; Pisano, A.; Profeta, R.; Runfola, P. La "Linea di Taormina": Residuo epidermico di una paleostruttura crostale del fronte cinematico maghrebide? *Acts 8° GNGTS Congr.* **1989**, *8*, 1197–2013.
76. Lentini, F.; Catalano, S.; Carbone, S. Note Illustrative Della Carta Geologica Della Provincia di Messina, Scala 1:50.000: S.EL.CA, Firenze. 2000; pp. 1–70. Available online: https://www.isprambiente.gov.it/Media/carg/note_illustrative/601_Messina_Reggio_Calabria.pdf (accessed on 1 July 2023).
77. Giunta, G.; Nigro, F.; Renda, P. Extensional tectonics during Maghrebides chain building since Late Miocene: Examples from Northern Sicily. *Ann. Soc. Geol. Polon.* **2000**, *70*, 81–89.
78. Giunta, G.; Nigro, F.; Renda, P.; Giorgianni, A. The Sicilian–Maghrebides Tyrrhenian Margin: A neotectonic evolutionary model. *Mem. Soc. Geol. Ital.* **2000**, *119*, 553–565.
79. Renda, P.; Tavarnelli, E.; Tramutoli, M.; Gueguen, E. Neogene deformations of Northern Sicily, and their implications for the geodynamics of the Southern Tyrrhenian Sea margin. *Mem. Soc. Geol. Ital.* **2000**, *55*, 53–59.
80. Nigro, F.; Renda, P. Plio-Pleistocene stike-slip deformation in NE Sicily: The example of the area between Capo Calavà and Capo Tindari. *Boll. Soc. Geol. Ital.* **2005**, *124*, 377–394.
81. Ogniben, L. Nota illustrativa dello schema geologico della Sicilia Nord-Orientale. *Riv. Min. Sic.* **1960**, *2*, 183–212.
82. Wezel, F.C. Geologia del flysch numidico della Sicilia nord-orientale. *Mem. Soc. Geol. Ital.* **1970**, *9*, 225–280.
83. Wezel, F.C. Flysch succession and the tectonic evolution of Sicily during the Oligocene and early Miocene. In *Geology of Italy*; Squires, C.H., Ed.; Earth Sciences Soc. Libyan Arabian Republic: Tripoli, Libya, 1974; pp. 105–127.
84. Giunta, G. Problematiche ed ipotesi sul bacino numidico nelle Maghrebidi siciliane. *Boll. Soc. Geol. Ital.* **1985**, *104*, 239–256.
85. Lentini, F. Carta Geologica della Provincia di Messina, Scala 1:50.000: S.EL.CA, Firenze. 2000. Available online: https://www2.regione.sicilia.it/beniculturali/dirbenicult/bca/ptpr/documentazione%20tecnica%20messina/CARTOGRAFIA/ANALISI/03_Geologia.pdf (accessed on 1 July 2023).
86. APAT. Carta Geologica d'Italia in scala 1:50.000—Foglio 598, S. Agata di Militello: ISPRA-Regione Siciliana-D.S.G.-uni., Palermo, SystemCart, Roma. 2013. Available online: https://www.isprambiente.gov.it/Media/carg/598_SANTAGATA_MILITELLO/Foglio.html (accessed on 1 July 2023).
87. Celico, P. *Prospezioni Idrogeologiche, Vol. I e Vol. II.*; Liguori Editore Napoli: Naples, Italy, 1988.
88. Pantaleone, D.V.; Vincenzo, A.; Fulvio, C.; Silvia, F.; Cesaria, M.; Giuseppina, M.; Ilaria, M.; Vincenzo, P.; Rosa, S.A.; Gianpietro, S.; et al. Hydrogeology of continental southern Italy. *J. Maps* **2018**, *14*, 230–241. [CrossRef]

89. Ferrara, V.; Amantia, A.; Pennisi, A. Lineamenti idrogeologici e vulnerabilità all'inquinamento degli acquiferi della fascia costiera tirrenica del messinese (versante settentrionale dei M. Peloritani—Sicilia NE). In Proceedings of the Congresso Biennale, Palermo Torre Normanna, Palermo, Italy, 21–23 September 1993.
90. Ferrara, V. Vulnerabilità all'inquinamento degli Acquiferi dell'area Peloritana (Sicilia Nord-Orientale), Studi Sulla Vulnerabilità Degli Acquiferi 14. In *Quaderni di Tecniche di Protezione Ambientale: Edizione Pitagora*; Pitagora Editrice Srl.: Bologna, Italy, 1999.
91. Civita, M. *Idrogeologia Applicata e Ambientale*; CEA Editore: Rozzano, Italy, 2005; p. 800. ISBN 8808087417.
92. Castany, G. Idrogeologia. In *Principi e Metodi: Edizione Flaccovio Dario, Palermo*; Pitagora Editrice Srl.: Bologna, Italy, 1985.
93. Hiscock, K.M.; Bense, V.F. *Hydrogeology: Principles and Practice*, 2nd ed.; Wiley-Blackwel: Oxford, UK, 2014.
94. Gorla, M. Pozzi per acqua. In *Manuale Tecnico di Progettazione: Edizione Flaccovio Dario*; Pitagora Editrice Srl.: Bologna, Italy, 2010.
95. Civita, M. *Geologia Tecnica*; ISED: Ottawa, ON, Canada, 1975.
96. Mouton, J.; Mangano, F.; Fried, J.J. *Studio Delle Risorse in Acque Sotterranee Dell'Italia*; Schafer Druckerei: Hannover, Germany, 1982; p. 515.
97. CASMEZ (CASsa per il MEZzogiorno)—Direzione Generale Progetti Speciali. Progetto Speciale n. 30: Utilizzazione delle acque degli schemi idrici intersettoriali della Sicilia. In *Indagini Idrogeologiche e Geofisiche per il Reperimento di Acque Sotterranee per L'approvvigionamento Idrico del Sistema IV Zona Nord Orientale Della Sicilia (Messinese)*; CASMEZ: Palermo, Italy, 1978.
98. Nogoshi, M.; Iragashi, T. On the propagation characteristic of the microtremors. *J. Seismol. Soc. Jpn.* **1970**, *24*, 24–40.
99. Forte, G.; Chioccarelli, E.; De Falco, M.; Cito, P.; Santo, A.; Iervolino, I. Seismic soil classification of Italy based on surface geology and shear-wave velocity measurements. *Soil Dyn. Earthq. Eng.* **2019**, *122*, 79–93. [CrossRef]
100. Harbaugh, A.W.; Banta, E.R.; Hill, M.C.; McDonald, M.G. MODFLOW2000, the U.S. Geological Survey modular ground-water model-User guide to modularization concepts and the Ground-Water Flow Process. *U.S. Geol. Surv. Open-File Rep.* **2000**, *92*, 121.
101. Guzman, J.A.; Moriasi, D.N.; Gowda, P.H.; Steiner, J.L.; Starks, P.J.; Arnold, J.G.; Srinivasan, R. A model integration framework for linking SWAT and MODFLOW. *Environ. Model. Softw.* **2015**, *73*, 103–116. [CrossRef]
102. McDonald, M.; Harbaugh, A. *A Three Dimensional Finite Difference Ground Water Flow Model*; U.S. Geological Survey: Reston, VA, USA, 1988.
103. Batu, V. Numerical Flow and Solute Transport Modeling in Aquifers. In *Applied Flow and Solute Transport Modeling in Aquifers: Fundamental Principles and Analytical and Numerical Methods*; CRC Press, Taylor & Francis Group: Boca Raton, FL, USA, 2006; pp. 215–344. ISBN 9780849335747.

water

MDPI

Article

Hydrogeophysical Investigation in Parts of the Eastern Dahomey Basin, Southwestern Nigeria: Implications for Sustainable Groundwater Resources Development and Management

Kehinde D. Oyeyemi [1], Ahzegbobor P. Aizebeokhai [1], Abayomi A. Olaojo [2], Emmanuel E. Okon [3], Divine V. Kalu [4] and Mohamed Metwaly [5,*]

1 Applied Geophysics Program, Department of Physics, Covenant University, Ota 112233, Nigeria; kdoyeyemi@yahoo.com (K.D.O.); apaizebeokhai@gmail.com (A.P.A.)
2 Department of Earth Sciences, Ajayi Crowther University, Oyo 200284, Nigeria; yomiolaojo@yahoo.com
3 Department of Geology, University of Calabar, Calabar 540004, Nigeria; etyboy911@yahoo.com
4 Department of Earth and Atmospheric Sciences, University of Houston, Houston, TX 77204-4020, USA; dckalu@cougarnet.uh.edu
5 Department of Archeology, College of Tourism and Archeology, King Saud University, Riyadh 11421, Saudi Arabia
* Correspondence: mmetwaly@ksu.edu.sa

Abstract: Geoelectrical resistivity measurements were conducted in five locations within the eastern portion of the Dahomey basin for the purpose of subsurface evaluation and detecting saturated zones. The locations are Covenant University (L1), Bells University (L2), Oju-Ore-Ilogbo Road (L3), Obasanjo-Ijagba Road (L4), and Iyana Iyesi (L5). The study was carried out to avert the common challenges of drilling low-yield groundwater boreholes in the area. A total of 30 Vertical Electrical Soundings (VES) and five two-dimensional Electrical Resistivity Tomography (ERT) data sets have been acquired along the study areas. The geoelectrical resistivity results were integrated with the borehole logs to generate the spatial distribution of the subsurface lithologies in the area. The delineated subsurface lithologies include the topsoil (lateritic clay), clayey sand, sandy clay, fine silty sand, coarse sand, and shale/clay units. The fine silty sand and coarse sand units were identified as the two main aquifer units within the area. The depths to the upper aquifer unit in the area include 31.7–96.7 m, 38.5–94.0 m, 30.7–57.5 m, 39.1–63.4 m, and 46.9–57.5 m for locations L1, L2, L3, L4, and L5, respectively. At the same time, the depths to the lower aquifer unit in the area include 43.4–112.7 m, 52.2–108.0 m, 44.2–72.5 m, 53.7–78.5 m, and 63.5–72.9 m for locations L1, L2, L3, L4, and L5, respectively. The estimated hydraulic parameters for both aquifers show they are highly productive with mean porosity, mean hydraulic conductivity, and mean transmissivity of 20–22%, 12.4–17.0 $\times$ 10^{-2} m/s, 1.56–2.18 m^2/s for the upper aquifer, and 48–50%, 371–478 $\times$ 10^{-2} m/s, 50.00–62.14 m^2/s for the lower aquifer. By focusing on these aquifer systems during exploration, sustainable groundwater resources can be secured, providing relief to homeowners within the study area who might otherwise face the frustration of drilling unproductive and low-yield boreholes. However, it is crucial to consider the presence of sub-vertical faults in the study area, as these faults can significantly impact groundwater development and management. These sub-vertical structural faults may lead to changes in the permeability, hydraulic conductivity, and transmissivity of the delineated aquifers, affecting their productivity across the divide and ultimately influencing the overall water availability in the area. Careful consideration of these geological factors is essential for effective aquifer management and sustainable groundwater utilisation.

Keywords: hydrogeological studies; groundwater; resources management; geoelectrical resistivity; sustainability

1. Introduction

Groundwater exploration and management are becoming essential research topics in arid and semi-arid regions due to the difficulty in gaining access to groundwater and the ongoing depletion of the water table. Areas with complex geology often bring about specific problems in the endeavour to correctly access groundwater resources, including the rampant drilling of dry boreholes and drilling into temporary productive aquifers. Thus, with a thorough understanding of subsurface geological properties, it is more effective to locate aquifers, regulate groundwater supplies, and develop for our collective gains. Although the earth's surface is about 71% covered in water, the ocean (saline water) holds approximately 96.5% of all earth's water, leaving aside 3.5% for water noticed in the air as water vapour, lakes, and streams, in the ice caps or glaciers, in the ground as soil moisture, and in aquifer beds [1–3]. Groundwater accounts for nearly 30.1% of all freshwater on and above the earth's surface, making it a critical source of freshwater for human life. Based on the vital role of groundwater in nature, the quantitative and qualitative characterisation of aquifers has turned out to be essential, intending to address a few hydrogeological parameters, for example, porosity, permeability, storativity, and hydraulic conductivity [4]. The most important parameter is the efficient permeability that allows a rock formation to store, transmit, and yield groundwater in reasonable quantities from the surface through its pore spaces, either by natural pressure as in a confined aquifer or through artificial pumping pressure like in an unconfined aquifer.

The rapid development and advancements in hydrogeophysical methods have made it a standout discipline within near-surface geophysics, offering innovative and sophisticated techniques for investigating subsurface hydrological processes. Hydrogeophysics can be depicted as the application of geophysical techniques for mapping subsurface structures and hydraulic parameters essential for groundwater evaluation and exploration [4]. It also connotes the assessment of subsurface hydrogeological properties and monitoring procedures vital for studying groundwater hydrology, either from ground surface measurements or throughout the well logs. These processes are linked with water resources, seepages along the vadose zone (the unsaturated portion of the near-surface), contaminant transport, and ecological and climate investigations of groundwater systems [4]. Several recent research studies have used geophysical methods to develop the field of hydrogeophysics and provide quantitative estimates of subsurface characteristics [5]. Generally, the productivity of subsurface aquifers, to a large extent, depends on their depth, thickness, rock physics parameters (such as permeability, porosity, mineral composition, and degree of water saturation), and hydraulic parameters [6–10]. Hydraulic parameters, such as porosity, hydraulic conductivity, transmissivity, specific yield, and storage coefficients, provide valuable insights into the behaviour and characteristics of aquifers [11–20]. They determine the flow patterns, storage capacity, and transport properties of groundwater within the subsurface [14–17]. Estimating these parameters enables hydrogeologists and water resource researchers to better comprehend the dynamics of aquifers, evaluate their potential, and make informed decisions regarding water supply, management, and protection [11,13,19,20]. Furthermore, hydraulic parameters play a vital role in aquifer characterization. They provide critical information about the physical properties of aquifers, such as porosity and permeability. Porosity describes the amount of void space available for water storage, while permeability relates to the ease with which water can flow through the aquifer. Accurate estimation of these parameters aids in delineating aquifer boundaries, identifying water-bearing zones, and evaluating the overall quality and suitability of the aquifer for various water supply purposes. Traditionally, these parameters were estimated through direct measurement techniques, such as pump tests and borehole logging, which can be time-consuming, costly, and limited in spatial coverage. Hydrogeophysical investigations offer an alternative approach, allowing for non-invasive and spatially extensive characterization of aquifer properties. The failure of most water boreholes and their subsequent abandonment during groundwater development are often due to an insufficient understanding of the complexity of the subsurface geology situation and hydraulic properties. Thus, the prime essence of conducting geophysical investigations for hydrogeophysical purposes is to provide detailed information on the subsurface geology that essentially helps in recommending suitable locations for productive wells. Hydrogeophysical investigations are valuable tools for estimating the hydraulic parameters of subsurface aquifers. Geophysical studies play a critical role in water management, providing valuable information for understanding subsurface conditions, groundwater resources, and hydrogeological processes [21]. In modern water management, geophysical evaluation strategies have become essential due to their non-invasive nature, cost-effectiveness, and ability to provide high-resolution data about the subsurface properties. In modern water management, geophysical studies are essential tools for acquiring comprehensive and reliable data on subsurface conditions and groundwater resources. They facilitate informed decision-making, efficient water allocation, and sustainable water management practices. The integration of geophysical data with other hydrological and geological information strengthens water resource planning and helps ensure the availability of clean and accessible water for current and future development [21]. Therefore, the current research focuses on the use of surface electrical resistivity measurements to identify and characterise the subsurface aquifers in some communities within the eastern part of the Dahomey Basin, southwestern Nigeria. This task was achieved through a detailed evaluation of the subsurface stratigraphic and structural features that control the area's hydrogeological setting. The geoelectrical resistivity measurements were used to evaluate the dynamic hydraulic parameters of the delineated aquifer systems to understand their heterogeneity and variability. Such applications can guide aquifer developments in arid and semiarid areas all over the world and introduce fast and reliable subsurface evaluation for groundwater research.

2. Geological Setting and Site Description

The study area is located in the Dahomey basin (or Benin), roughly between latitudes 6°37′ N–6°44′ N and longitudes 3°9′ E–3°15′ E, close to the mainland Gulf of Guinea margin (Figure 1). Generally, the area has predominant dry and wet climatic seasons, and the terrain has a mild slope. The dry season runs from November to March, and the rainy season runs from April to October [22]. However, because of its proximity to the Atlantic Ocean, this region frequently sees sporadic rainfall throughout the dry season. The primary source of groundwater recharge comes from the average annual rainfall, which is roughly 2000 mm [23]. Moreover, the two significant rivers (Atura and Yewa) drain the study area (Figure 1b) and recharge groundwater resources within the eastern part of the Dahomey basin [24–26].

Geologically, the Abeokuta Group, which is separated into the Ise, Afowo, and Araromi Formations, makes up the Cretaceous stratigraphy gathered from outcrops and drilling records [27]. The basement complex is unconformably covered by the Abeokuta Group, which is then followed in that order by the Ewekoro, Akinbo, Oshosun, Ilaro, and Benin Formations. These lithostratigraphic units have been discussed by a number of authors [28–31]. It is known that the Abeokuta Group is primarily composed of shale-clay layers and poorly sorted ferrous grit, siltstone, and mudstone. The Abeokuta Group is underlain by the Ewekoro Formation, a primarily Paleocene shallow marine limestone [32]. A predominantly shale unit of the Late Paleocene to Early Eocene Akinbo Formation lies on top of the Ewekoro Formation. Pure white, coarse sand and a trace of clay make up the Akinbo formation's upper layer. The Oshosun Formation, which is normally marine and is Eocene in age, is deposited on top of the Akinbo Formation, and it laterally extends into thick mud. According to descriptions, Oshosun is made up of heavily laminated, glauconitic, and phosphate-containing Eocene shale. The Ilaro Formation primarily consists of a sequence of coarse, sandy estuary, deltaic, and continental layers with dramatic lateral facies alterations. The Benin Formation, which mostly consists of Tertiary alluvial deposits and coastal plain sands, lies beneath the Ilaro Formation. Most of the area is covered with coastal plains and recent deposits, which are mainly poorly graded sand and clayey deposits.

Figure 2 presents the subsurface lithologic units identified from the drilled boreholes in the area. The dominant lithology consists of sandy materials of different sizes. The uppermost layer is composed of a lateritic unit, which acts as an impermeable layer. Below this unit, several sandy horizons allow for the transmission of groundwater. These clay units and sandy clay horizons restrict further infiltration of groundwater from likely contaminants at the surface. These impermeable units, namely lateritic clayey sand, and compacted sandy clay, confine the aquifers within the study area. The findings from the geoelectric sequence align with the deductions made from the borehole lithostratigraphic units, confirming that the aquifer systems in the area are confined.

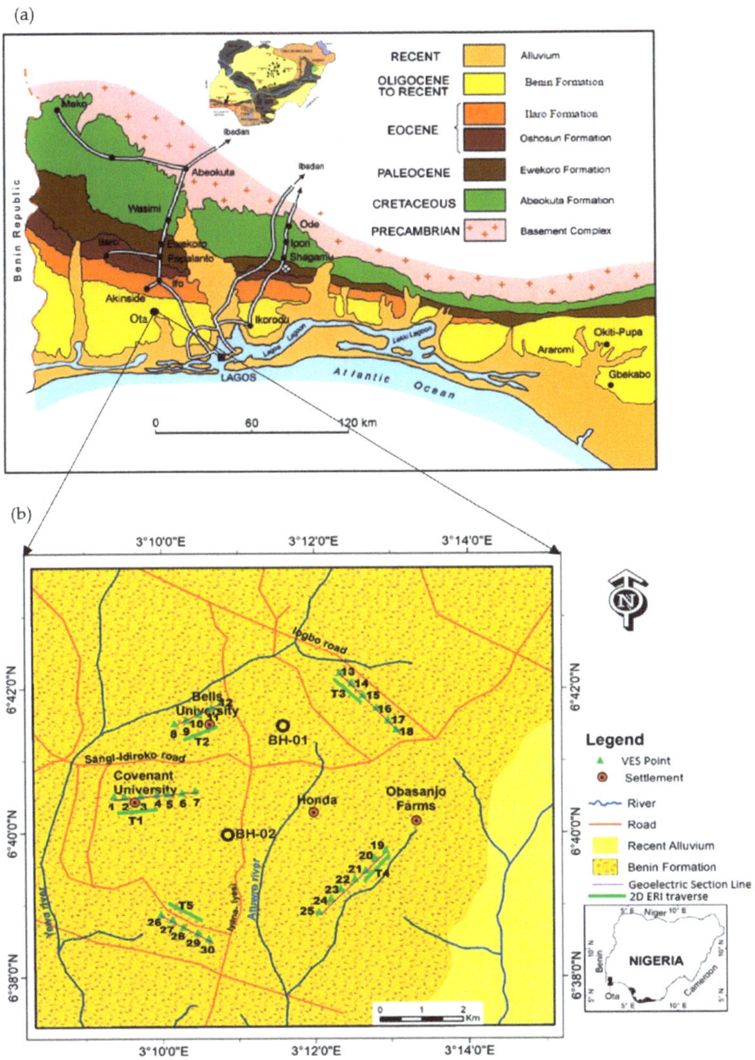

Figure 1. (**a**) Geological map of the Dahomey basin (modified after [33]) (**b**) A local geology map for the study area showing the locations of measured VES stations and 2D ERI profiles.

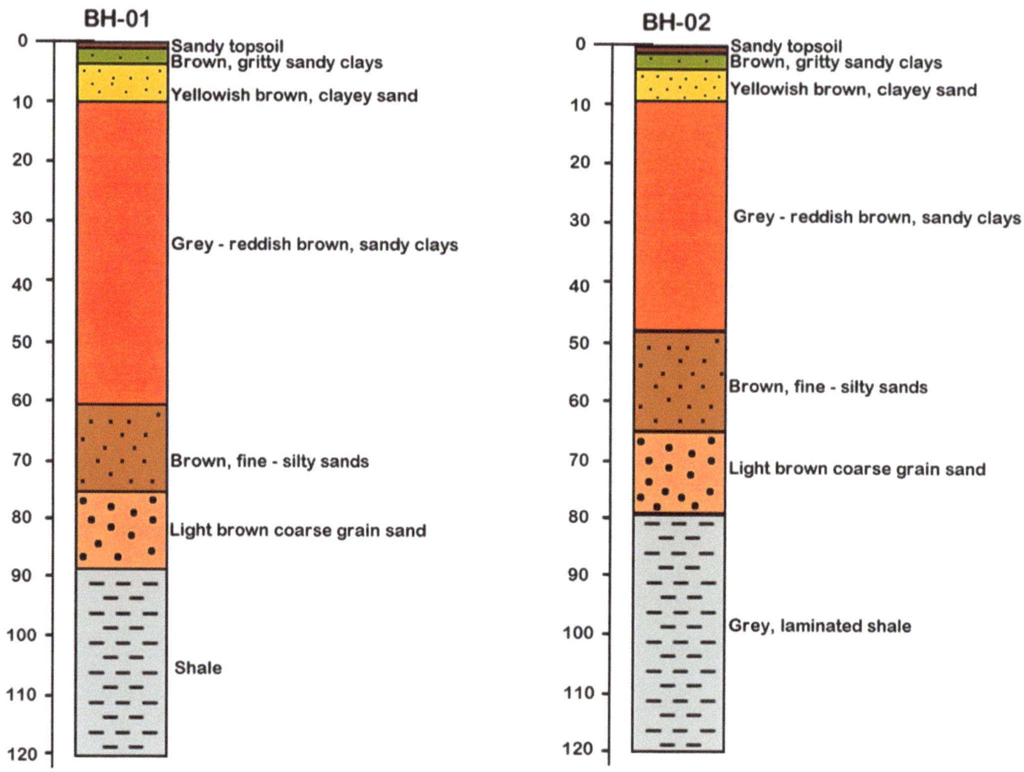

Figure 2. The dominant lithological succession along the study area deduced from the borehole log data for the location refers to Figure 1.

3. Materials and Methods

3.1. Data Acquisition Procedure

To study the subsurface geological setting and characterise the more promising groundwater-saturated layers, Vertical Electrical Soundings (VES) and two-dimensional Electrical Resistivity Tomography (ERT) data sets have been acquired along the study area (Figure 1b). A total of thirty (30) VESs were conducted across the study area, with seven (7) VES in location L1 (Covenant University), five VES in location L2 (Bells University), six VES in location L3 (Oju-Ore-Ilogbo road), seven VES in location L4 (Obasanjo-Ijagba road), and five VES in location L5 (Iyana-Iyesi). The Schlumberger array was utilised for the survey with a maximum half-current electrode spacing (AB/2) of 240 m, utilising the ABEM Terrameter (SAS 4000). The geoelectrical resistivity soundings were conducted to determine the vertical distributions of the aquifer units and the subsurface lithologic stratigraphy along the study area. Five 2D ERT traverse profiles were also conducted along the five investigated locations. The survey was taken manually along the five traverses of a 500 m long roll-along technique with 51 electrode positions for each traverse line. Wenner electrode configuration has been applied with electrode offsets of 10 m to trace the lateral resistivity distribution along the measured profiles and then interpret the results in the expected subsurface geological settings. The directions of the traverses T1, T2, T3, T4, and T5 are west-east, northeast-southwest, northwest-southeast, and northeast-southwest (Figure 1b). The Terrameter system displayed the acquired resistivity values three times before showing the fourth value, which is an average of the previous values. To ensure acquiring good data sets, the electrode coupling with the ground was crosschecked, and in cases where the ground was dry, electrode contact was improved by watering the ground.

3.2. Data Processing and Inversion

On a log-log graph sheet, the measured apparent resistivity values were plotted against the half current spacing (AB/2) to analyse the field datasets for each 1-D VES. The results were used to create field curves that were matched with theoretical master curves for the Schlumberger array in order to calculate the thickness and resistivity of the geoelectric layers. To generate geoelectric model parameters for the demarcated strata, WINRESIST software version (1.0) was using the estimated geoelectric parameters as initial models. The 2D ERT data sets were inverted using the RES2DINV inversion code based on the principle of inversion, which aims to estimate the subsurface resistivity distribution from the measured apparent resistivity data. Inversion is a mathematical process that involves solving an inverse problem where the unknown resistivity distribution is inferred from the observed data. Using the finite difference or finite element method, the software calculates the expected apparent resistivity values for a given subsurface resistivity model [34]. The forward modelling process involves solving the governing partial differential equations that describe the flow of electrical current through the subsurface. By comparing the calculated apparent resistivity values with the observed data, the algorithm seeks to minimise the difference, or misfit, between them. In the inversion process, RES2DINV aims to find the resistivity model that best explains the observed apparent resistivity data. It starts with an initial resistivity model, which can be based on prior geological information or assumed resistivity values. The inversion algorithm iteratively adjusts the resistivity values within the model grid to minimise the misfit between the observed and calculated apparent resistivity values.

3.3. Hydraulic Parameter Estimation

The estimation of hydraulic parameters holds significant importance in aquifer studies. These parameters provide essential information for understanding groundwater flow, aquifer characteristics, groundwater modelling, water resource planning, and groundwater remediation. Accurate estimation of these parameters enhances our ability to effectively manage and protect this vital natural resource, ensuring its sustainable use for present and future generations. The fundamental equations for geoelectrical exploration assume a porous medium with an insulating matrix where electrical currents pass through the water present within the pore spaces. The electrical resistivity of an aquifer is primarily influenced by the porosity and fluid resistivity within the pores. The geoelectrical data collected at the surface holds valuable information about the aquifer, which can be interpreted by experienced geophysicists for hydrogeological investigations [35,36]. In an ideal scenario, the physical factors governing electric current flow, such as tortuosity and porosity, also control water flow within a porous medium. Building on this analogy, numerous empirical equations have been reported in the literature, establishing correlations between electrical resistivity and hydraulic conductivity [37,38]. These equations offer valuable insights into the hydraulic properties of aquifers based on geoelectrical data, further enhancing our understanding of groundwater systems. Equation (1) is the relation used to compute porosity and hydraulic conductivity from the geoelectrical measurements [39,40].

$$\rho = a\rho_w\phi^{-m} \tag{1}$$

where a and m represent the electrical tortuosity parameter [41] and cementation factor, while ρ_w and ϕ represent the resistivity of groundwater and aquifer porosity. For a clean, unconsolidated sandy aquifer with no interbedding clays, a and m are assumed to be 1.0 and 1.3, respectively. The hydraulic conductivity was estimated using the Kozeny-Carman method [42,43] presented in Equation (2). The Kozeny-Carman method is widely accepted as one of the primary formulas for calculating hydraulic conductivity. The Kozeny-Carman equation offers a convenient and widely used approach to estimating hydraulic conductivity by considering porosity and grain size diameter. This equation incorporates important parameters, such as water density (ρ_w) in grammes per cubic centimetre (g/cm^3), porosity (ϕ), viscosity (η), acceleration due to gravity (g) in centimetres per second squared (cm/s^2), and the dominant grain size (d) in centimeters (cm). Referred to as the Kozeny-Carman method, it conceptualises a rock with primary porosity as a network of capillaries, satisfying the Navier-Stokes equation. The resulting hydraulic conductivity (K) can be expressed in different units, such as centimetres per second (cm/s), metres per second (m/s), or meters per day (m/day), depending on the chosen unit system. This versatility allows the method to suit various applications and scenarios. The Kozeny-Carman method relies on rock sampling and analysis, enabling the determination of the dominant grain size (d) from the grain size distribution curve using Equation (2). In this context, d_{10} and d_{60} denote the grain diameter at 10% and 60% cumulative frequency, respectively, obtained through sieve analysis. The method's ability to estimate hydraulic conductivity based on readily available data makes it a valuable tool in hydrogeological investigations and groundwater studies.

$$K = \frac{\rho_w g}{\mu} \frac{d^2}{180} \frac{\phi^3}{(1-\phi)^2} \tag{2}$$

$$d = \frac{d_{10} + d_{60}}{2} \sqrt{\frac{d_{10}}{d_{60}}} \tag{3}$$

Equation (2) is simplified to give Equation (4). The constant "C" in the simplified Kozeny-Carman Equation (4) incorporates factors such as the shape and arrangement of sediment particles as well as the tortuosity of flow paths within the porous medium. The value of "C" can vary depending on the specific characteristics of the sediment or soil being analysed [44]. The following approximate values are adopted for the following sedimentary grain attributes: well-sorted, rounded sands: $C \approx 5$–15, moderately sorted sands: $C \approx 10$–30, poorly sorted sands and silts: $C \approx 20$–50, and clayey sediments: $C \approx 50$–100 or higher.

$$K = \frac{C\phi^3}{(1-\phi)^2} \tag{4}$$

Hydraulic conductivity (K) and transmissivity (T) are related through Equation (5):

$$T = K \times b \tag{5}$$

where T is the transmissivity of the aquifer, K is the hydraulic conductivity of the aquifer, and b is the thickness of the aquifer perpendicular to the direction of flow. Transmissivity represents the ability of an aquifer to transmit water under a hydraulic gradient. It is calculated by multiplying the hydraulic conductivity (K) by the thickness of the aquifer (b) in the direction perpendicular to the flow. The relationship between hydraulic conductivity and transmissivity is valuable in groundwater studies and resource evaluations, as transmissivity indicates the potential for water movement within an aquifer under a given hydraulic gradient.

4. Results

4.1. Vertical Electrical Sounding

The interpretation of the VES data and the geoelectric sections in all the study locations (L1–L5) revealed around eight geoelectrical layers. The estimated geoelectric parameters for the identified geoelectric layers are uniform among all the VES curves, and an example is shown in Table 1. The interpretation of the subsurface lithology from the geoelectric layers at the five locations was established based on the inhomogeneity of electrical resistivity properties and the information from the drilled boreholes and wells integrated with the known local geological setting.

Table 1. Example of the estimated parameters for VES data sets along the Covenant University (L1) that were utlised to construct the Geoelectric resistivity sections.

VES		Layer 1	Layer 2	Layer 3	Layer 4	Layer 5	Layer 6	Layer 7	Layer 8
	Lithology	Lateritic Clay	Clayey Sand	Sandy Clay	Sandy Clay		Fine Silty-Sand (Upper aquifer)	Coarse Sand (Lower aquifer)	Shale/Clay
1	Resistivity	89	142.8	1039.3	1543.9	3107.6	347.4	125.4	86.6
	Thickness	1.3	2.4	4.6	8.4	16	12.8	12.5	
	Depth	1.3	3.8	8.3	16.8	32.7	45.5	58	
2	Resistivity	83.5	241.2	782.7	1034.5	3209.5	386.7	121	48.5
	Thickness	1.1	2	4	6.7	41.2	12.3	12.1	
	Depth	1.1	3.1	7.1	13.8	55	67.3	79.3	
3	Resistivity	108.9	214.2	915.2	2628.8	10,341.50	378.5	119.6	45.7
	Thickness	1	2.1	3.4	3.9	23	13.1	13	
	Depth	1	3.1	6.5	10.4	33.4	46.5	59.5	
4	Resistivity	51.7	222.7	874.1	980.2	5994.2	389.7	120.8	47.4
	Thickness	0.9	1.6	5.6	8.3	45	13.5	13.5	
	Depth	0.9	2.6	8.2	16.5	61.5	75	88.1	
5	Resistivity	24.5	376.7	399.5	971.8	3102	356.2	128	34.1
	Thickness	0.9	6.5	19.9	8	22	12	13.9	
	Depth	0.9	7.4	27.2	35.3	57.3	69.2	83.2	
6	Resistivity	59.3	287.2	1109.2	1355.6	9784.1	390.1	132	83.6
	Thickness	1	2.8	7.3	20.1	65.5	15.9	14.2	
	Depth	1	3.8	11.1	31.2	96.7	112.7	126.9	
7	Resistivity	132.8	119.4	508.5	2804.8	5817.9	361.6	119.9	36.7
	Thickness	1	3.9	2.9	5.6	18.3	11.8	11.9	
	Depth	1	4.9	7.8	13.4	31.7	43.4	55.3	

Figure 3 reveals the representative of the inverted VES numbers (1–7) conducted within Covenant University (Location L1). The first geoelectric layer is topsoil, adjudged to be a lateritic clay soil with a resistivity range of 24.5–132.8 Ωm, and a thickness range of 0.9–1.3 m. The topsoil resistivity values are low because it is evident that the layer contains some lateritic clay. The second geoelectric layer shows resistivity values ranging from 119.4–376.7 Ωm, and a thickness range of 1.6–6.5 m that can be interpreted as clayey sand deposits. The third delineated layer with inverse model resistivity values of 399.5–1109.2 Ωm is interpreted as a sandy clay unit with a thickness range of 2.9–19.9 m. The fourth and fifth delineated layers had resistivity ranges of 971.8–2804.8 Ωm and 3102.0–10,341.5 Ωm, and layer thickness ranges of 3.9–8.4 m and 16.0–65.5 m, respectively. This zone is interpreted as a sandy clay layer, which seems to be the confinement of the underlying saturated units. The sixth geoelectric layer has a resistivity range of 347.4–390.1 Ωm, and a layer thickness range of 12.0–15.9 m, and is interpreted as the upper saturated silty sand layer. The seventh delineated layer shows a resistivity range of 119.6–132.0 Ωm, and a thickness range of 11.9–14.2 m and is represented as a lower saturated layer of coarse sand. The last delineated resistivity layer has a resistivity range of 45.7–86.6 Ωm and is interpreted as a shale or clay unit. The summary of the estimated geoelectric parameters from the interpreted VESs is presented in Table 1, and the corresponding geoelectric sections constructed are presented in Figure 4. Two anticipated sub-vertical faults were mapped in the area, as displayed in the geoelectric section.

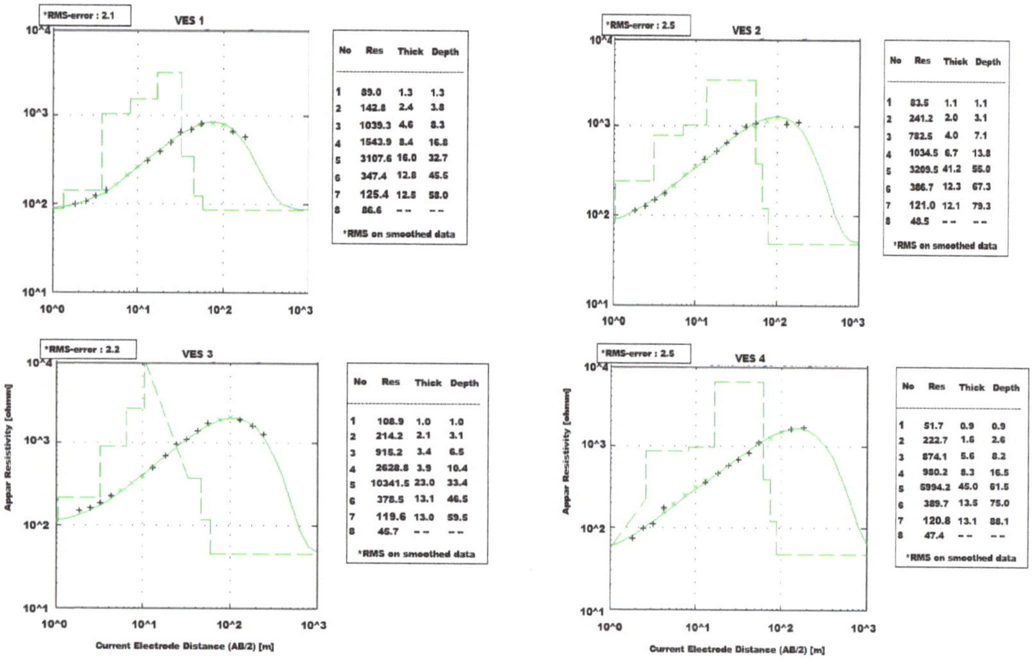

Figure 3. Representative of the inverted VES curves within Covenant University (L1) and the resulting digital subsurface models.

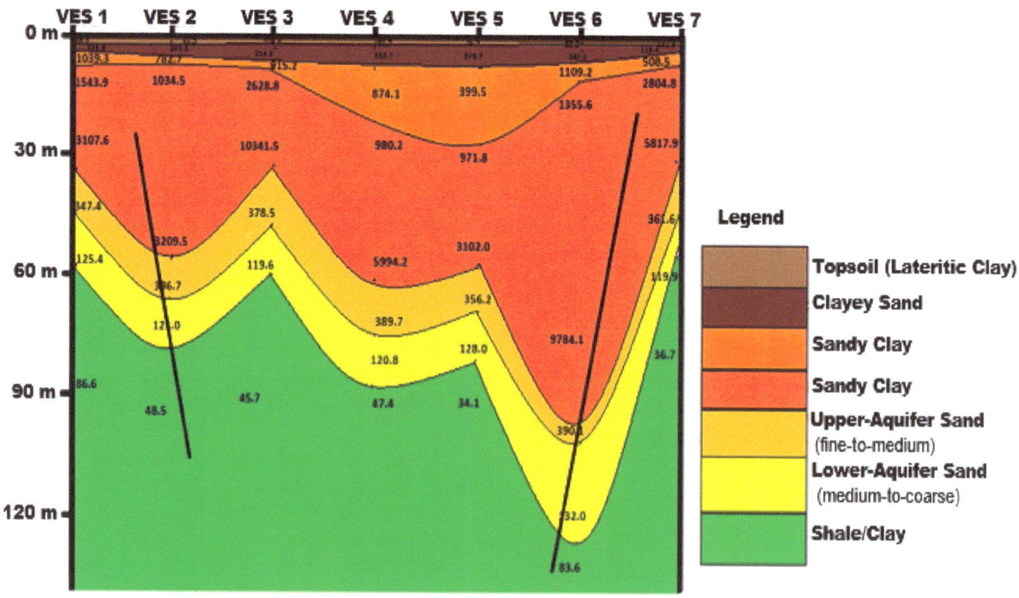

Figure 4. Geoelectric subsurface section from sounding results carried out along the Covenant University, showing the inferred faults.

A total of five VESs were conducted within the Bells University (Location L2) campus, and the interpretation of the VES data equally revealed eight geoelectric strata within the subsurface. The representative inverted VES curves and the resulting model parameters are presented in Figure 5. The first unit of the inverse model shows variable resistivity values in the range 73.4–322.1 Ωm, which is represented by the topsoil of lateritic clay with a thickness range of 0.9–1.9 m. The high resistivity characteristics of the topsoil at some VES points may be attributed to the compaction due to surface activities. The second layer has a resistivity range of 99.5–276.6 Ωm, and a thickness range of 4.4–8.0 m, which can be interpreted as a clayey sand unit. The third layer shows resistivity values of 570.3–1088.9 Ωm and a thickness range of 4.4–8.0 m. This layer is considered a sandy clay unit. The fourth and fifth mapped layers have resistivity ranges of 852.3–1831.4 Ωm and 1914.0–8177.0 Ωm respectively, and the thickness range is 31–83.3 m. This layer can be interpreted as sandy clay and represents the confining bed for the underlying saturated units. The delineated sixth layer has a resistivity range of 363.3–408.0 Ωm and a thickness of 13.1–14.3 m, which can be interpreted as sandy clay. The seventh layer shows a resistivity range of 120.9–143.3 Ωm and a thickness range of 13.0–14.0 m and is interpreted as a medium-to-coarse sand unit. These two zones are expected to be saturated based on the inverted resistivity ranges and the previous geological and hydrological information. This layer overlies a basal shale unit with a resistivity range of 43.8–236.6 Ωm. The constructed geoelectric sections (Figure 6) show a sub-vertical fault based on the sharp changes in the layers' thicknesses and resistivities.

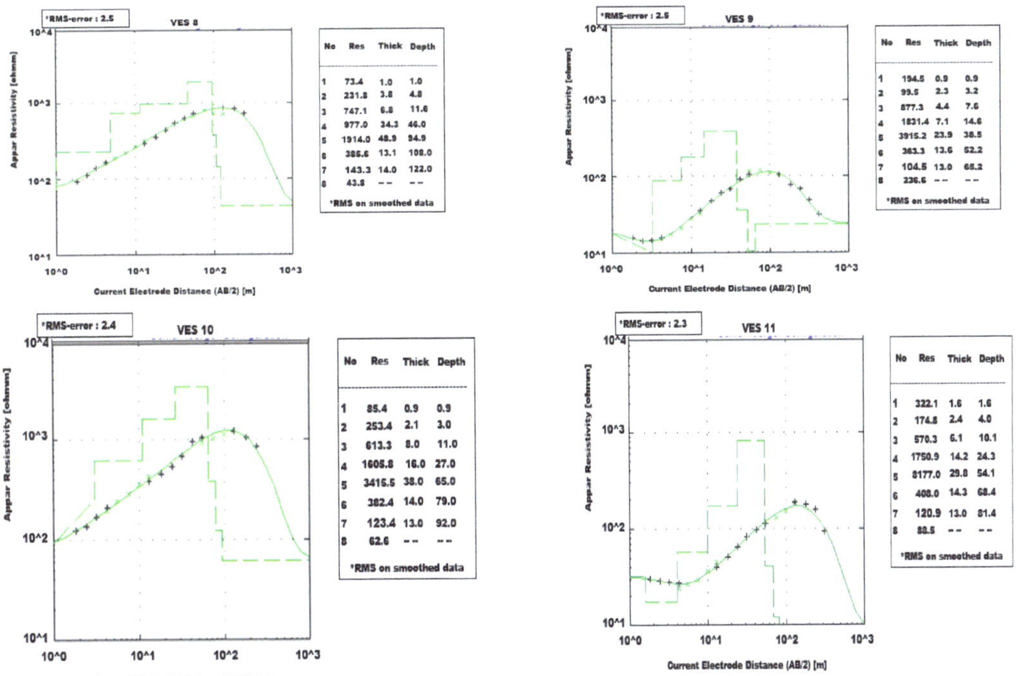

Figure 5. Representative of the inverted VES curves within Bells University (L2) and the resulting digital subsurface models.

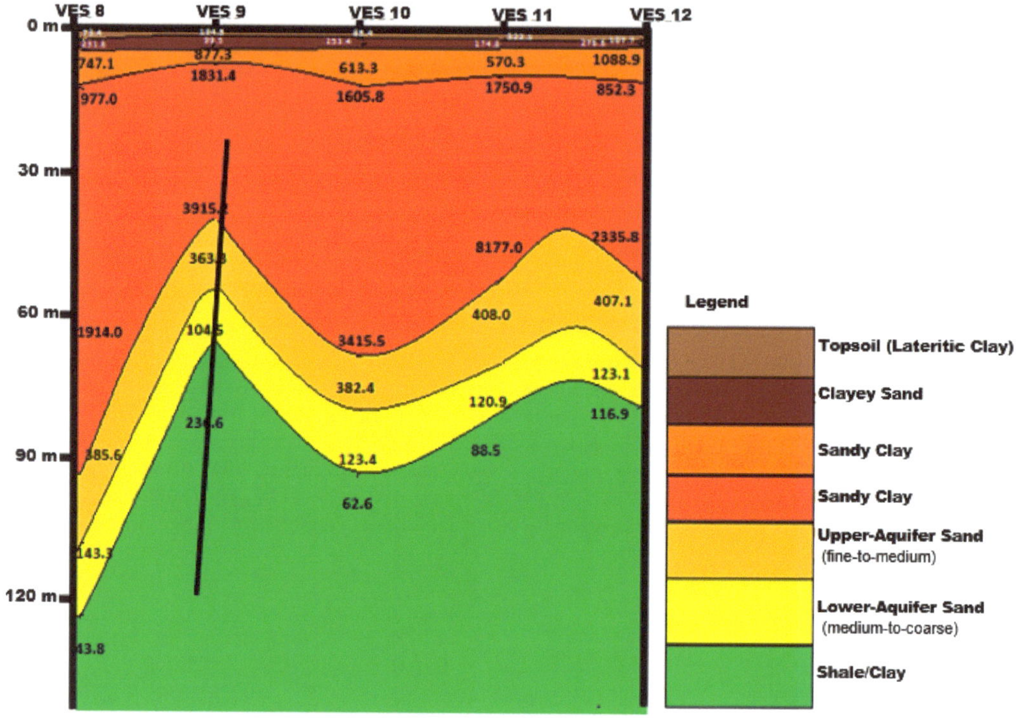

Figure 6. Geoelectric subsurface section from sounding results carried out along the Bells University, showing the inferred faults.

The representatives of inverted sounding curves for the six VESs conducted along the Oju-Ore-Ilogbo road (location L3) are presented in Figure 7. Like the other locations, eight geoelectric layers were attained for each VES station. The first geoelectric layer is the topsoil, which has a resistivity range of 50.0–138.0 Ωm, and a thickness range of 0.7–1.5 m. The topsoil zone is composed of lateritic clay. The undelayed second layer has a resistivity value range of 39.2–209.1 Ωm and a thickness range of 1.0–2.5 m, which can be interpreted as a clayey sand layer. The third delineated layer is characterised by resistivity values of 187.4–711.6 Ωm and is revealed to be a sandy clay unit with a thickness range of 1.8–3.5 m. The fourth and fifth layers show resistivity ranges of 841.5–3064.5 Ωm and 1119.7–3727.45 Ωm and thickness ranges of 8.4–26.2 m and 17.6–31.7 m, respectively. This zone is interpreted as sandy clay confining the undelayed saturated units. The sixth geoelectric layer is the upper saturated zone in the area, which has a resistivity range of 359.5–404.8 Ωm, and a thickness of 13.6–16.3 m and can be interpreted as a sand saturated layer. The seventh delineated layer is the lower saturated zone, with a resistivity range of 116.3–126.7 Ωm and a thickness of 14.0 m, and represents a saturated sand unit. The mapped basal layer has a resistivity range of 27.7–236.0 Ωm and can be interpreted as the shale layer (Figure 8).

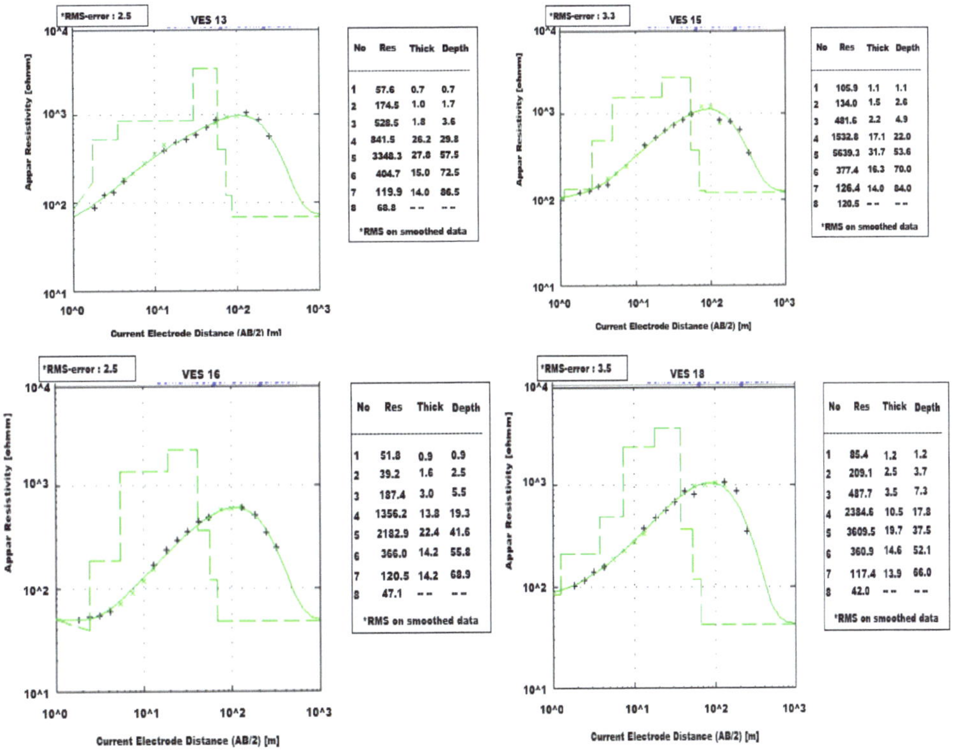

Figure 7. Representative of the inverted VES curves within Oju-Ore-Ilogbo Road (L3) and the resulting digital subsurface models.

Figure 9 shows the representative of the inverted VES numbers 19–25 conducted along the Obasanjo-Ijagba road (Location L4). Eight subsurface geoelectric layers were interpreted from the sounding data in the area and used to construct the subsurface geoelectric resistivity section (Figure 9). The first layer is the topsoil, which has a resistivity range of 22.5–223.6 Ωm, a thickness range of 0.6–1.7 m, and is interpreted as lateritic clay. The delineated second layer shows resistivity values ranging from 103.7–335.5 Ωm and a thickness of 1.6–3.8 m. It is interpreted as a clayey sand unit. The third delineated layer has a high resistivity value of 414.8–1375.5 Ωm with a thickness range of 4.4–9.8 m, which represents a sandy clay unit. The fourth and fifth delineated layers have high resistivity values of 1187.3–1842.6 Ωm and 2162.5–4064.5 Ωm, and a thickness of 8.5–23.6 m and 15.4–30.4 m, respectively. This unit is represented as highly compacted sandy clay, which is confining the underlying saturated units. The sixth layer shows a resistivity range of 364.5–411.4 Ωm and a thickness range of 13.8–16.1 m, which can be interpreted as the upper saturated silty sand unit. The seventh delineated layer has a resistivity range of 117.5–122.1 Ωm and a thickness ranging from 13.7–14.1 m and represents a lower saturated sand unit. The delineated basal layer shows a resistivity range of 36.0–81.8 Ωm, which can be interpreted as the shale unit (Figure 10).

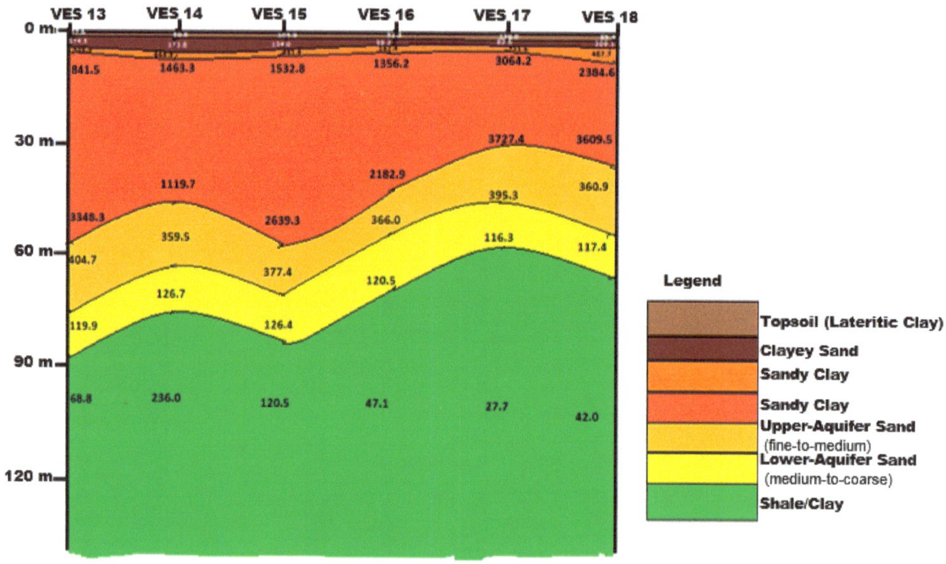

Figure 8. Geoelectric subsurface section from sounding results carried out along Oju-Ore-Ilogbo Road.

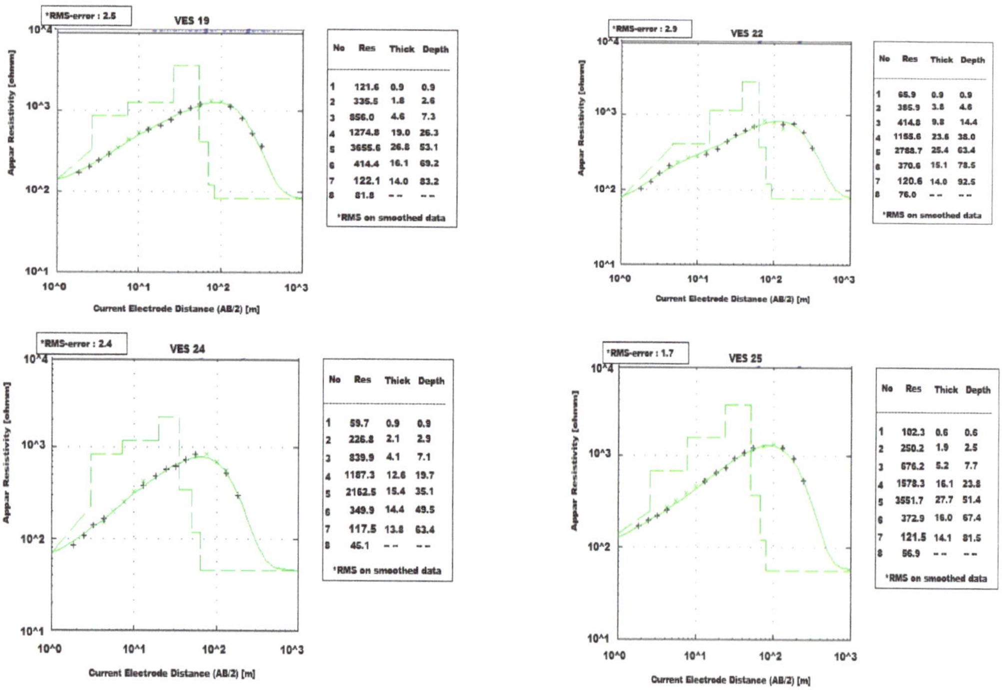

Figure 9. Representative of the inverted VES curves within Obasanjo-Ijagba Road (L4) and the resulting digital subsurface models.

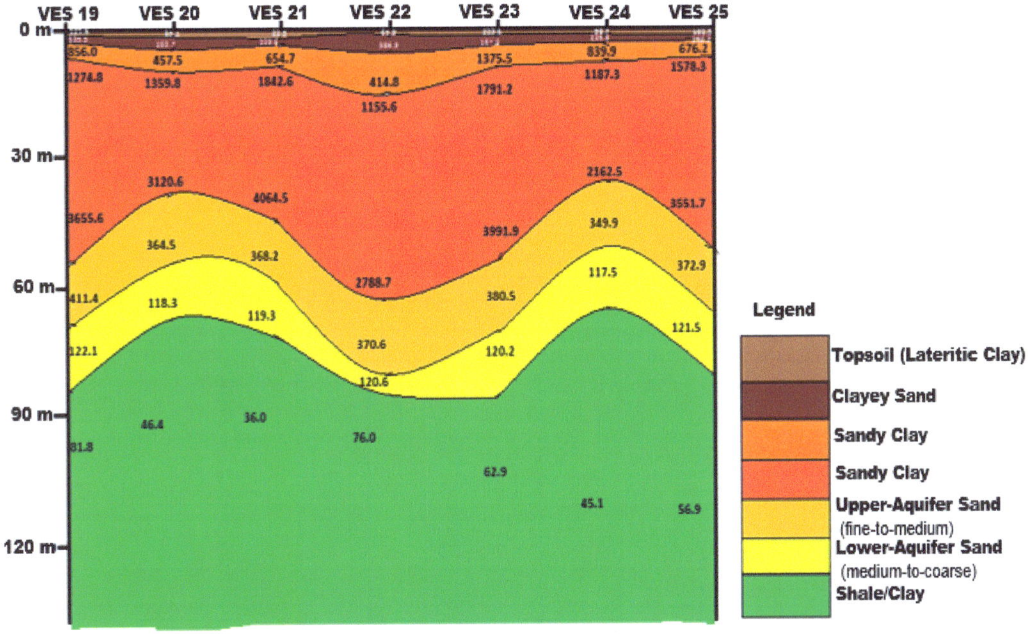

Figure 10. Geoelectric subsurface section from sounding results carried out along Obasanjo-Ijagba road.

Five VESs were conducted at Iyana-Iyesi (Location L5), and the interpretation of these soundings data equally revealed eight subsurface geoelectric layers that can be discussed as follows: The representative inverted VES numbers 26–30 curves for the area are presented in Figure 11. Similarly, eight geoelectric layers were delineated, starting with the topsoil, which has variable resistivity values of 53.5–185.4 Ωm with a thickness range of 1.0–1.4 m, representing the lateritic clay layer. The second layer shows resistivity values of 150.8–720.7 Ωm and a thickness of 2.5–3.4 m; this layer represents a clayey sand unit. The third layer has high resistivity values of 796.6–1289.8 Ωm and a thickness range of 5.4–7.4 m, it is interpreted as a sandy clay unit. The fourth and fifth delineated layers have higher resistivity, ranging from 1365.9 to 2179.0 Ωm and 2713.7 to 3885.2 Ωm, and the thickness ranges are 13.5–16.7 m and 21.3–30.8 m, respectively. These layers represent sandy clay units, which are considered the confining beds for the underlying saturated units. The delineated sixth layer has a resistivity range of 368.9–372.6 Ωm and a thickness range of 15.1–15.4 m which represents the silty sand unit, which is considered the upper saturated zone. The seventh layer shows a resistivity range of 120.2–121.3 Ωm and a thickness range of 14.0–14.1 m and is interpreted as a saturated sand unit. This layer overlies a basal shale unit with a resistivity range of 50.1–63.3 Ωm (Figure 12).

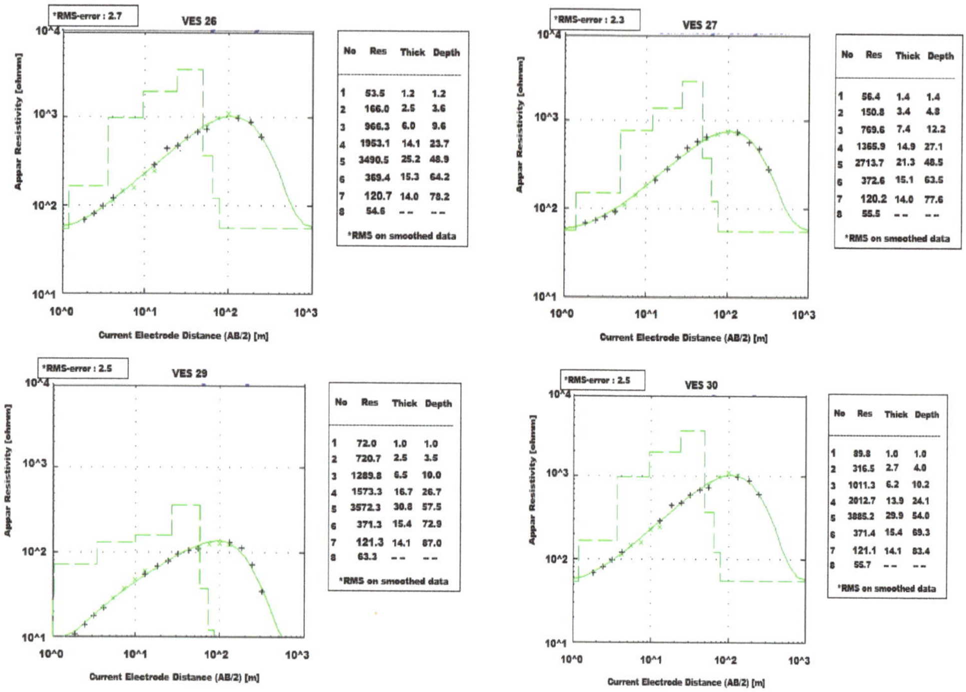

Figure 11. Representative of the inverted VES curves within Iyana-Iyesi (L5) and the resulting digital subsurface models.

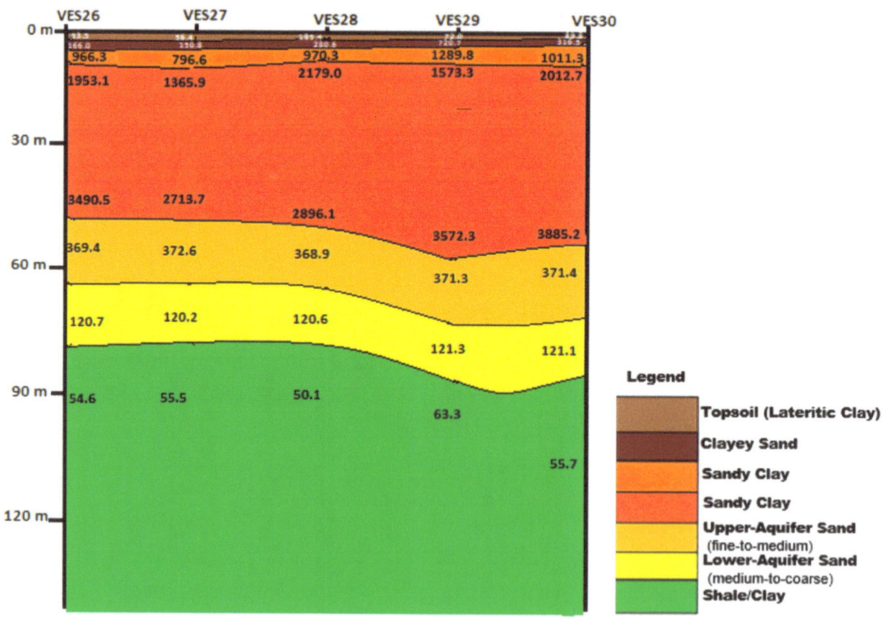

Figure 12. Geoelectric subsurface section from sounding results carried out along the Obasanjo-Ijagba road.

4.2. 2D Electrical Resistivity Imaging

Though both L_1-norm and L_2-norm inversion techniques were tested for the acquired 2D ERT datasets, only the inverse models using the L2-norm are presented because they better represent the subsurface. The 2D ERT inversion generally reveals the geoelectric layers in more detail and is equivalent to the estimated results using the VESs techniques (Figures 13–17). A few iterations were adequate to achieve a good match between the measured and modelled resistivity, and the resulting 2D resistivity section revealed the subsurface layer distribution in relation to the previous geological and VES results. The colour code represents the 2D resistivity distributions, with emphasis on the dry and saturated zones. The inversion of the ERT traverse for location L1 is presented in Figure 13, which reveals the lateral resistivity distribution up to 70 m depth. The 2D ERT profile for traverse T1 shows the presence of the shallow low resistivity layer with resistivities <250 Ωm which represents the clayey topsoil with variable resistivity values according to the surface activities and whose thickness is a maximum of 20 m. This layer is followed by a layer of high resistivity values >800 Ωm with a thickness of 70 m or more. This layer represents the dry sandy clay cap unit, and it is extended along the measured section. The most important unit that appears at elevation 30 m has a low resistivity character, which represents the saturated sandy clay layer. It appears at distinctive locations: at the beginning of the measured profile at 320 m. The sub-vertical sharp resistivity boundaries between the low and high values appear at the beginning and at 330 m horizontal distance, which may be attributed to the presence of a fault (Figure 13).

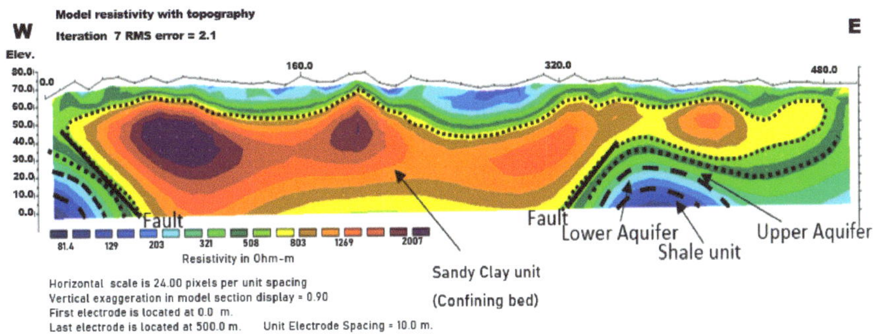

Figure 13. 2D subsurface geoelectrical resistivity image for traversing T1 Covenant University; dashed lines delineate the boundaries between the different layers. The solid lines are the fault locations.

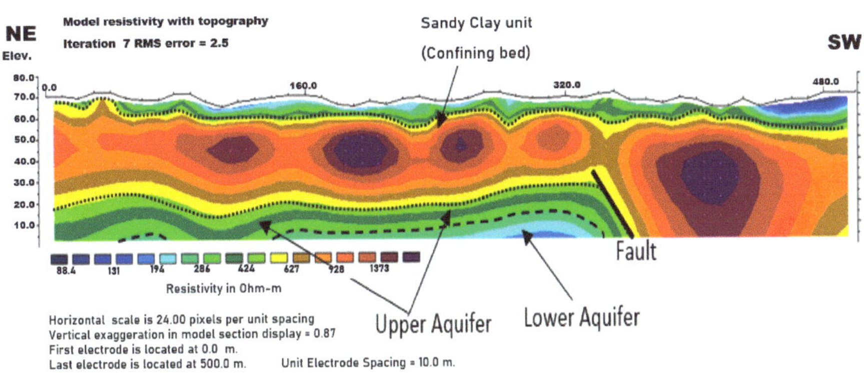

Figure 14. 2D subsurface geoelectrical resistivity image for traverse T2 at Bells University.

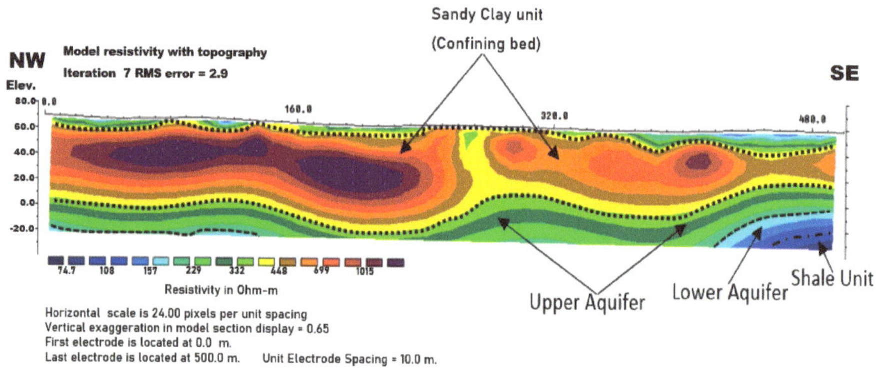

Figure 15. Two–dimensional geoelectrical resistivity image for traverse T3 along the Oju-Ore-Ilogbo road (L3).

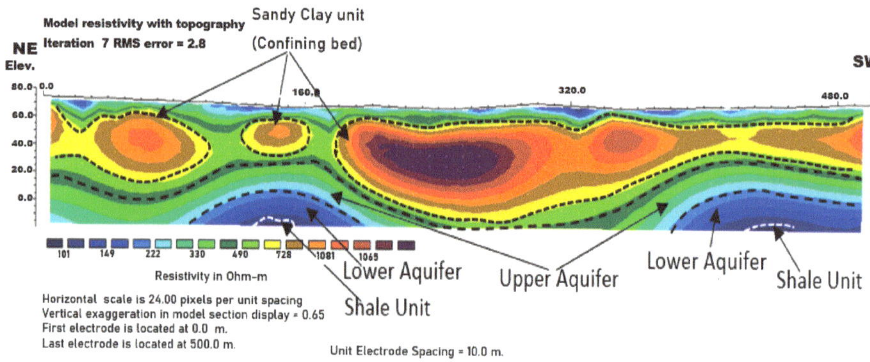

Figure 16. Two-dimensional geoelectrical resistivity image for traverse T4 along the Obasanjo-IJagba road (L4).

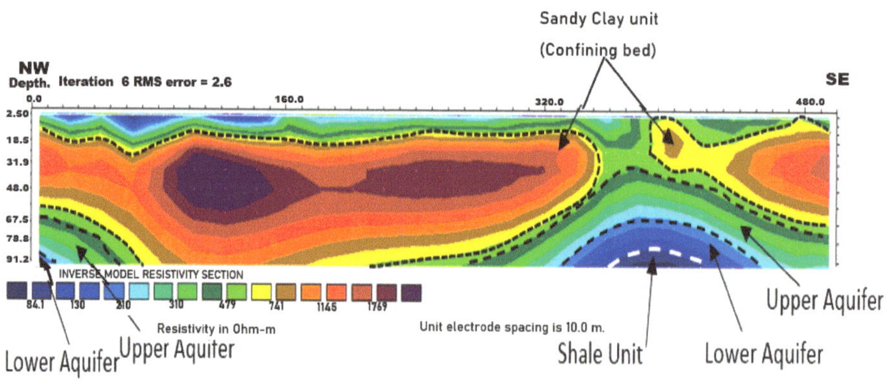

Figure 17. Two-dimensional geoelectrical resistivity image for traverse T5 at Iyana-Iyesi (L5).

Similarly, the 2D ERT profile for traverse T2 within Bells University is presented in Figure 14 with the same resistivity value distributions as in T1. The shallow low resistivity layer with resistivities <250 Ωm represents the clayey topsoil with variable resistivity values according to the surface activities, and its thickness is a maximum of 10 m. This layer is followed by a layer of high resistivity values >700 Ωm with a thickness of 60 m and more along the SW direction. This layer represents the dry sandy clay cap unit, and it is extended along the measured section. The most important units that appear at elevation 20 m have a low resistivity character, which represents the saturated sandy clay layer. It appears horizontally along the measured profile and laterally in contact with the high resistivity layer at a horizontal distance of 340 m. This probably reflects the presence of a subvertical fault (Figure 14).

The 2D ERT profile for traverse T3 is presented in Figure 15 and shows similar subsurface resistivity distributions to the previous profiles. Inspecting the inverted resistivity profiles reveals similar subsurface lithological successions in the other two locations. The shallow and thin low resistivity layer with resistivities <250 Ωm represents the clayey topsoil with variable resistivity values. This layer is followed by a layer of high resistivity values >600 Ωm with a thickness of 60 m, thinning in the SE direction. This layer represents the dry sandy clay cap unit, and it is extended along the measured section. The most important units that appear at elevation 0 m have low resistivity, which represents the saturated sandy clay layer. It appears horizontally and smoothly along the measured profile. At the end of the profile, the low resistivity zone (<100 Ωm) appears underneath the saturated zone. This zone represents the unit bounding the upper saturated layers. (Figure 15). The inverse model of the 2D ERT profile T4 conducted along the Oju-ore-Ilogbo road (location L4) is displayed in Figure 16. Inspecting the inverted resistivity profiles reveals similar subsurface lithological successions to the other 2D profiles. The shallow and thin low resistivity layer with resistivities <220 Ωm represents the clayey topsoil. The second layer has a relatively high resistivity value of >600 Ωm along most of the measured profile with a thickness of about 60–70 m, reflecting the old topography of the underlain sand clay layer. This layer represents the dray sandy clay cap unit, and it is extended along the measured profile. The most important units that appear at elevation 30 m with low resistivity characters represent the saturated sandy clay layer with an irregular bottom topographic surface. At the bottom of the saturated layer, a low resistivity zone (<100 Ωm) appears, which is interpreted as a shale unit, bounding the upper saturated layers (Figure 16).

The fifth 2D ERT profile is presented in Figure 17, which shows the main subsurface resistivity layer. The shallow and thin low resistivity layer with resistivities <210 Ωm represents the clayey topsoil. This is followed by a high resistivity value > 950 Ωm along most of the measured profile, with a thickness of about 70–80 m along most of the profile. This layer represents the dray sandy clay cap unit, and it is extended along the measured profile except at distances of 230–260 m, where the layer is dissected. The low resistivity character layer appears at elevations of 70 m and 30 m, which represents the saturated sandy clay layer with an irregular topographic surface. At the bottom of the saturated layer, a low resistivity zone (<100 Ωm) appears, which is interpreted as a shale unit bounding the upper saturated layers (Figure 17).

4.3. Aquifers Hydraulic Parameters

Hydraulic parameters play a crucial role in understanding and managing aquifers, which are vital sources of groundwater. To effectively utilise and sustainably manage these invaluable resources, it is essential to accurately estimate hydraulic parameters. Based on the Archie law (Equation (1)), the upper and lower aquifer porosities can be calculated from the interpreted resistivity values deduced from the measured VES stations. Then the Dar Zarrouk parameters (example: Table 1) have been used to calculate the two aquifers' hydraulic conductivity (Equations (2) and (4)). Using the aquifer thickness, the transmissivity values can be estimated. Table 2 shows the calculated hydraulic parameters for the two aquifers along the five investigated sites. The calculated hydraulic values for the two aquifers at each VES station are presented in Figures 18 and 19. The prediction models for the estimated hydraulic parameters of both upper and lower aquifers are presented in Figures 20 and 21. Equations (6)–(8) are the linear prediction models relating the estimated transmissivity, hydraulic conductivity, and porosity with the mean true resistivity of the upper aquifer. The lower aquifer hydraulic estimated parameters versus mean resistivity values can be expressed in Equations (9)–(11).

$$Mean\ Transmissivity = 6.1993 - (0.012) Mean\ R_f \quad (6)$$

$$Mean\ K(10^{-2}) = 54.46 - (0.011) Mean\ R_f \quad (7)$$

$$Mean\ Porosity = 0.37 - (4.37 \times 10^{-4}) Mean\ R_f \quad (8)$$

Table 2. Estimated hydraulic parameters for upper and lower aquifers in the study area.

	Upper Aquifer				Lower Aquifer			
Location	**Mean R_F Ω-m**	**Mean Porosity**	**Mean K (m/s)10^{-2}**	**Mean T (m^2/s)**	**Mean R_F Ω-m**	**Mean Porosity**	**Mean K (m/s)10^{-2}**	**Mean T (m^2/s)**
L1(V1–V7)	347.4	0.2182	17.0	2.18	125.4	0.4777	400	50.00
L2(V8–V12)	386.7	0.2009	12.7	1.56	121.0	0.4910	457	55.30
L3(V13–V18)	378.5	0.2042	13.5	1.76	119.6	0.4955	478	62.14
L4(V19–V25)	389.7	0.1997	12.4	1.68	120.8	0.4917	460	62.10
L5(V26–V30)	356.2	0.2140	15.9	1.90	128.0	0.4703	371	51.57

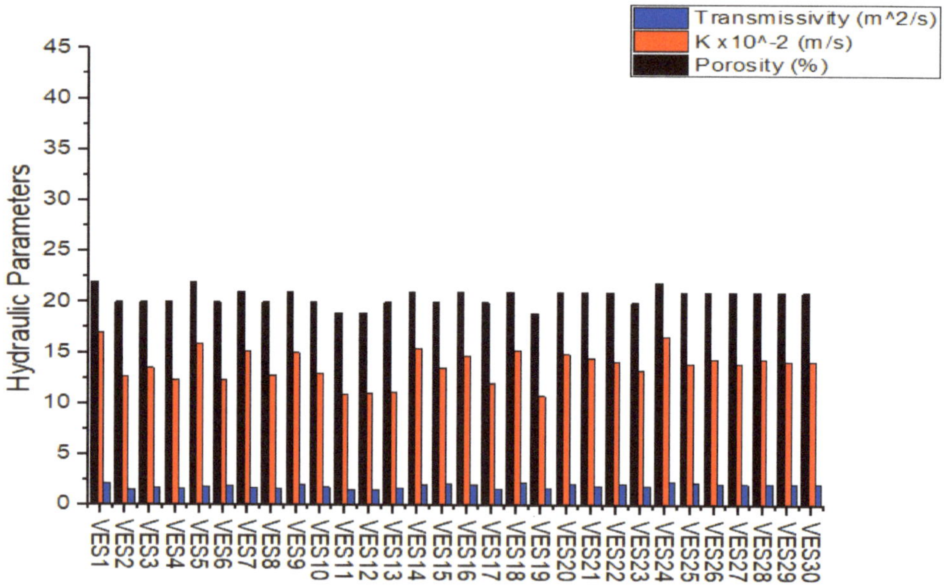

Figure 18. Estimated porosity, hydraulic conductivity, and transmissivity for the upper aquifer were deduced from the measured- VES stations.

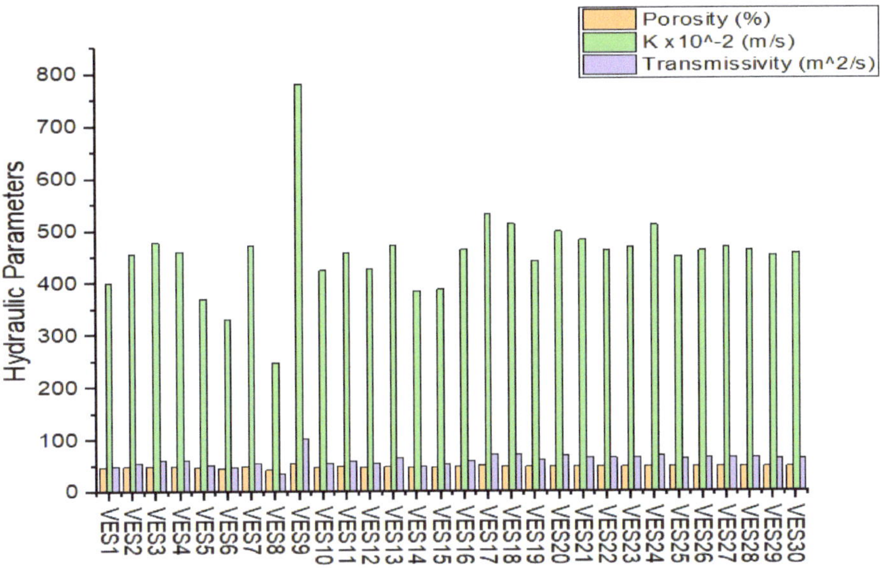

Figure 19. Estimated porosity, hydraulic conductivity, and transmissivity for lower aquifer were deduced from the measured- VES stations.

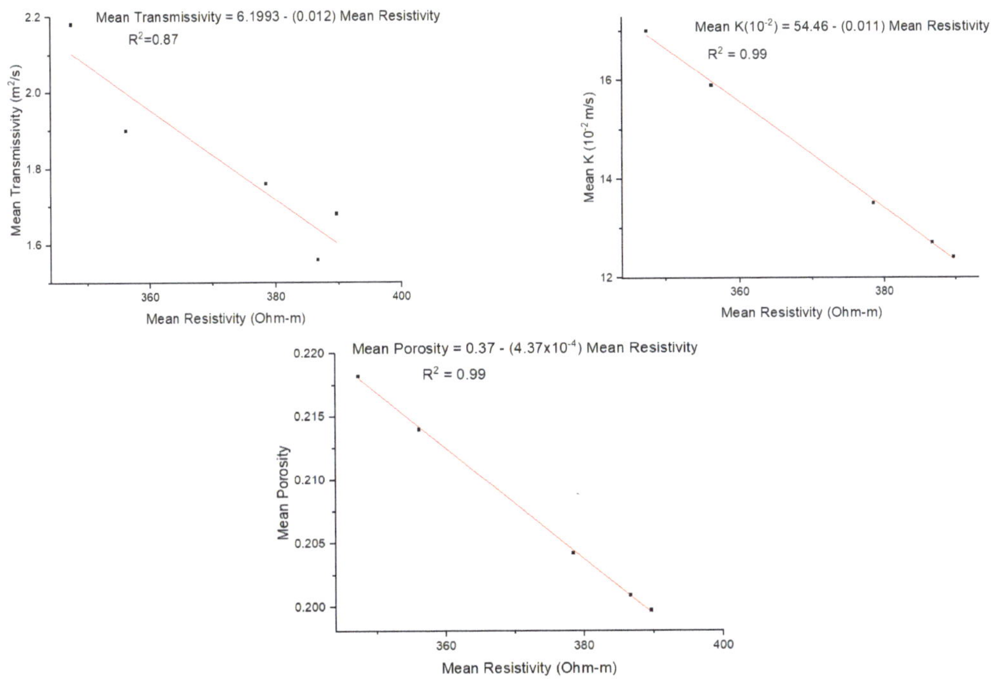

Figure 20. Predicted model for mean transmissivity of upper aquifer from the true mean resistivity.

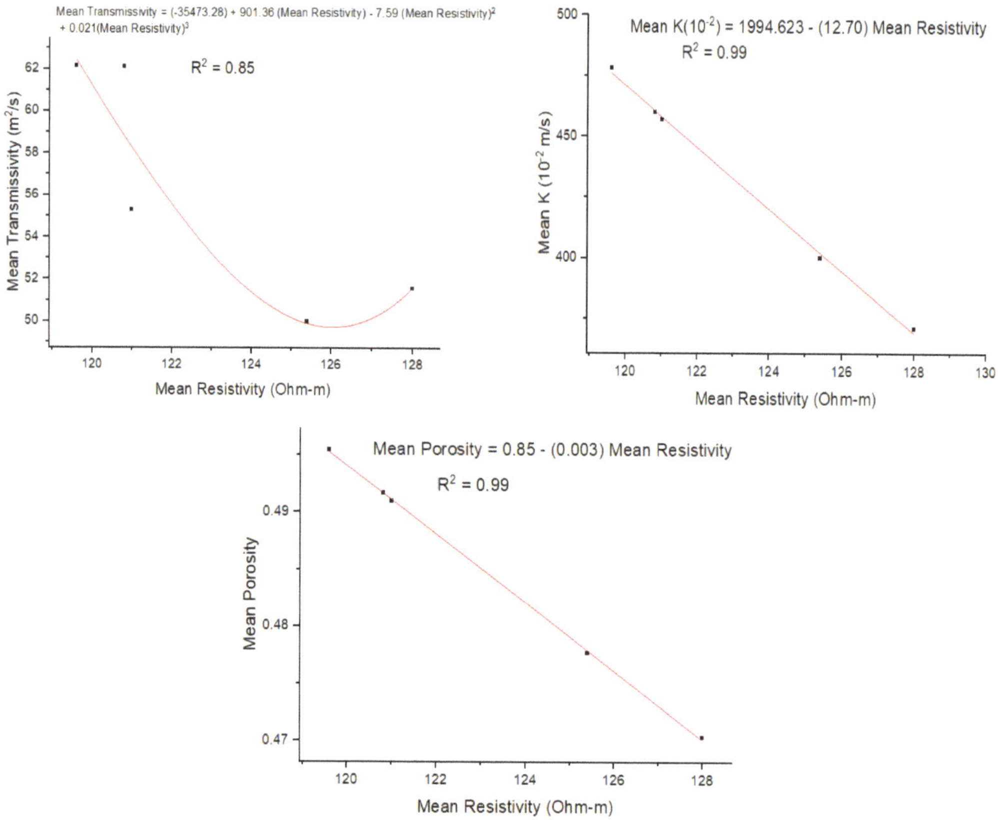

Figure 21. Predicted model for mean transmissivity of the lower aquifer from the true mean resistivity.

The prediction models for hydraulic conductivity and porosity are linear, while the mean transmissivity model is non-linear (3rd-order polynomial). This confirms that the estimated hydraulic values are acceptable for both aquifers.

$$Mean\ Transmissivity = (-35473.28) + 901.36\,(Mean\ R_f) - 7.59\,(Mean\ R_f)^2 + 0.021\,(Mean\ R_f)^3 \quad (9)$$

$$Mean\ K(10^{-2}) = 1994.623 - (12.70) Mean\ R_f \quad (10)$$

$$Mean\ Porosity = 0.85 - (0.003) Mean\ R_f \quad (11)$$

5. Discussion

5.1. Subsurface Characterisation and Aquifer Delineation

The VESs and the 2D electrical resistivity images were integrated for subsurface evaluation up to depths ranging from 90–120 m and delineated the dry and saturated subsurface zones. The geological and borehole information has been considered in the inversion of the VES data sets. Then the borehole and VES models were considered in the inversion and interpretation of the 2D ERT profiles, which show more details about the lateral extension of subsurface layer successions. The delineated geoelectric layers started with the topsoil (mostly lateritic clay), clayey sand, sandy clay, sand, and shale. Most of the interpreted layers are laterally extended along the investigated areas in the same order. On the inverted-sounding data and the 2D ERT sets, the topsoil unit is thin and dominant along most of the surveyed areas. The second dominant layer is the sandy clay unit, which has relatively high resistivity characters across both data sets, with more details about the thickness and continuation appearing in the 2D ERT inverted profiles. The high resistivity character of the sandy clay unit is referred to as the compaction and the clay content, which have been described as being rich in kaolin and intercalated with phosphates. The intercalated phosphates are thought to be part of the reported thin bands of phosphate belonging to the Ilaro Formation [29,32,45]. Despite the resistivity data indicating that this layer is dry, many homeowners tried to hand-dig wells to extract water from this layer with very limited success. Underlain by the high-resistivity clayey sand layer, the low-resistivity characters appear with more details about the layer's thickness and depth, which are considered saturated with groundwater. This is the sandy clay layer, and based on the geological and borehole information, this unit can be classified into two saturated zones with different grain sizes.

The high resistive unit is confining the unconsolidated sand unit, which forms the upper saturated zone along the study area. It is composed mainly of silty sand deposits. This upper saturated layer is thought to be part of the tertiary alluvium deposits of the Benin Formation [29,32,45]. The lower saturated zone is a coarse-grained sand formation that is perhaps more porous and permeable compared to the upper aquifer system, which is the reason for its lower resistivity characteristics. The lower aquifer is interpreted to be part of the coastal plain sands of the Benin Formation as well. Moreover, many faults were identified by the vertical and sub-vertical sharp contact between the low and high resistivity units. These expected faults penetrate both upper and lower saturated zones at different depths and could scientifically affect sustainable groundwater exploration, development, and management within the two aquifer systems [9,23,46,47].

5.2. Implications for Groundwater Resource Development and Management

The spatial distributions of the true stratigraphic thickness of both the lower and upper saturated zones are presented in Figures 22 and 23. The thickness of an aquifer plays a significant role in groundwater development and management. It determines the storage capacity, sustainability of extraction, water quality, well yield, recharge potential, hydrological dynamics, and adaptability to changing conditions. The thicknesses of both lower and upper saturated zones increase south-westward up to the Iyana-Iyesi area and decrease north-westward down to the Canaan land area (Figures 22 and 23). It denotes the vertical extent of the saturated zone within the aquifer, indicating the depth from which water can be extracted [48]. Aquifer thickness influences its water storage capacity, as a thicker aquifer can store more water, ensuring a larger volume of groundwater available for various uses. The thickness of the aquifer directly influences its sustainable yield, providing a reliable and steady supply of water over an extended period. The estimated hydraulic parameters reveal the productive capacity of the delineated aquifers. Porosity measurements are fundamental to characterising aquifers and understanding their hydrogeological properties. Accurate porosity data are used in groundwater models and simulations to predict aquifer behaviour under different scenarios. It plays a vital role in determining the volume of water that an aquifer can store and transmit. The estimated porosity values range between 20 and 22% within the upper aquifer, while the values range from 48 to 50% for the lower aquifer. The estimated porosity values for both aquifers show their high capacity to store water.

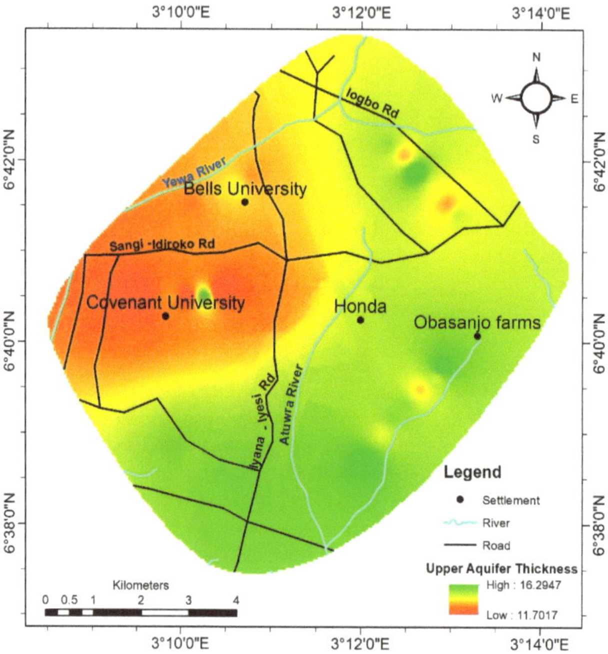

Figure 22. Spatial distribution of the Upper aquifer thickness in meters.

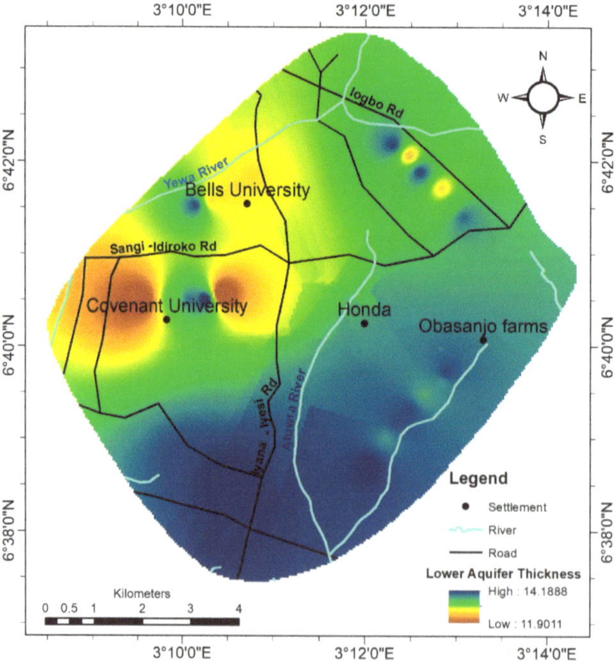

Figure 23. Spatial distribution of the Lower aquifer thickness in metres.

The estimated hydraulic conductivity values of both delineated upper and lower aquifers are high, with values ranging from 12.4–17.0 × 10^{-2} m/s for the upper aquifer unit and 371–478 × 10^{-2} m/s for the lower aquifer unit. Aquifers with high hydraulic conductivity can transmit water more easily, resulting in higher groundwater flow rates. The estimated hydraulic conductivity values of both delineated upper and lower aquifers are high, with values ranging from 1.56–2.18 m^2/s for the upper aquifer unit and 50.00–62.14 m^2/s for the lower aquifer unit. Understanding the hydraulic conductivity and transmissivity of the subsurface aquifer units is essential for effective groundwater management, water resource planning, and environmental protection. The estimated porosity, hydraulic conductivity, and transmissivity have high values for both delineated major aquifers in the study area.

Moreover, the structural faults within some parts of the study area cut across the delineated aquifers and can have significant impacts on groundwater resource development and management. The faults can serve as conduits for water to enter the aquifer, resulting in increased recharge rates. Groundwater wells may need to be carefully sited to avoid faults, and additional measures may be required to prevent contamination through faults. Thus, it is important to carefully consider the presence of faults when planning groundwater projects and to take steps to mitigate any negative impacts that may result.

6. Conclusions

Groundwater resources have many advantages over surface water, first in terms of comprehensive applications and usability in agriculture, domestic, and manufacturing industries. Thus, there is a need for these natural resources to be appropriately managed and protected to ensure their sustainability. Hydrogeophysical investigations have been employed within the eastern Dahomey basin to provide subsurface information and characterise the multi-layer aquifers within the subsurface. The subsurface lithologic units include the topsoil (Lateritic clay), clayey sand, sandy clay (confining bed), fine-to-medium sand (upper aquifer system), medium-to-coarse sand (lower aquifer system), and shale or clay belonging to the Akinbo Formation, which was delineated in all the locations (L1–L5). The shallower clayey sand and sandy clay formations serve as potential low-yield aquifers that are useful only for hand-dug wells in the study area. Two major aquifers were delineated within the area. The upper aquifer is a fine-silty sand unit with a mean thickness range and mean resistivity range of 13.0–15.3 m and 347.4–389.7 Ωm in the entire area. The estimated hydraulic parameters for the upper aquifer reveal that it is highly productive. The mean porosity range is 20–22%, the mean hydraulic conductivity range is 12.4 × 10^{-2} m/s–17.0 × 10^{-2} m/s, and the mean transmissivity range is 1.56–2.18 m^2/s. The delineated lower aquifer is coarse sand, with mean resistivity ranges of 119.6–128.0 Ωm and a mean thickness range of 13.0–14.1 m. The estimated hydraulic parameters for the lower coarse sand aquifer unit have a mean porosity range of 48–50%, a hydraulic conductivity range of 371–478 × 10^{-2} m/s, and a mean transmissivity range of 50.0–62.14 m^2/s. Targeting these aquifers for sustainable groundwater resources during exploration will save the homeowners within the study area from the pain of drilling unproductive and low-yield boreholes. Also, some sub-vertical faults within the study area will affect groundwater resource development and management in the area since the occurrence of these sub-vertical structural faults within the subsurface will result in changes in the permeability of the delineated aquifers across the divide, thereby affecting the productivity of the aquifer units in the area. There is a need to establish an hydrogeophysics observatory in the study area to obtain time-lapse hydrogeological data such as groundwater level data, pumping test data, groundwater recharge rate, and groundwater quality data. Integrating these data with the findings of this research would enable the building of effective groundwater models of the delineated, multi-layered aquifers in the area. With robust groundwater models of the aquifers, managed aquifer recharge (MAR) projects will be effective in the area to sustain groundwater supply. Moreover, there is a need to understand groundwater flow and transport processes within the delineated aquifer systems in the area to predict the movement and fate of contaminants in the aquifer. This is crucial for managing and mitigating the impacts of groundwater pollution, protecting water quality, and safeguarding public health in the area.

Author Contributions: Conceptualization, K.D.O., A.P.A. and M.M.; methodology, K.D.O., A.P.A., A.A.O. and D.V.K.; software, E.E.O., D.V.K. and M.M.; validation, A.P.A., K.D.O., A.A.O. and M.M.; formal analysis, K.D.O., A.A.O. and A.P.A.; field investigation, K.D.O., E.E.O., A.A.O. and A.P.A.; resources, D.V.K. and M.M.; data curation, K.D.O. and A.P.A.; writing—original draft preparation, K.D.O., A.A.O. and E.E.O.; writing—review and editing, K.D.O. and M.M.; supervision, A.P.A. and M.M.; project administration, A.P.A. and K.D.O.; funding acquisition, M.M. All authors have read and agreed to the published version of the manuscript.

Funding: This work was funded by the Deputyship for Research and Innovation, "Ministry of Education" in Saudi Arabia (IFKSUOR3-093-1).

Data Availability Statement: Not applicable.

Acknowledgments: The authors wish to express their appreciation to the Deputyship for Research and Innovation, "Ministry of Education" in Saudi Arabia for funding this research (IFKSUOR3-093-1). Also, the authors appreciate the Covenant University Center for Research, Innovation, and Development (CUCRID) for the research support. The first author is thankful to both UNESCO-TWAS and CSIR, India, for the CSIR-TWAS Postdoctoral Fellowship at the National Geophysical Research Institute, India.

Conflicts of Interest: The authors declare no conflict of interest.

References

1. Barlow, J.R.B.; Clark, B.R. *Simulation of Water-Use Conservation Scenarios for the Mississippi Delta Using an Existing Regional Groundwater Flow Model: U.S. Geological Survey Scientific Investigations Report 2011–2019*; U.S. Geological Survey: Reston, VA, USA, 2011; 14p.
2. UNESCO. *Water a Shared Responsibility. The United Nations World Water Development Report 2*; UNESCO: New York, NY, USA, 2006; p. 601.
3. WHO/UNICEF. *Water for Life. Making it Happen. A Decade for Action 2005–2015*; WHO: Geneva, Switzerland, 2005; p. 44.
4. Rubin, Y.; Hubbard, S.S. *Hydrogeophysics, Water Science and Technology Library*; Springer: New York, NY, USA, 2005.
5. Nwankwoala, H.O.; Udom, G.J. Influence of land reclamation on the status of groundwater in Borokiri area of Port Harcourt, Niger Delta, Nigeria. *Int. J. Nat. Appl. Sci.* **2008**, *4*, 431–434. [CrossRef]
6. Kaya, M.A.; Ozürlan, G.S.; Engül, E. Delineation of soil and groundwater contamination using geophysical methods at a waste disposal site in çanakkale, Turkey. *Environ. Monit. Assess.* **2007**, *135*, 441–446. [CrossRef] [PubMed]
7. Balkaya, Ç.; Kaya, M.A.; Goktürkler, G. Delineation of shallow resistivity structure in the city of Burdur, SW Turkey by vertical electrical sounding measurements. *Environ. Geol.* **2009**, *57*, 571–581. [CrossRef]
8. Olaojo, A.A.; Oladunjoye, M.A.; Sanuade, O.A. Geoelectrical assessment of polluted zone by sewage effient in University of Ibadan campus southwestern Nigeria. *Environ. Monit. Assess.* **2018**, *190*, 24. [CrossRef] [PubMed]
9. Oyeyemi, K.D.; Aizebeokhai, A.P.; Ndambuki, J.M.; Sanuade, O.A.; Olofinade, O.M.; Adagunodo, T.A.; Olaojo, A.A.; Adeyemi, G.A. Estimation of aquifer parameters from surficial geophysical methods: A case study of Ota, Southwestern Nigeria. *IOP Conf. Ser. Earth Environ. Sci.* **2018**, *173*, 012028. [CrossRef]
10. Oyeyemi, K.D.; Abuka-Joshua, J.; Rotimi, O.J.; Dieppos, B.; Gomo, M.; Olaojo, A.A.; Falae, P.O.; Metwaly, M. Geoelectrical Characterization of Coastal Aquifers in Agbado Ijaye, Lagos, Southwestern Nigeria; Implications for groundwater resources. *Sustainability* **2023**, *15*, 3538. [CrossRef]
11. Sendrós, A.; Himi, M.; Lovera, R.; Rivero, L.; Garcia-Artigas, R.; Urruela, A.; Casas, A. Geophysical Characterization of Hydraulic Properties around a Managed Aquifer Recharge System over the Llobregat River Alluvial Aquifer (Barcelona Metropolitan Area). *Water* **2020**, *12*, 3455. [CrossRef]
12. Mohammed, M.A.A.; Abdelrahman, M.M.G.; Szabó, N.P.; Szűcs, P. 2023. Innovative hydrogeophysical approach for detecting the spatial distribution of hydraulic conductivity in Bahri city, Sudan: A comparative study of Csókás and Heigold methods. *Sustain. Water Resour. Manag.* **2023**, *9*, 107. [CrossRef]
13. Mohammed, M.A.A.; Szabó, N.P.; Szűcs, P. Characterization of groundwater aquifers using hydrogeophysical and hydrogeochemical methods in the eastern Nile River area, Khartoum State, Sudan. *Environ. Earth Sci.* **2023**, *82*, 219. [CrossRef]
14. Szabó, N.P.; Kormos, K.; Dobróka, M. Evaluation of hydraulic conductivity in shallow groundwater formations: A comparative study of the Csókás' and Kozeny–Carman model. *Acta Geod. Geophys.* **2015**, *50*, 461–477. [CrossRef]
15. Hasan, M.; Shang, Y.; Jin, W.; Akhter, G. Estimation of hydraulic parameters in a hard rock aquifer using integrated surface geoelectrical method and pumping test data in southeast Guangdong, China. *Geosci. J.* **2021**, *25*, 223–242. [CrossRef]

MDPI AG
Grosspeteranlage 5
4052 Basel
Switzerland
Tel.: +41 61 683 77 34

Water Editorial Office
E-mail: water@mdpi.com
www.mdpi.com/journal/water

www.ingramcontent.com/pod-product-compliance
Lightning Source LLC
Chambersburg PA
CBHW041415210926
51850CB00013B/3277
* 9 7 8 3 7 2 5 8 4 3 5 2 7 *